INTRODUCTION TO COMPOSITE MATERIALS

Stephen W. Tsai
U.S. Air Force Materials Laboratory

H. Thomas Hahn
Washington University

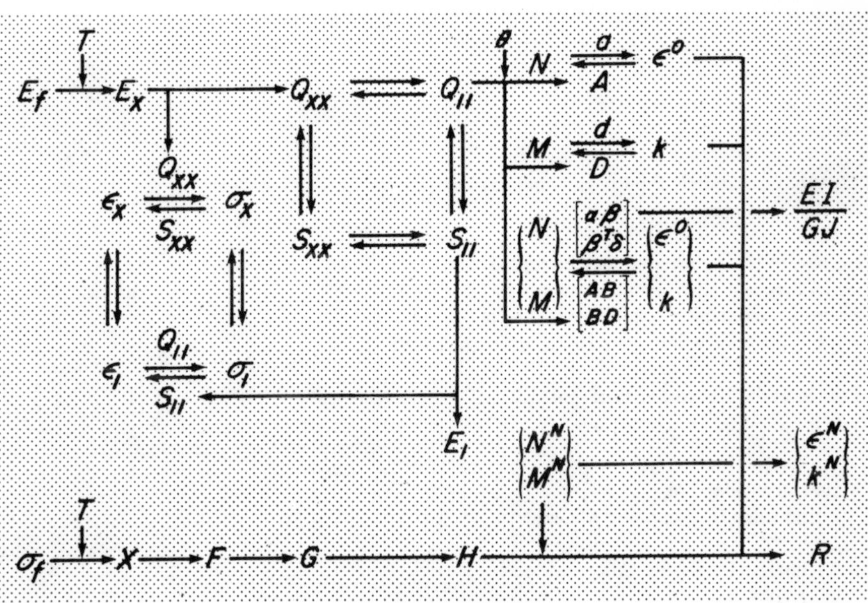

LANCASTER · BASEL

Published in the Western Hemisphere by
Technomic Publishing Company, Inc.
851 New Holland Avenue
Box 3535
Lancaster, Pennsylvania 17604 U.S.A.

Distributed in the Rest of the World by
Technomic Publishing AG

Printed in the United States of America

10 9 8 7 6 5 4

Main entry under title:
Introduction to Composite Materials

A Technomic Publishing Company book
Bibliography: p.
Includes index p. 453

Library of Congress Card No. 80-51965
ISBN No. 0-87762-288-4

contents

preface

The objective of this book is to introduce the governing principles of the stiffness and strength of uni- and multi-directional composite materials. It is intended as a textbook at the undergraduate and graduate levels. It can also serve as a reference for the engineers in industry. A course in strength of materials provides the desired background.

We hope to show that composite materials are conceptually simple. They offer unique opportunities in design beyond being lighter substitutes of conventional materials. The structural performance offered by composite materials is much more versatile than can be realized with conventional materials. We can put such versatility to our advantage in design through a full comprehension of the principles governing their structural behavior. An elucidation of these principles is the underlying theme of our book.

Our approach relies heavily on a unified and recurring formulation. Closed form solutions and simplified formulas are presented whenever possible. Figures, tables and charts of numerical results for one typical composite material throughout the book are given. Since our book is meant to be self-contained, references are given only when a set of pertinent data and established concepts are cited, or when a detailed derivation is omitted in the text. We wish to apologize for any inadvertent omissions. We are happy to see the rule-of-mixtures equations applicable in many situations. We feel that the governing principles can be presented more clearly if we separate geometric factors from material properties whenever possible. Eventually, the reader will find that transformation equations and the state of combined stresses are the key concepts in the study of composite materials.

In addition to the basic text presented in nine chapters, a special section on notation and terminology follows this preface. At the end of each chapter, a principal nomenclature for that chapter is listed. Transformation equations are summarized in Appendix A; unit conversion tables, in Appendix B; and listing of general references in the English language, in Appendix C.

This book is based in part on several United States Air Force reports written by us. We are indebted to our colleagues for their invaluable help. The opportunities to teach and to learn from the students at the Composite Materials Computation Workshops at the University of California at Berkeley, and at similar ones in Stuttgart, Tokyo, Osaka, Peking, and other places are gratefully acknowledged. The formulas in the book can be conveniently solved by programmable pocket calculators. Such programs are available from the authors.

STEPHEN W. TSAI
H. THOMAS HAHN

Dayton and St. Louis
May 1980

notation and terminology

The choice of notation and terminology can be a source of confusion. Definitions and explanations of our choice and format are listed as follows:

- The contracted notation is used which calls for the use of engineering shear strain. (See Table 1.3)
- Poisson's ratios are defined in Equations 1.5 and 1.6 for the on-axis unidirectional composites; in Table 3.15 for the off-axis. They are different from those in existing literature.
- The angles of coordinate transformation and ply orientation have the same sign. Proper signs are incorporated in the equations such that the transformation of stress and strain (the behavioral quantities) goes from the reference axes to the material symmetry axes. The transformation of modulus, compliance, expansion coefficients and other properties of material goes from the material symmetry axes to the reference axes; the opposite of the behavioral quantities. Separate equations are listed in Appendix A to show the differences.
- The laminate code in Equations 4.1 and 5.1 follows ascending order from the bottom to the top ply.
- A balanced laminate means that each off-axis ply or ply group is matched by one with opposite ply orientation. This is meaningful only for the in-plane modulus. A balanced laminate will be orthotropic in its in-plane behavior, but is not orthotropic in its flexural behavior.
- There are coupling coefficients beyond the traditional Poisson's ratio. The 61 and 62 components are the shear coupling coefficients; the 16 and 26, the normal coupling. Such coefficients can be applied to off-axis unidirectional composites as in Table 3.15, and to symmetric laminates in Equations 4.18 and 5.23 et al. These coefficients are treated as engineering constants. Comparable coefficients are not defined for general laminates. These laminates have unique couplings between the in-plane and flexural behavior defined by the B or β matrix.

- The curvature-displacement (k-w) relations in Equation 5.9 must have negative signs. The twisting curvature must have a factor of 2 to be consistent with the engineering shear strain.
- Symmetry is used for numerous situations.

  a) Material property symmetry or reciprocity:

  $$Q_{ij} = Q_{ji},\ A_{ij} = A_{ji}, \text{ etc.}$$

  b) Material symmetry in terms of structures:

  Anisotropy, Orthotropy, Square Symmetry, Isotropy

  c) Odd and even symmetry of material property transformation:

  $$Q_{11}(\theta) = Q_{11}(\theta + n\pi),\ Q_{66}(\theta) = Q_{66}\left(\theta + \frac{n\pi}{2}\right)$$

  $$Q_{16}(\theta) = -Q_{16}(-\theta), \text{ etc.}$$

  d) Midplane symmetry of a laminate:

Symmetric laminate:	$\theta(z) = \theta(-z)$
Asymmetric laminate:	$\theta(z) \neq \theta(-z)$
Antisymmetric laminate:	$\theta(z) = -\theta(-z)$

- Shear is a source of ambiguity.

Longitudinal shear:	σ_s or ϵ_s
Longitudinal shear modulus:	E_s
Longitudinal shear strength:	S
In-plane shear:	N_6

  Interlaminar or transverse shears (σ_{xz} and σ_{yz}) are not covered or discussed in this book.
- Strength ratio is defined as the allowed over the applied stress or strain, as in Equation 7.48. This should not be confused with the stress ratio used in design handbooks. Stress ratio is the reciprocal of our strength ratio. Both use R as the symbol.
- Subscripts and superscripts are omitted from symbols if their meaning is self-evident; e.g., U_1 means U_{1Q} or U_{1S} in Chapter 3; V_i means V_{1A} in Chapter 4, V_{1D} in Chapter 5.
- Inconsistency may exist in the last digit of numerical results due to round-off. Intermediate steps in calculations do not always carry the correct exponent or units. The final step, however, should be the correct answer with the correct units.

chapter 1
stiffness of unidirectional composites

The stiffness of unidirectional composites, like any other structural material, can be defined by appropriate stress-strain relations. We will show that the coefficients or material constants of these relations can be packaged in a set of engineering constants, compliance components, or modulus components. The components of any one set are directly expressible in terms of the components of the other sets. The stiffness of unidirectional composites is governed by the same stress-strain relation that is valid for conventional materials. Only the number of independent constants are four for composites and two for conventional materials.

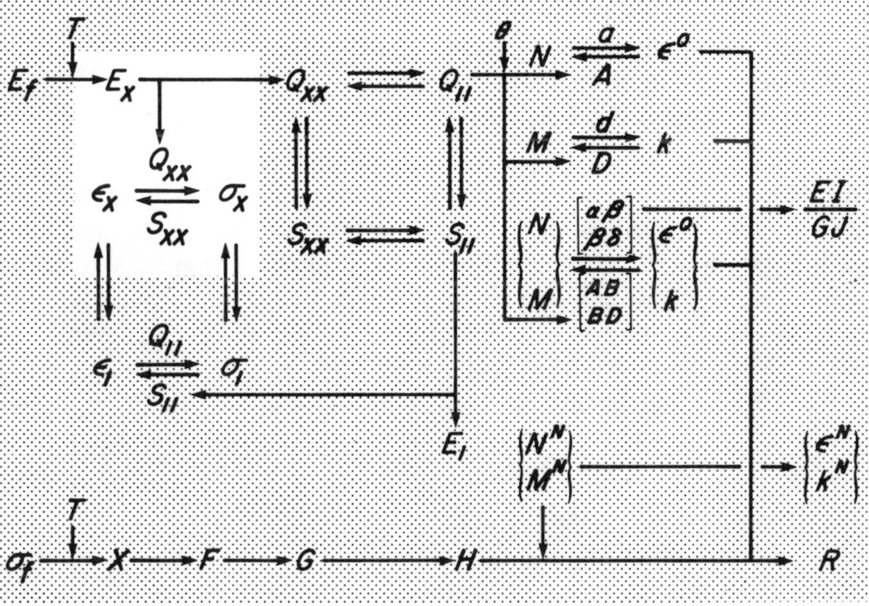

1. stress

Stress is a measure of internal forces within a body. This together with strain are the key variables for the determination of stiffness and strength of a material. The mechanisms of deformation and failure are also interpreted in terms of the state of stress and strain. They are the fundamental variables for the mechanical behavior of materials similar to temperature and heat flux for heat conduction; or pressure, volume and temperature for gas.

There is no direct measurement for stress. Instead, stress is inferred or derived from the following:

- Applied forces using stress analysis.
- Measured displacements also using stress analysis.
- Measured strains using stress-strain relations.

When we talk about stress we usually mean the average stress over some physical dimension. This is similar to population measured over a city, county or state. In our study of composites we deal with three levels of average stress:

- Micromechanical or local stress is that calculation based on distinct, continuous phases of fiber, matrix and, in some cases, the interface and voids.
- Ply stress is that calculation based on assumed homogeneity within each ply or ply group where the fiber and matrix are smeared and no longer recognized as distinct phases.
- Laminate stress resultant N or moment M is an average of ply stresses across the thickness of a laminate. The individual plies are smeared.

In Figure 1 we show two levels of this idealization of average stresses. On the micromechanical level in (*a*) the fiber and matrix stresses vary from point to point within each constituent phase. The average of these stresses is the ply stress. In a laminate or on the macromechanical level, each ply or ply group has its own ply stress. The average of several ply stresses is the laminate stress or stress resultant N.

We will use contracted notation in this book. Single subscripts for stress and strain, and double subscripts for compliance and modulus will be followed. The conversion from the conventional or tensorial notation to the contracted notation is shown in Table 1.1.

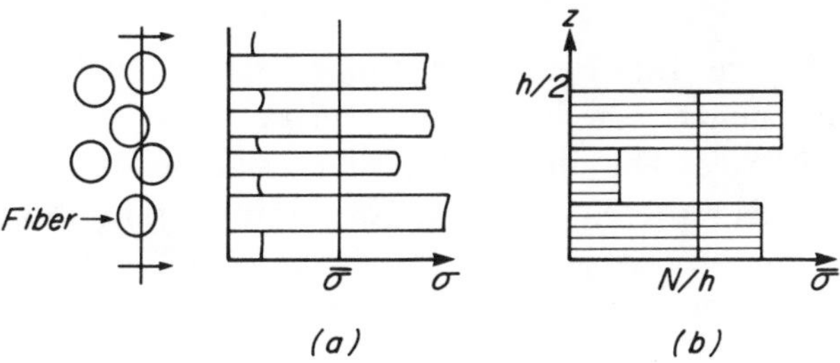

Figure 1.1 Schematic relations between local and average stresses:
(*a*) Micromechanical level where stresses in fiber and matrix are recognized. This average is the ply stress.
(*b*) Macromechanical level where stresses in plies and ply groups are recognized. This average is the laminate stress.

table 1.1
stress components in contracted notation

Conventional or tensorial notation				Contracted notation		
σ_x	σ_{xx}	σ_1	σ_{11}	σ_x	or	σ_1
σ_y	σ_{yy}	σ_2	σ_{22}	σ_y	or	σ_2
σ_{xy}	σ_{xy}	σ_{12}	σ_{12}	σ_s	or	σ_6

The single subscript system can be readily extended to the index notation to be introduced later. The subscript s or 6 is therefore used to designate the shear component in the x-y plane. The use of subscript 6 for the shear stress component is derived from the 6 components in 3-dimensional stress. Although subscript 3 has occasionally been used for this shear component, it is a source of confusion since 3 can also be used for the 3rd normal stress component in 3-dimensional problems. Subscript 6 is used to avoid this confusion.

The state of stress in a ply or ply group is predominantly plane stress. The nonzero components of plane stress are those listed in Table 1.1. The remaining three components are of secondary and local nature and will not be treated in this book. It is convenient to represent the state of plane stress in a 3-dimensional stress-space where the three orthogonal axes correspond to the three stress components. The stress-space is shown in Figure 1.2. Here each applied stress, represented by three stress components, can be readily portrayed as a vector in this 3-dimensional space. The unit vector which signifies the direction of the applied stress is represented by the conventional notation of

$$(i,j,k)$$

where the components of the unit vectors are directional cosines. All three unit vectors are shown in Figure 1.2. Typical unit vectors for simple states of stress will be shown in the following table.

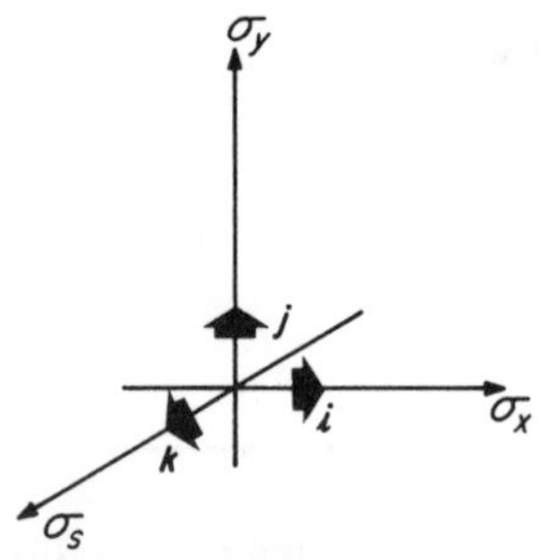

Figure 1.2 Stress components in 3-dimensional stress-space. Unit stress vectors are also shown as arrows.

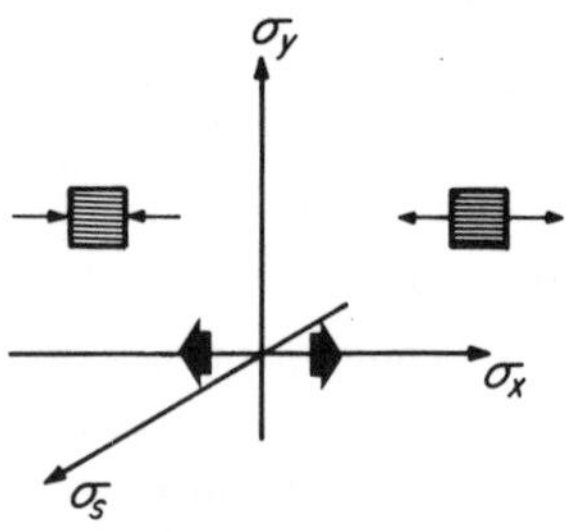

Figure 1.3 Longitudinal uniaxial stresses in tension and compression. The respective unit vectors are (±1,0,0).

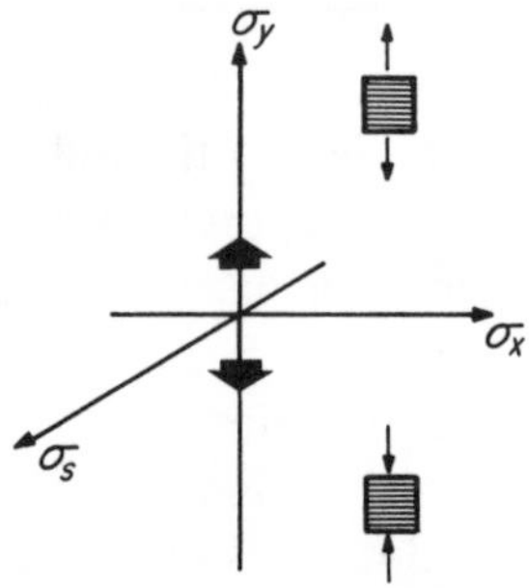

Figure 1.4 Transverse uniaxial stress in tension and compression. The respective unit vectors are (0,±1,0).

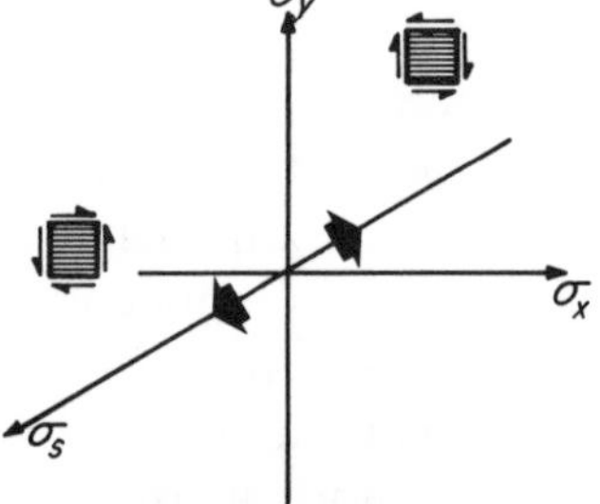

Figure 1.5 Positive and negative longitudinal shears. The respective unit vectors are (0,0,±1).

table 1.2
unit vectors for simple stress states

Type of stress	Unit vector	Figure no.
Longitudinal tension	(1,0,0)	1.3
Longitudinal compression	(−1,0,0)	1.3
Transverse tension	(0,1,0)	1.4
Transverse compression	(0,−1,0)	1.4
Positive longitudinal shear	(0,0,1)	1.5
Negative longitudinal shear	(0,0,−1)	1.5

The sign convention must be observed faithfully when we deal with composites. The difference between tensile and compressive strengths may be several hundred percent. Moreover, there can be an even greater difference between positive and negative shear strengths in composites. For conventional materials signs are often immaterial, but here this attitude can be fatal. We must be precise and accurate about signs. This is a necessary discipline when we work with composites.

In Figure 1.6 the sign convention is shown in detail. All components in (*a*) are positive; in (*b*) negative. For the normal components, signs are no problem. Shear, however, is more difficult. The rule is that a shear is positive if the shear is acting on a positive face and directed toward a positive axis; or the shear is positive if it is acting on a negative face and directed toward a negative axis. Thus, two positives or two negatives would make a positive shear. If we have a mixture of positive and negative the shear is negative.

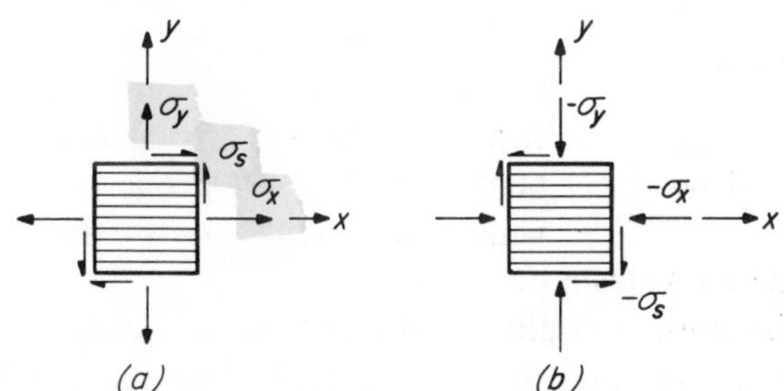

Figure 1.6 Sign convention for stress components:
(*a*) All components shown are positive.
(*b*) All components shown are negative.

2. strain

Relative displacements in a plane will induce 2-dimensional strain. If the displacements do not vary from point to point within a material, there will only be rigid body motion and no strain. Thus, strain is simply the spatial variation of the displacements. There is no material property involved. Strain is related to displacement. Both are geometric quantities.

From the definition of strain, we can establish the stress-strain relation. The constants in this relation govern the stiffness of composites. This process is the same for conventional materials.

Let Δu = Relative infinitesimal displacement along the x-axis
Δv = Relative infinitesimal displacement along the y-axis

From Figure 1.7 we can define:

$$\epsilon_x = \lim_{\Delta x \to 0} \frac{\Delta u}{\Delta x} = \frac{\partial u}{\partial x}$$

$$\epsilon_y = \lim_{\Delta y \to 0} \frac{\Delta v}{\Delta y} = \frac{\partial v}{\partial y} \tag{1.1}$$

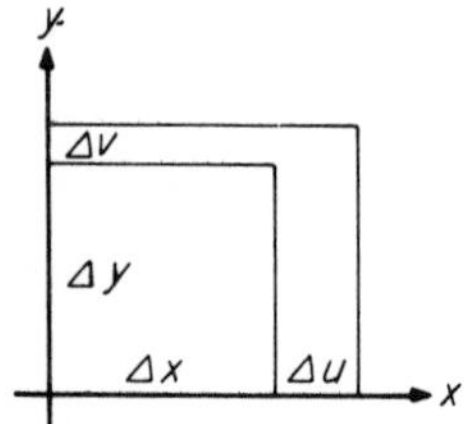

Figure 1.7 Normal strain and displacement relations.

The partial differentiation is used because the displacements are functions of both x and y coordinates. Strain, like stress, is a local property. In general it varies from point to point in a material. Only in special cases is the state of strain or stress uniform; we call this homogeneous strain or stress. This special case is pertinent to testing for property determination where we deliberately try to create a simple, homogeneous strain or stress.

Note that the normal strain components are associated with changes in the lengths of an infinitesimal element. The rectangular element before deformation remains rectangular although its length and width may change. There is no distortion produced by the normal strain components. Distortion is measured by the change of angles. The original rectangular element would be distorted into a parallelogram. Geometrically this is equivalent to stretching one diagonal and compressing the other. This combined action will produce distortion which is measured by shear strain. Figure 1.8 shows the combined action produced by the same displacements that produced the normal strain components in Equation 1.1. The desired shear strain is:

$$\epsilon_s = a + b \tag{1.2}$$

$$\text{where } a = \tan a \cong \frac{\partial v}{\partial x}, \qquad b = \tan b \cong \frac{\partial u}{\partial y} \tag{1.3}$$

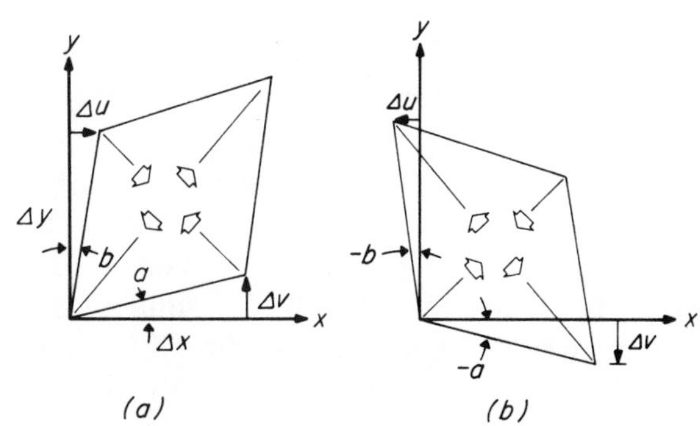

Figure 1.8 The strain-displacement relation for shear strain. The arrows show the stretching and compressing of the diagonals. This shear strain is positive in (*a*); and negative in (*b*).

The resulting strain displacement relation is

$$\epsilon_s = \frac{\partial v}{\partial x} + \frac{\partial u}{\partial y} \tag{1.4}$$

This is the engineering shear strain which is twice the tensorial strain. Engineering shear strain is used because it measures the total change in angle, or the total angle of twist in the case of a rod under torsion. This factor of 2 is often a source of confusion. When in doubt, the strain-displacement relation, in Equation 1.4, is the best place for clarification.

As with stress, contracted notation will be used for strain components. The conversion table between the components of the conventional or tensorial strain and the contracted strain is shown in Table 1.3.

table 1.3
strain components in contracted notation

Conventional or tensorial notation				Contracted notation
ϵ_x	ϵ_{xx}	ϵ_1	ϵ_{11}	ϵ_x or ϵ_1
ϵ_y	ϵ_{yy}	ϵ_2	ϵ_{22}	ϵ_y or ϵ_2
$2\epsilon_{xy}$	$2\epsilon_{xy}$	$2\epsilon_{12}$	$2\epsilon_{12}$	ϵ_s or ϵ_6

Strain vectors can also be portrayed in strain-space. Because of the coupling between the normal strain components, known as the Poisson's effect, the response to a uniaxial stress creates a biaxial strain state. For example, for conventional as well as unidirectional materials an extension is coupled with a lateral contraction if the applied uniaxial stress is tensile. In Figure 1.9 we will show the unit strain vectors as the result of uniaxial longitudinal and transverse tensile stresses in (*a*) and (*b*), respectively. If the applied stress is compressive, the direction of all the stress and strain unit vectors will be reversed.

3. stress-strain relations

We will limit the composites of this book to the linearly elastic materials. The response of materials under stress or strain follows a straight line up to failure. With assumed linearity we can use superposition which is a very powerful tool. For example, the net result of combining two states of stress is precisely the sum of the two states—no more and no less. The sequence of the stress application is immaterial. We can assemble or disect components of stress and strain in whatever pattern

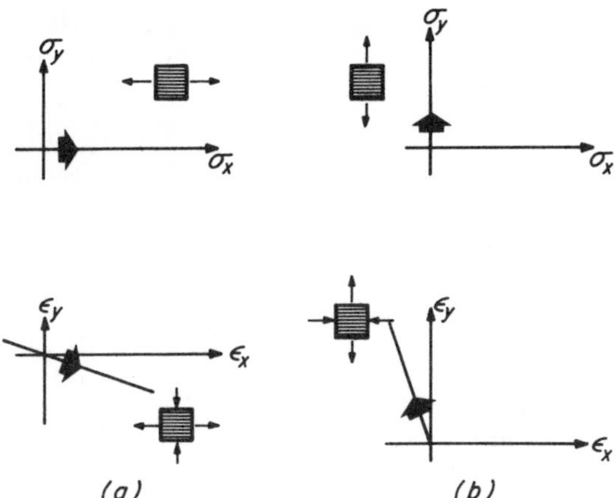

Figure 1.9 Unit strain vectors resulting from uniaxial stresses.
(*a*) Biaxial strain $(1,-\nu,0)$ resulting from uniaxial stress $(1,0,0)$.
(*b*) Biaxial strain $(-\nu,1,0)$ resulting from uniaxial stress $(0,1,0)$.

we choose without affecting the result. Combined stresses are the sum of simple, uniaxial stresses. The addition is done component by component.

Secondly, elasticity means full reversibility. We can load, unload and reload a material without incurring any permanent strain or hysteresis. Elasticity also means that the material's response is instantaneous. There is no time lag, no time or rate dependency.

Experimentally observed behavior of composites follows closer to linear elasticity than nearly all metals and nonreinforced plastics. The assumed linear elasticity for composites appears to be reasonable. If we are to go beyond the linearity assumption, such as the incorporation of nonlinear elasticity, plasticity and viscoelasticity, the increased complexity is beyond the scope of this book.

For unidirectional composites, the stress-strain relations can be derived by the superposition method. We must recognize that two orthogonal planes of symmetry exist for unidirectional composites: one plane is parallel to the fibers; and the other is transverse to the fibers. Symmetry exists when the structure of the material on one side of the plane is the mirror image of the structure on the other side. The two orthogonal planes are shown in Figure 1.10, where the x-axis is along the longitudinal direction of the fiber while the y-axis is in the

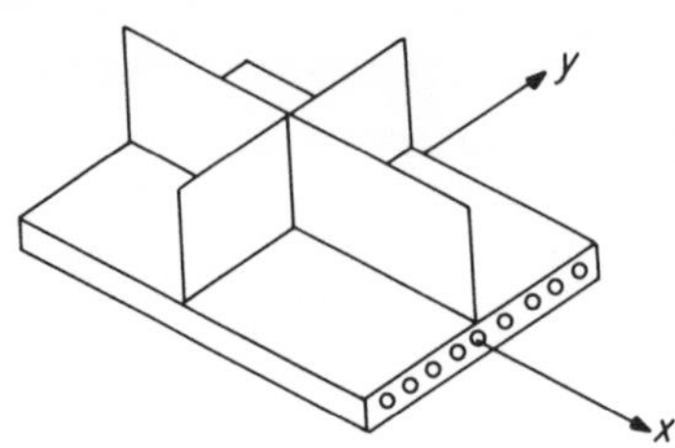

Figure 1.10 Two orthotropic planes of symmetry of unidirectional composites. Axes x-y coincide with the longitudinal and transverse directions. This material symmetry is called orthotropic and on-axis.

transverse direction. When the reference axes x-y coincide with the material symmetry axes, we call this the on-axis orientation. The stress-strain relation in this chapter is limited to this special case. The off-axis orientation will be discussed in Chapter 3.

The on-axis stress-strain relation can be derived by superpositioning the results of the following simple tests:

a. uniaxial longitudinal tests

The applied uniaxial stress and the resulting biaxial strain were shown in Figure 1.9(a). The stress-strain curves for this test are shown in Figure 1.11, from which we can establish the following stress-strain relations:

$$\epsilon_x = \frac{1}{E_x}\sigma_x$$

$$\epsilon_y = -\frac{\nu_x}{E_x}\sigma_x = -\nu_x \epsilon_x \qquad (1.5)$$

where E_x = Longitudinal Young's modulus, also designated E_L

ν_x = Longitudinal Poisson's ratio $= -\dfrac{\epsilon_y}{\epsilon_x}$

(This is also called the major Poisson's ratio, and designated by ν_{LT}, ν_{12}, or sometimes ν_{21}.)

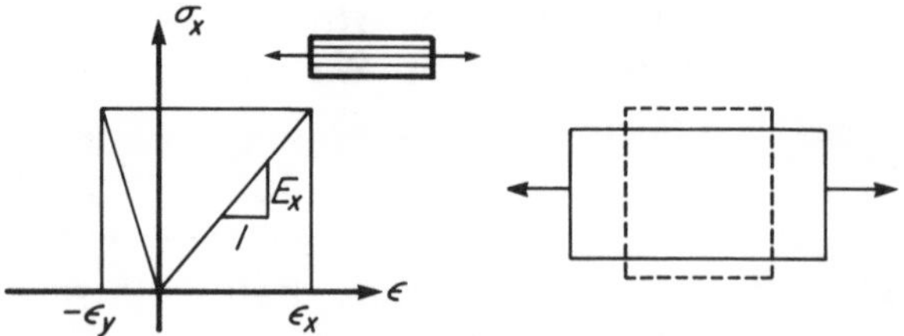

Figure 1.11 Uniaxial longitudinal tensile test. A square will be deformed into a rectangle.

b. uniaxial transverse tests

The applied uniaxial stress and the resulting biaxial strain were shown in Figure 1.9(*b*). The stress-strain curves for this test were shown in Figure 1.12, from which the following stress-strain relations can be established:

$$\epsilon_y = \frac{1}{E_y}\sigma_y$$
$$\epsilon_x = -\frac{\nu_y}{E_y}\sigma_y = -\nu_y\,\epsilon_y \qquad (1.6)$$

where E_y = Transverse Young's modulus, also designated E_T

ν_y = Transverse Poisson's ratio $= -\dfrac{\epsilon_x}{\epsilon_y}$

(This is also called the minor Poisson's ratio, and designated by ν_{TL}, or ν_{21} or sometimes ν_{12}.)

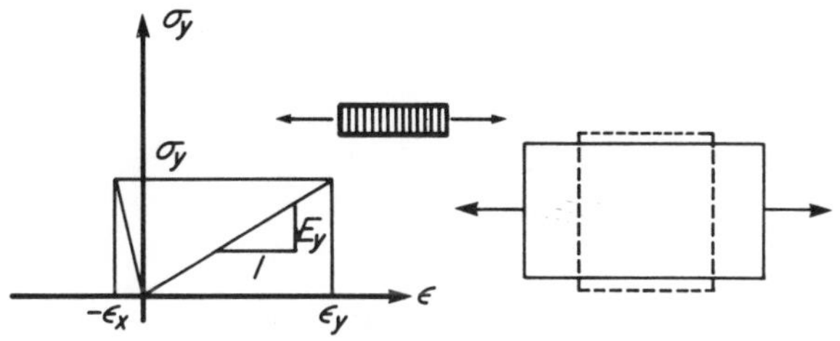

Figure 1.12 Uniaxial transverse tensile test.

c. longitudinal shear test

We apply another simple state of stress, the pure shear, to our unidirectional composite. This is shown in Figure 1.13. The resulting stress-strain relation is:

$$\epsilon_s = \frac{1}{E_s}\sigma_s \qquad (1.7)$$

where E_s = Longitudinal shear modulus
(This is also called longitudinal-transverse shear modulus and designated by G_{LT} or G_{12}.)

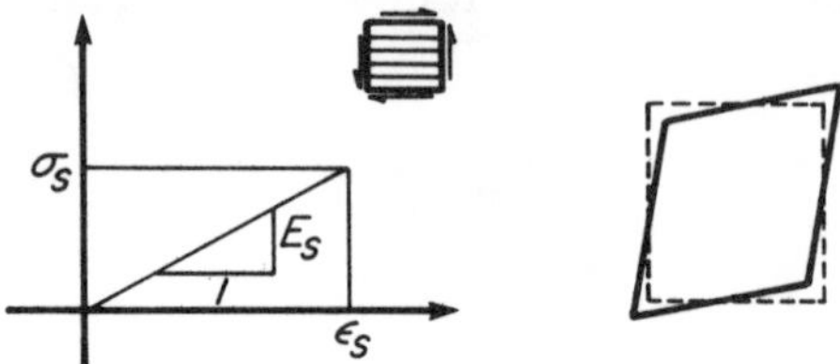

Figure 1.13 Longitudinal shear test. A square is distorted into a parallelogram.

By applying the principle of superposition, we can sum up the contribution of each stress component in Equation 1.5, 1.6 and 1.7 to the resulting strain components. The final stress-strain relation for our unidirectional composite is:

$$
\begin{aligned}
\epsilon_x &= \frac{1}{E_x}\sigma_x - \frac{\nu_y}{E_y}\sigma_y \\
\epsilon_y &= -\frac{\nu_x}{E_x}\sigma_x + \frac{1}{E_y}\sigma_y \\
\epsilon_s &= \frac{1}{E_s}\sigma_s
\end{aligned}
\tag{1.8}
$$

This is the on-axis stress-strain relation of a unidirectional composite; i.e., the material is in its orthotropic symmetry orientation. Conventional materials have the same functional relations.

These simultaneous equations can be repackaged in a matrix multiplication table, wherein each row in the table is equal to the sum of products from each column and its column heading. This rule should be self-evident if we compare the first of Equation 1.8 with the first row of Table 1.4. This and all subsequent tables will be drawn in italics when matrix multiplication is in force.

table 1.4
on-axis stress-strain relation for unidirectional composites in terms of engineering constants

	σ_x	σ_y	σ_s
ϵ_x	$\frac{1}{E_x}$	$-\frac{\nu_y}{E_y}$	
ϵ_y	$-\frac{\nu_x}{E_x}$	$\frac{1}{E_y}$	
ϵ_s			$\frac{1}{E_s}$

All the material constants of the stress-strain relation shown in this table are called engineering constants. They are the familiar constants used for conventional materials with subscripts added to denote the directionality of properties. Many design formulas for structural elements are written in terms of engineering constants. Thus the use of engineering constants will often facilitate the use of composites for structural applications. This concession to the state-of-the-art design methodology, however, can lead to an unnecessarily complicated design procedure. In fact, engineering constants for composites can be clumsy and should be replaced by the components of compliance and stiffness. A change of notation from engineering constants in Table 1.4 to components of compliance in Table 1.5 can be done by direct substitution.

table 1.5
on-axis stress-strain relation for unidirectional composites in terms of compliance

	σ_x	σ_y	σ_s
ϵ_x	S_{xx}	S_{xy}	
ϵ_y	S_{yx}	S_{yy}	
ϵ_s			S_{ss}

The relations between these two sets of elastic constants are:

$$S_{xx} = \frac{1}{E_x} \qquad S_{yy} = \frac{1}{E_y} \qquad S_{ss} = \frac{1}{E_s}$$

$$S_{yx} = -\frac{\nu_x}{E_x} \qquad S_{xy} = -\frac{\nu_y}{E_y} \tag{1.9}$$

or conversely,

$$E_x = \frac{1}{S_{xx}} \qquad E_y = \frac{1}{S_{yy}}$$

$$\nu_x = -\frac{S_{yx}}{S_{xx}} \qquad \nu_y = -\frac{S_{xy}}{S_{yy}} \qquad (1.10)$$

$$E_s = \frac{1}{S_{ss}}$$

From Equation 1.8 we can solve for stress in terms of strain for which we have the following equations:

$$\sigma_x = m E_x [\epsilon_x + \nu_y \epsilon_y]$$

$$\sigma_y = m E_y [\nu_x \epsilon_x + \epsilon_y] \qquad (1.11)$$

$$\sigma_s = E_s \epsilon_s$$

where $m = [1 - \nu_x \nu_y]^{-1}$

To eliminate the clumsiness of engineering constants in this stress-strain relation, we introduce components of stiffness in Table 1.6.

table 1.6
on-axis stress-strain relation for unidirectional composites in terms of stiffness

	ϵ_x	ϵ_y	ϵ_s
σ_x	Q_{xx}	Q_{xy}	
σ_y	Q_{yx}	Q_{yy}	
σ_s			Q_{ss}

The following relations exist between engineering constants and the components of stiffness.

$$
\begin{aligned}
Q_{xx} &= mE_x & Q_{yy} &= mE_y \\
Q_{yx} &= m\nu_x E_y & Q_{xy} &= m\nu_y E_x \\
Q_{ss} &= E_s
\end{aligned}
\tag{1.12}
$$

or conversely

$$
\begin{aligned}
E_x &= \frac{Q_{xx}}{m} & E_y &= \frac{Q_{yy}}{m} \\
\nu_x &= \frac{Q_{yx}}{Q_{yy}} & \nu_y &= \frac{Q_{xy}}{Q_{xx}} \\
E_s &= Q_{ss}
\end{aligned}
\tag{1.13}
$$

where $m = \left[1 - \frac{Q_{xy}}{Q_{xx}} \frac{Q_{yx}}{Q_{yy}}\right]^{-1}$

We have seen three sets of material constants, any of which can completely describe the stiffness of on-axis unidirectional composites. The characteristics of each set is summarized in the following:

- Stiffness is used to calculate the stress from strain. This is the basic set needed for the stiffness of multidirectional laminates.
- Compliance is used to calculate the strain from stress. This is the set needed for the calculation of engineering constants. This is not needed for the stiffness of multidirectional laminates.
- Engineering constants are the carryover from the conventional materials. Old designers feel more comfortable working with the engineering constants.

As stated earlier, from one set of constants we can readily find the other sets. They are all equivalent. There is a direct relationship between the stiffness and compliance. One is the inverse of the other. We will discuss the process of inversion later.

4. symmetry of compliance and stiffness

We wish to show that the coupling components of compliance and those of stiffness are equal; or, in the terminology of matrix algebra, that the compliance and stiffness matrices are symmetric. Since the only coupling that we have seen thus far is the Poisson coupling, the symmetry condition states that the Poisson coupling components are equal, as follows:

$$S_{xy} = S_{yx}, \qquad Q_{xy} = Q_{yx} \tag{1.14}$$

We can demonstrate the validity of these equalities from the stored elastic energy in a body subjected to stress and strain. Let the stored energy at a point in the orthotropic body be

$$W = \frac{1}{2}[\sigma_x \epsilon_x + \sigma_y \epsilon_y + \sigma_s \epsilon_s] \tag{1.15}$$

Substituting the stress-strain relation in terms of compliance from Table 1.5 into Equation 1.15,

$$W = \frac{1}{2}[S_{xx}\sigma_x{}^2 + (S_{xy} + S_{yx})\,\sigma_x\sigma_y + S_{yy}\sigma_y{}^2 + S_{ss}\sigma_s{}^2] \tag{1.16}$$

We will recover the stress-strain relation by differentiation of this energy term:

$$\begin{aligned} \epsilon_x &= \frac{\partial W}{\partial \sigma_x} = S_{xx}\,\sigma_x + \frac{1}{2}[S_{xy} + S_{yx}]\,\sigma_y \\ \epsilon_y &= \frac{\partial W}{\partial \sigma_y} = \frac{1}{2}[S_{xy} + S_{yx}]\,\sigma_x + S_{yy}\,\sigma_y \end{aligned} \tag{1.17}$$

Matching the like constants between this set and those in Table 1.5, the only condition that satisfies both sets is

$$S_{xy} = S_{yx} \tag{1.18}$$

By substituting the stiffness relations in Table 1.6 into Equation 1.15 we can also show that

$$Q_{xy} = Q_{yx} \tag{1.19}$$

The last two equations state the symmetry or reciprocal conditions of the Poisson coupling. A similar symmetry condition can be applied to engineering constants. From Equation 1.19, for example, we have

$$\nu_x E_y = \nu_y E_x \tag{1.20}$$

or

$$\frac{\nu_x}{\nu_y} = \frac{E_x}{E_y} \tag{1.21}$$

With these symmetry conditions, the number of independent constants for the on-axis, orthotropic unidirectional composite are reduced by one, from five to four in Tables 1.4 to 1.6. If additional symmetry conditions exist, the number of constants can be further reduced. Specifically, two such cases exist:

- Square Symmetric Materials

  If the longitudinal and transverse properties are equal, i.e.,

$$\begin{aligned} Q_{xx} &= Q_{yy} \\ S_{xx} &= S_{yy} \\ E_x &= E_y \end{aligned} \tag{1.22}$$

  we have a square symmetric material. But because of the additional relation in Equation 1.22, the number of independent constants are three, one less than the orthotropic material. A cross-ply laminate is a square symmetric material in the plane of the laminate. Many woven fabrics are also square symmetric.

- Isotropic Materials

  We know that isotropic materials have only two independent constants because there is another relation among the three remaining constants, i.e.,

$$Q_{ss} = \frac{Q_{xx} - Q_{xy}}{2}$$

$$S_{ss} = 2\,[S_{xx} - S_{xy}] \qquad (1.23)$$

$$G = \frac{E}{2\,(1+\nu)}$$

This relation is derived from the equivalence between the state of pure shear and that of equal tension-compression. This equivalence is only valid for isotropic materials. The derivation of this relationship will be discussed later.

In summary, the stress-strain relations which govern the stiffness of all materials have the identical form for unidirectional composites as for conventional materials. There is no additional terms or more complex relationship. The only difference is the number of independent constants; four for composites versus two for conventional materials. But there are no conceptual and operational barriers that would make composites intrinsically difficult to work with. In fact, once we understand composites, we automatically will understand conventional materials as special cases of composites.

5. stiffness data for typical unidirectional composites

a. Measured engineering constants for a number of unidirectional composites are listed in Table 1.7. The fiber volume fraction and specific gravity are also included. These constants are normally derived directly from simple tests. They are not coefficients of the stress-strain relations. The unit ply thickness is 125×10^{-6} meter.

b. The compliance components for the same composites in Table 1.7 are listed in Table 1.8. These components are computed from Table 1.7 using the formulas in Equation 1.9. The compliance components are the coefficients of the stress components in the stress-strain relation. We need this relation to go from stress to strain.

c. The stiffness components for the same composites are listed in Table 1.9. These components are calculated using the formulas in Equation 1.12. These components are needed to go from strain to stress. They are also needed to calculate the stiffness of laminated composites.

table 1.7
engineering constants, fiber volume and specific gravity of typical unidirectional composites

Type	Material	E_x GPa	E_y GPa	ν_x	E_s GPa	v_f	Specific gravity
T300/5208	Graphite /Epoxy	181	10.3	0.28	7.17	0.70	1.6
B (4)/5505	Boron /Epoxy	204	18.5	0.23	5.59	0.5	2.0
AS/3501	Graphite /Epoxy	138	8.96	0.30	7.1	0.66	1.6
Scotchply 1002	Glass /Epoxy	38.6	8.27	0.26	4.14	0.45	1.8
Kevlar 49 /Epoxy	Aramid /Epoxy	76	5.5	0.34	2.3	0.60	1.46

table 1.8
compliance components of typical unidirectional composites $(TPa)^{-1}$

Type	S_{xx}	S_{yy}	S_{xy}	S_{ss}
T300/5208	5.525	97.09	−1.547	139.5
B (4)/5505	4.902	54.05	−1.128	172.7
AS/3501	7.246	111.6	−2.174	140.8
Scotchply 1002	25.91	120.9	−6.744	241.5
Kevlar 49/Epoxy	13.16	181.8	−4.474	434.8

table 1.9
stiffness components of typical unidirectional composites (GPa)

Type	m	Q_{xx}	Q_{yy}	Q_{xy}	Q_{ss}
T300/5208	1.0045	181.8	10.34	2.897	7.17
B (4)/5505	1.0048	205.0	18.58	4.275	5.79
AS/3501	1.0059	138.8	9.013	2.704	7.1
Scotchply 1002	1.0147	39.16	8.392	2.182	4.14
Kevlar 49/Epoxy	1.0084	76.64	5.546	1.886	2.3

6. sample problems

a. find strain from stress

$$\text{Given stress vector: } (400, 60, 15) \quad \text{MPa} \tag{1.24}$$

For compliance of T300/5208 from Table 1.8:

$$\begin{aligned} S_{xx} &= 5.525 \quad (\text{TPa})^{-1} \\ S_{yy} &= 97.09 \quad (\text{TPa})^{-1} \\ S_{xy} &= -1.547 \quad (\text{TPa})^{-1} \\ S_{ss} &= 139.5 \quad (\text{TPa})^{-1} \end{aligned} \tag{1.25}$$

Using stress-strain relation in terms of compliance, such as that in Table 1.5:

$$\begin{aligned} \epsilon_x &= (5.525 \times 400 - 1.547 \times 60) \times 10^{-6} \\ &= 2.117 \times 10^{-3} \\ \epsilon_y &= 5.206 \times 10^{-3} \\ \epsilon_s &= 139.5 \times 15 \times 10^{-6} = 2.092 \times 10^{-3} \end{aligned} \tag{1.26}$$

If a different material is used, we only need to replace the compliance components in Equation 1.25 with different data. If the new material is Scotchply 1002, we can get the compliance from Table 1.8.

$$
\begin{aligned}
S_{xx} &= 25.91 \quad (\text{TPa})^{-1} \\
S_{yy} &= 120.9 \quad (\text{TPa})^{-1} \\
S_{xy} &= -6.744 \quad (\text{TPa})^{-1} \\
S_{ss} &= 241.5 \quad (\text{TPa})^{-1}
\end{aligned}
\tag{1.27}
$$

With the same applied stress as Equation 1.24, the resulting strain is:

$$
\begin{aligned}
\epsilon_x &= (25.91 \times 400 - 6.744 \times 60) \times 10^{-6} = 9.959 \times 10^{-3} \\
\epsilon_y &= 4.556 \times 10^{-3} \\
\epsilon_s &= 241.5 \times 15 \times 10^{-6} = 3.623 \times 10^{-3}
\end{aligned}
\tag{1.28}
$$

Since the glass composite is less stiff than the graphite composite, the strain produced by the same applied stress is expected to be larger in the glass composite. If we compare the strain components by components between Equation 1.26 and 1.28, the strain in the glass composite is larger in two components, and smaller in one. The moral of the story is that biaxial stress and strain states are complex. Disciplined, analytic approach is straightforward and is definitely preferred over guesswork. Guessing is not reliable because it is difficult to guess the result of a matrix multiplication.

b. find stress from strain

This process is the inverse of the previous example. If we are given the strain in Equation 1.26 and apply it to a T300/5208 composite, the resulting stress must be calculated by using

- Stress-strain relation in terms of stiffness, such as that in Table 1.6, and
- Stiffness components in Table 1.9,

$$
\begin{aligned}
Q_{xx} &= 181.8 \quad \text{GPa} \\
Q_{yy} &= 10.34 \quad \text{GPa} \\
Q_{xy} &= 2.897 \quad \text{GPa} \\
Q_{ss} &= 7.17 \quad \text{GPa}
\end{aligned}
\tag{1.29}
$$

The resulting stress is:

$$
\begin{aligned}
\sigma_x &= 181.8 \times 2.117 + 2.897 \times 5.206 = 400 \text{ MPa} \\
\sigma_y &= 2.897 \times 2.117 + 10.34 \times 5.206 = 60 \text{ MPa} \\
\sigma_s &= 7.17 \times 2.092 = 15 \text{ MPa}
\end{aligned}
\tag{1.30}
$$

Note that the original stress of Equation 1.24 has been recovered. If our composite is Scotchply 1002, we should use the stiffness components listed in Table 1.9 for this material:

$$
\begin{aligned}
Q_{xx} &= 39.16 \quad \text{GPa} \\
Q_{yy} &= 8.392 \quad \text{GPa} \\
Q_{xy} &= 2.182 \quad \text{GPa} \\
Q_{ss} &= 4.14 \quad \text{GPa}
\end{aligned}
\tag{1.31}
$$

The resulting stress from the applied strain in Equation 1.28 is:

$$
\begin{aligned}
\sigma_x &= 39.16 \times 9.959 + 2.182 \times 4.556 = 400 \text{ MPa} \\
\sigma_y &= 2.182 \times 9.959 + 8.392 \times 4.556 = 60 \text{ MPa} \\
\sigma_s &= 4.14 \times 3.623 = 15 \text{ MPa}
\end{aligned}
\tag{1.32}
$$

Note again that the original stress of Equation 1.24 has been recovered.

7. conclusions

We have shown that the stiffness of unidirectional composites relative to the material symmetry axes (axes parallel and transverse to the fibers) are dictated by four elastic constants. These constants are the coefficients of the various forms of the stress-strain relations.

When the strains are the independent variables the stress-strain relations in terms of the components of the stiffness shall be used. When the stresses are the independent variables the stress-strain relations in terms of the components of the compliance shall be used. There is a one-to-one relation that exists between the stiffness and the compliance. We can calculate the components of stiffness from those of the compliance; or we can just as easily compute the components of compliance from those of the modulus.

There is another set of elastic constants which we call the engineering constants. These constants are derived from measurements of simple tests. These constants are more familiar to the users of composites because these constants possess exact counterparts in isotropic materials. Again, of the engineering constants only four are independent. The symmetry condition that exists for the stiffness and compliance components is not applicable to the engineering constants. Again, a one-to-one relation or complete interchangeability exists among the engineering constants and the components of stiffness and compliance.

The important issue is the functional form of the stress-strain relation for unidirectional composites. The form is exactly the same as that for the conventional isotropic materials. It is for this reason we believe that composite materials are conceptually as simple as conventional materials.

For composite materials whose stiffness properties are not listed in this chapter, four independent constants must be obtained either by direct measurements or from appropriate sources. The process of determining the stresses and strains remain. If a woven fabric is used instead of the unidirectional composites comparable constants must be obtained. The fabric is treated as a homogeneous material in the same fashion as the unidirectional composites are treated. In general, four constants are needed. If the fabric has a square weave; i.e., the properties along two orthogonal directions are identical, we will have a square symmetric material for which there are only three independent constants.

Finally, the directionally dependent material property is a unique feature of composite materials. All four independent material constants must be known. The stiffness of conventional materials, on the other hand, can be represented by the Young's modulus alone because the Poisson's ratio for isotropic materials is approximately 0.3. Furthermore, Poisson's ratios often appear with unity, such as m in Equation 1.11 and G in Equation 1.23; small variations in Poisson's ratios often have insignificant effect. For composite materials, Poisson's ratios are not bounded and can have very significant effect. Young's modulus alone is not sufficient to describe the stiffness of composite materials.

8. homework problems

a. Find the components of modulus and compliance for an aluminum with the following independent engineering constants:

$$E = 69 \text{ GPa}; \nu = 0.3. \tag{1.33}$$

What is the effect on the components if the Poisson's ratio is changed to 0.25?

b. Find the components of modulus and compliance of a square symmetric material with the following independent engineering constants:

$$E_x = E_y = 96 \text{ GPa}; \nu_x = \nu_y = 0.03; E_s = 7.17 \text{ GPa}. \tag{1.34}$$

c. Find the resulting strain in the aluminum above from an imposed stress vector (400,60,15) MPa in Equation 1.24. Apply the calculated strain to the aluminum and see if the original stress vector is recovered.

d. Repeat Problem (*c*) for the square symmetric material described in Problem (*b*).

e. Use the symmetry condition of Equation 1.20 and interchange the two non-zero off-diagonal terms in Table 1.4. Show the condition for apparent infinite stiffness in the x-direction under biaxial stress is:

$$\nu_x = \sigma_x / \sigma_y \tag{1.35}$$

Show the condition for infinite areal stiffness under plane hydrostatic pressure p is:

$$\nu_x = \nu_y = 1; \sigma_x = \sigma_y = -p. \tag{1.36}$$

f. Find direct expressions for the compliance components in terms of the modulus components for an orthotropic material. Write down the modulus components in terms of the compliance components.

g. Given two pieces of a unidirectional composite material joined in a manner shown in Figure 1.14 determine the deformed shape under uniaxial stress. Is the deformed shape *a*, *b* or *c*? What principle is involved?

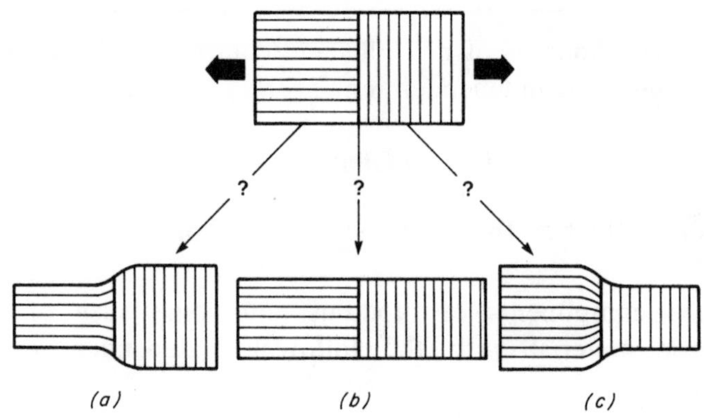

Figure 1.14 Possible deformed shapes of a 0/90 composite.

h. The maximum stress criterion of a unidirectional composite states that failure occurs when one of the equalities is met. The data are those for T300/5208, with the assumption that tensile and compressive strengths are equal.

$$\begin{array}{lll} \sigma_x \leqslant X & \sigma_y \leqslant Y & \sigma_s \leqslant S \\ \leqslant 1500 \text{ MPa} & \leqslant 40 \text{ MPa} & \leqslant 68 \text{ MPa} \end{array} \tag{1.37}$$

This criterion appears as a rectangle in stress space in Figure 1.15(*a*). It is not drawn to scale. Similarly, the maximum strain criterion states

$$\begin{array}{lll} \epsilon_x \leqslant \dfrac{X}{E_x} & \epsilon_y \leqslant \dfrac{Y}{E_y} & \epsilon_s \leqslant \dfrac{S}{E_s} \\ \leqslant 8.28 \times 10^{-3} & \leqslant 3.85 \times 10^{-3} & \leqslant 9.42 \times 10^{-3} \end{array} \tag{1.38}$$

This is shown in strain space in Figure 1.15(*b*). Use the linear orthotropic stress-strain relations to draw to scale: (1) the maximum strain criterion in stress space, and (2) the maximum stress criterion in strain space.

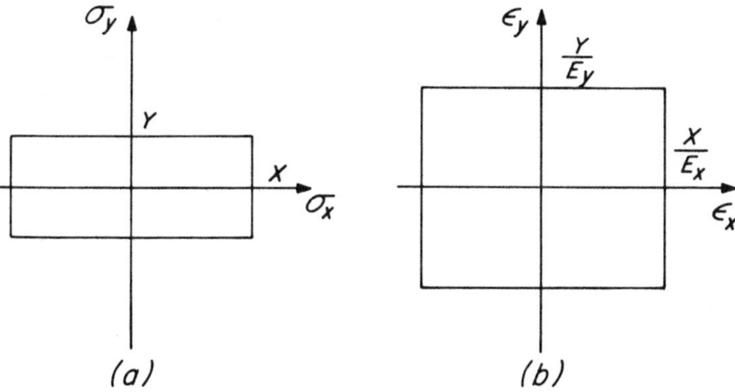

Figure 1.15 Maximum stress and maximum strain failure criteria.

i. The longitudinal shear stress/shear strain curve for most unidirectional composites is nonlinear. A typical curve (not to scale) is shown in Figure 1.16. The maximum stress S and strain (ϵ_s') are indicated. The linear approximation of the maximum strain can be based on the tangent modulus in Equation (1.38). Show to scale for T300/5208 the maximum strain criterion in stress space similar to Figure 1.17(*a*), and the maximum stress criterion in strain space similar to Figure 1.17(*b*) for both the tangent and the maximum shear strains, assuming the maximum shear is three times the tangent shear strain.

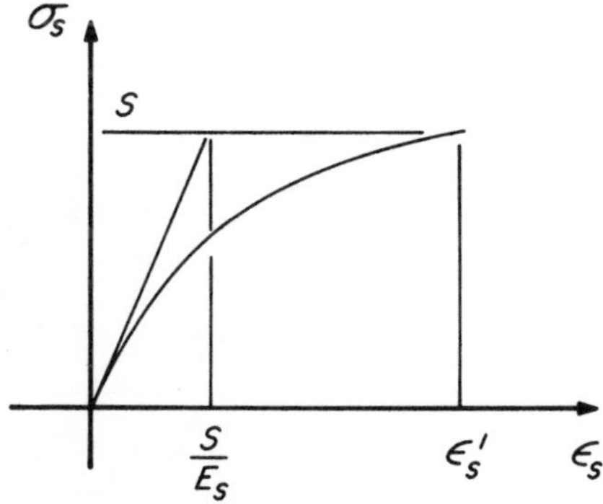

Figure 1.16 Longitudinal shear curve.

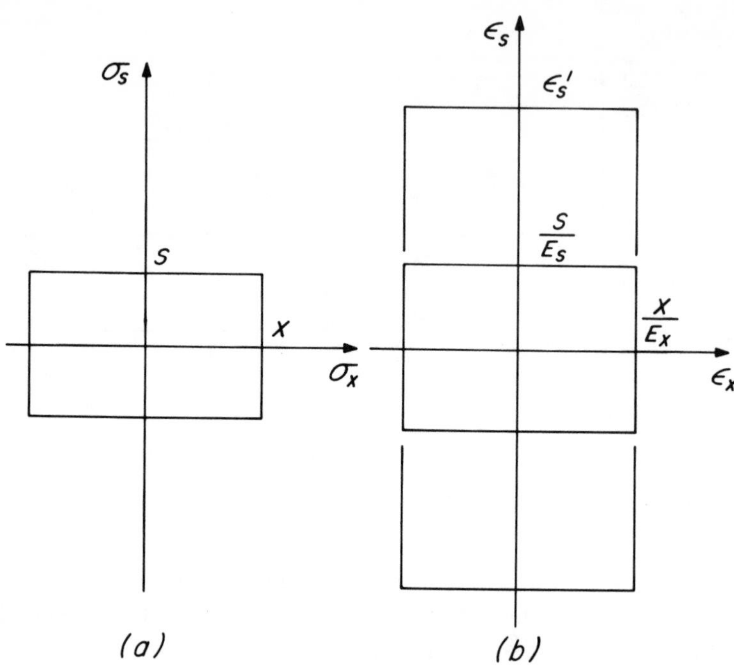

Figure 1.17 Maximum stress and maximum strain failure criterion in the $\sigma_x - \sigma_s$ plane.

nomenclature

E	= Young's modulus for isotropic materials
E_x	= Longitudinal Young's modulus of unidirectional components
E_y	= Transverse Young's modulus of unidirectional components
G	= Shear modulus for isotropic materials
E_s	= Longitudinal shear modulus of unidirectional components
m	= Dimensionless multiplying constant = $[1 - \nu_x \nu_y]^{-1}$
Q_{ij}	= Stiffness components; $i,j = x,y,s$
S_{ij}	= Compliance components; $i,j = x,y,s$
u	= Displacement along the x-axis
v	= Displacement along the y-axis
W	= Stored elastic energy
ϵ_i	= Strain components; $i,j = x,y,s$
σ_i	= Stress components; $i,j = x,y,s$
ν	= Poisson's ratio for isotropic materials
ν_x	= Longitudinal Poisson's ratio
ν_y	= Transverse Poisson's ratio
Sub x	= Normal component along the x-axis
Sub y	= Normal component along the y-axis
Sub s	= Shear component in the x-y plane

chapter 2
transformation of stress and strain

The change of stiffness of unidirectional composites as a function of ply orientation is a unique feature of composites. This change can be related to the orientational variations of stress and strain. We will derive the relations that govern these variations; namely, the transformation equations. There are three formulations for the transformation; viz., the conventional power functions, the double angle functions, and the invariant functions. Each formulation has its unique characteristic and is useful for special purposes. All three formulations are equivalent and will yield the same answer.

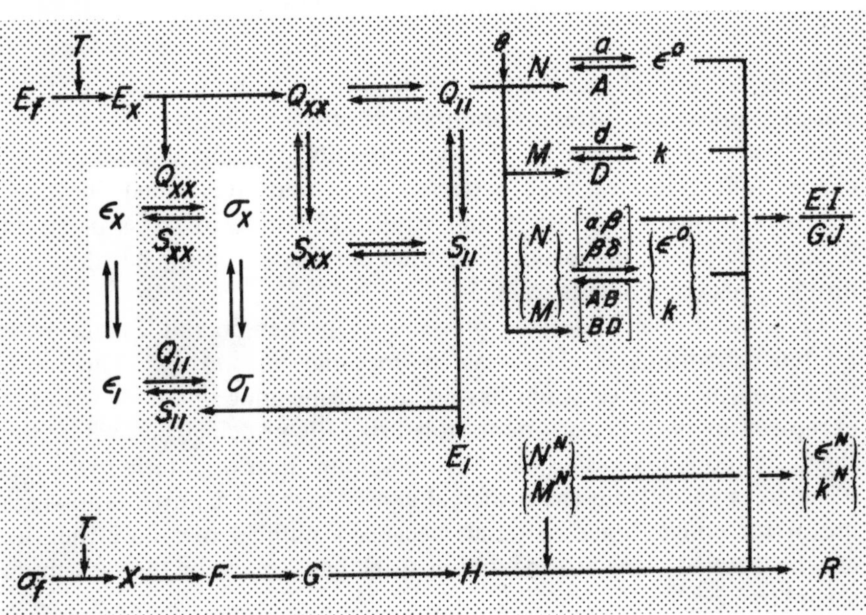

1. background

Up to this point, we have dealt only with the stiffness of unidirectional composites in their material symmetry axes, as shown in Figure 2.1. In this reference coordinate system, we call this type of symmetry orthotropic; see Figure 1.10 for graphic illustration. General orthotropic configuration occurs when the ply orientation is different from 0 or 90 degrees. This is shown in Figure 2.2. We also call the latter configuration the off-axis as distinguished from the on-axis in Figure 2.1.

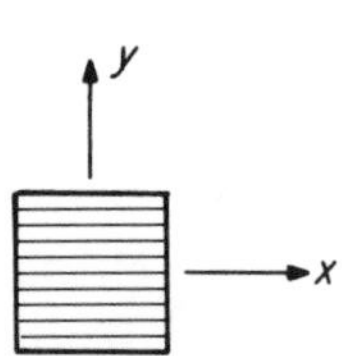

Figure 2.1 Material symmetry axes of a unidirectional composite. The x-axis is along the fiber and is in the longitudinal direction. This on-axis configuration is called orthotropic.

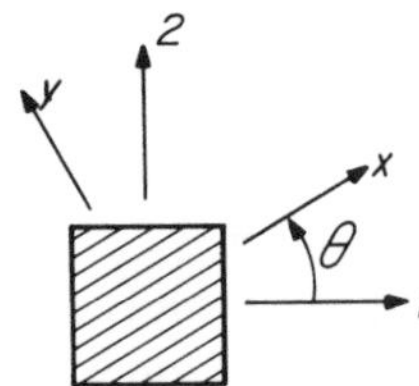

Figure 2.2 Off-axis or generally orthotropic configuration of a unidirectional composite. Counterclockwise rotation of the ply-orientation is positive; clockwise rotation, negative.

There are several reasons that we need to know how stress and strain can be expressed in different orientations of the coordinate axes. As the angle varies in Figure 2.2, the components of stress and strain will change following prescribed patterns. This variation is called the transformation equation of stress and strain. The state of stress or strain remains the same, independent of the coordinate system, but the magnitude of its components change.

In conventional materials, physical properties do not change with reference coordinates. This class of materials is isotropic. The transformation of stress or strain has no special meaning or utility with the exception of two special orientations; viz., the principal axes where the shear component vanishes, and the maximum shear orientation which is 45 degrees away from the principal axes. These special orientations are useful for conventional materials because a number of failure theories can be applied using the maximum principal stresses or the maximum shear stresses.

In composite materials, we need to know the transformation equations of stress and strain for a number of reasons.

First, the properties of composites are not isotropic. The state of stress or strain existing relative to the on-axis configuration is important in determining the stiffness and strength of composite materials. We can then use the transformation equations, for example, to find the on-axis stress from the applied stress in an off-axis orientation or vice versa.

Secondly, transformation equations are needed to determine the principal stresses or strains. The same transformation equations also define the invariants of stress and strain. The concepts of principal axes and invariants are fundamental for the understanding of composite materials. The same concepts can be extended to those for stiffness and strength. They will be explained later.

Finally, transformation equations for stress and strain, together with the on-axis stress-strain relations of Chapter 1, can be used to determine the off-axis compliance and modulus of unidirectional composites. The sequence of operations is illustrated in Figure 2.3 for the compliance and described as follows:

- The originally applied stress to an off-axis composite is shown in (*a*), expressed in components 1,2,6.
- If we apply stress transformation to the components of (*a*) in the 1-2 system, we will get (*b*), the same state of stress but expressed in the different components of the on-axis, *x-y* system; i.e., in components *x,y,s*. This is a positive stress transformation.
- Since we know the relations between the on-axis stress and strain from Table 1.5, we can determine the induced strain in the on-axis, *x-y* system, which is shown in (*c*).
- We can then apply inverse strain transformation to get the strain components in the off-axis, 1-2 system from the on-axis, *x-y* system; i.e., from (*c*) to (*d*) in Figure 2.3. This is a negative strain transformation. Then we have the induced strain in (*d*) as the result of the applied stress in (*a*), both of which are in the off-axis, 1-2 system.

We can go from (*a*) to (*d*) directly if we know the off-axis stress-strain relation. This can be derived by merging the three steps in this figure into one.

If the imposed strain is given in Figure 2.3(*a*) instead of the stress, the induced off-axis stress can be determined by a very analogous method. The off-axis modulus of a unidirectional composite can then

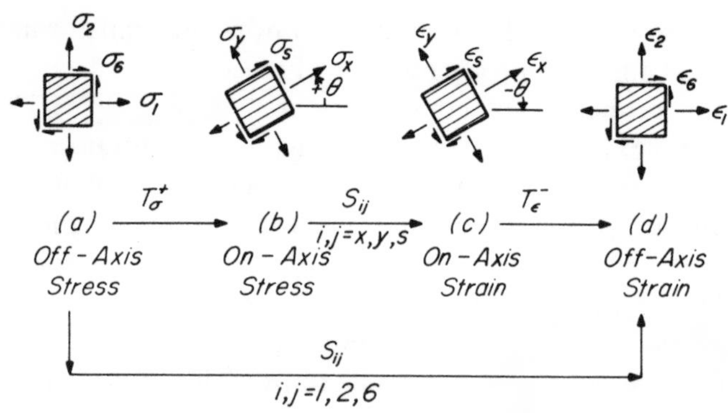

Figure 2.3 Determination of the off-axis compliance:
From (*a*) to (*b*): use positive stress transformation.
From (*b*) to (*c*): use the on-axis stress-strain relation in compliance.
From (*c*) to (*d*): use negative strain transformation.

be derived from this process. The sequence of operations is illustrated in Figure 2.4.

- From off-axis strain to on-axis strain, use positive strain transformation. This is the operation from (*a*) to (*b*).
- From on-axis strain to on-axis stress, use the on-axis stress-strain relation in modulus, as in Table 1.6. This is the operation from (*b*) to (*c*).
- From on-axis stress to off-axis stress, use negative stress transformation. This operation is from (*c*) to (*d*).

Alternatively, we can go from (*a*) to (*d*) directly if we know the off-axis modulus.

The scope of this chapter is to show stress and strain transformation. The formulas of the transformation are simple and easy to use, but the most critical part of the operation is the sign convention. As we have repeatedly mentioned, signs are critical for the study of composites. Such emphasis is not called for in the case of conventional materials because their behavior is often insensitive to signs and directions.

The notations associated with coordinate transformation are arbitrary. The components of the original versus the transformed, the old versus the new, the 1-2 versus the *x*-*y* systems, or the on-axis versus the off-axis are based on a matter of judgment, and certainly vary from author to author and from situation to situation. Only the definitions

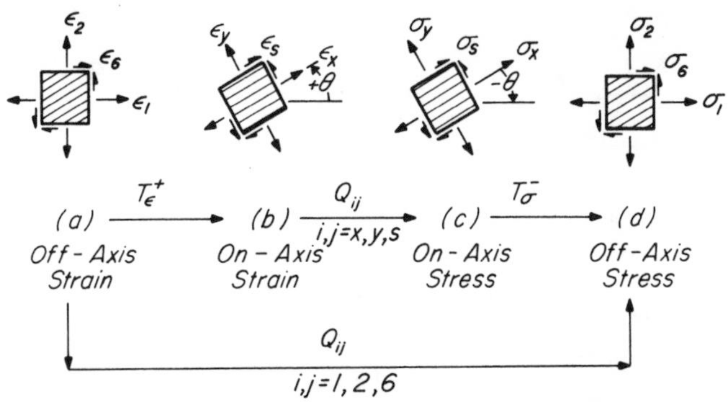

Figure 2.4 Determination of the off-axis modulus:
From (*a*) to (*b*): use positive strain transformation.
From (*b*) to (*c*): use the on-axis stress-strain relations in modulus.
From (*c*) to (*d*): use negative or inverse stress transformation.

of the on-axis and the off-axis are normally fixed. The key issue is the initial definition such as that shown in Figure 2.2, where the reference coordinates and the ply orientation are illustrated. This choice is made for convenience because most transformations for the stress and strain in composite materials go from the off-axis to the on-axis orientation. But these are exceptions, such as the negative or inverse transformation between step (*c*) and step (*d*) in Figure 2.3 and 2.4.

2. transformation of stress

Now we would like to derive the relations between two sets of stress components; one set expressed in the 1-2 system, and the other in the *x-y* system. The latter is rotated from the former by a positive angle as shown in Figure 2.2 and repeated in Figure 2.5(*a*). In Figure 2.5(*b*) and (*c*), the two sets of stress components, one with numerical subscripts, the other with letter subscripts, are also shown. The arrows indicate the direction of the positive component, following the sign convention in Figure 1.6.

The transformation of stress can be derived from the balance of forces. Consider a free-body diagram shown in Figure 2.6(*a*) which is a wedge slicing across fibers in a typical infinitesimal unit area like that in Figure 2.5(*b*). The sides of this wedge have the following lengths

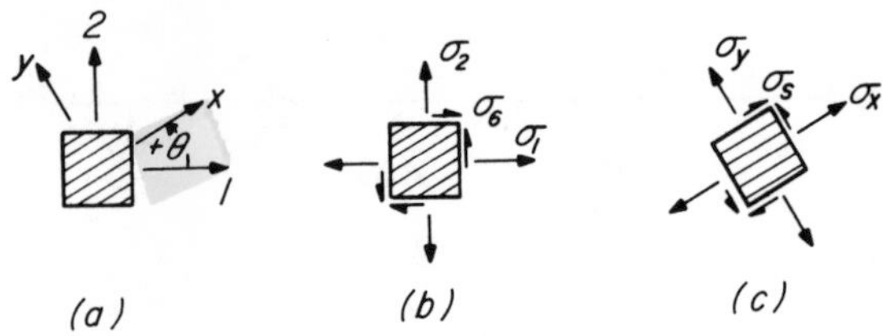

Figure 2.5 Stress transformation: changes in stress components due to coordinate rotation or transformation.

(*a*) Relation between the 1-2 and *x*-*y* systems. Counterclockwise rotation is positive.
(*b*) The off-axis or old stress components, with numerical subscripts.
(*c*) The on-axis or new stress components, with letter subscripts.

All arrows for the components are pointing in a positive direction.

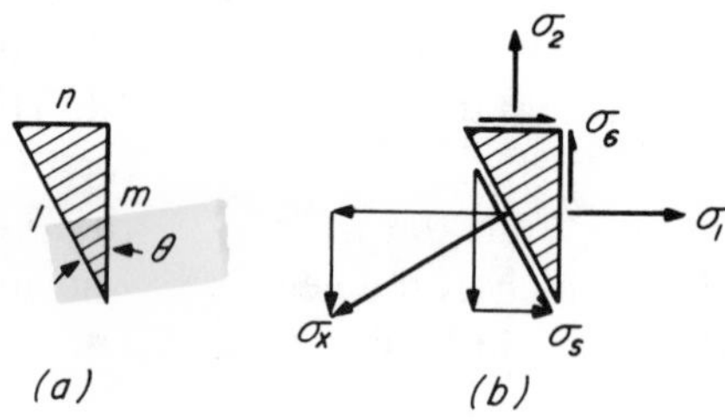

Figure 2.6 Free-body diagram for the balance of stress components. The components of the on-axis, *x*-*y* coordinates can be expressed in terms of those of the off-axis, 1-2 coordinates. All stress components shown are derived from Figure 2.5(*b*) and (*c*).

relative to unity hypotenuse; also shown in Figure 2.6(*a*):

$$m = \cos\theta, \qquad n = \sin\theta \tag{2.1}$$

The forces exerted on the sides of this triangular free body, shown only schematically in Figure 2.6(*b*), are the products of the stress components multiplied by the appropriate lengths of the sides.

- Balance of horizontal (along the 1-axis) forces yields:

$$m\sigma_x - n\sigma_s = m\sigma_1 + n\sigma_6 \tag{2.2}$$

- Balance of vertical (along the 2-axis) forces yields:

$$n\sigma_x + m\sigma_s = n\sigma_2 + m\sigma_6 \tag{2.3}$$

Keeping in mind that the original 1-2 components are given, and we are looking for the new *x-y* components, we can find the unknown components by solving simultaneously Equations 2.2 and 2.3 as follows:

$$\sigma_x = m^2\sigma_1 + n^2\sigma_2 + 2mn\sigma_6 \tag{2.4}$$

$$\sigma_s = -mn\sigma_1 + mn\sigma_2 + [m^2 - n^2]\sigma_6 \tag{2.5}$$

It is assumed here that the off-axis stress components are normally given and the on-axis components are desired. This is usually the case when we study composite materials. It is important to know that Equations 2.4 and 2.5 are applicable independent of material properties. The description of the on-axis and off-axis is made for sake of convenience and is not intended to restrict the transformation equations to a specific material.

We can now repeat the process by slicing a wedge parallel to fibers in the unit area as in Figure 2.7. On this plane the normal stress component is acting transversely to the fibers. The free body diagram for this wedge is shown in Figure 2.7(*b*), from which the following relations can be established:

- Balance of horizontal forces yields:

$$n\sigma_y - m\sigma_s = n\sigma_1 - n\sigma_6 \tag{2.6}$$

- Balance of vertical forces yields:

$$m\sigma_y + n\sigma_s = m\sigma_2 - n\sigma_6 \tag{2.7}$$

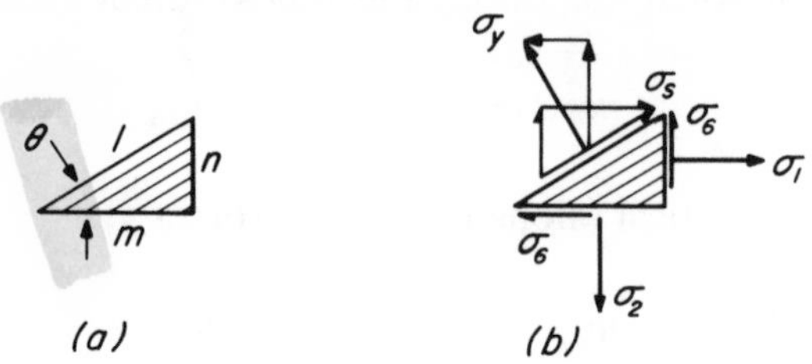

Figure 2.7 Free-body diagram for the balance of stress components. This is the same as Figure 2.6 except the new plane is sliced along the fibers. Positive components are derived from Figure 2.5(*b*) and (*c*).

If we solve the last two equations simultaneously, we get

$$\sigma_y = n^2\sigma_1 + m^2\sigma_2 - 2mn\sigma_6 \tag{2.8}$$

$$\sigma_s = -mn\sigma_1 + mn\sigma_2 + [m^2 - n^2]\sigma_6 \tag{2.9}$$

Note that the shear stress expressed in Equation 2.9 is the same as that in Equation 2.5 as it should be. Thus, the three equations for stress transformation are Equations 2.4, 2.8 and 2.9. These equations can be packaged in a matrix multiplication table as follows:

table 2.1
stress transformation equations in power functions

	σ_1	σ_2	σ_6
σ_x	m^2	n^2	$2mn$
σ_y	n^2	m^2	$-2mn$
σ_s	$-mn$	mn	m^2-n^2

$m = \cos\theta$, $n = \sin\theta$

The transformation equations above are expressed in terms of second power of sines and cosines. We can rewrite these equations using double angle trigonometric identities as follows:

$$m^2 = \cos^2\theta = \frac{1}{2}+\frac{1}{2}\cos 2\theta$$

$$n^2 = \sin^2\theta = \frac{1}{2}-\frac{1}{2}\cos 2\theta \qquad (2.10)$$

$$2mn = \sin 2\theta$$

$$m^2 - n^2 = \cos 2\theta$$

When we substitute these identities into the equations in Table 2.1, we get

$$\sigma_x = \frac{1}{2}[\sigma_1 + \sigma_2] + \frac{1}{2}[\sigma_1 - \sigma_2]\cos 2\theta + \sigma_6 \sin 2\theta$$

$$\sigma_y = \frac{1}{2}[\sigma_1 + \sigma_2] - \frac{1}{2}[\sigma_1 - \sigma_2]\cos 2\theta - \sigma_6 \sin 2\theta \qquad (2.11)$$

$$\sigma_s = -\frac{1}{2}[\sigma_1 - \sigma_2]\sin 2\theta + \sigma_6 \cos 2\theta$$

Introducing a notation commonly used in photoelasticity, we have

$$p = \frac{1}{2}[\sigma_1 + \sigma_2], \; q = \frac{1}{2}[\sigma_1 - \sigma_2], \; r = \sigma_6$$

or (2.12)

$$\bar{p} = \frac{1}{2}[\sigma_x + \sigma_y], \; \bar{q} = \frac{1}{2}[\sigma_x - \sigma_y], \; \bar{r} = \sigma_s$$

where super bars refer to the on-axis orientation. We can now express the stress transformation equations in terms of double angles and the notation in Equation 2.12. This new formulation is shown in a matrix multiplication table as follows:

table 2.2
stress transformation in double angle function — I

	p	q	r
σ_x	1	$\cos 2\theta$	$\sin 2\theta$
σ_y	1	$-\cos 2\theta$	$-\sin 2\theta$
σ_s		$-\sin 2\theta$	$\cos 2\theta$

θ is positive in counter-clockwise rotation

There is an alternative arrangement for Table 2.2 where the column headings and the trigonometric functions are interchanged. This arrangement is useful for certain ply orientation such as 45 degrees, in which case the column with the cosine function vanishes.

table 2.3
stress transformation in double angle functions — II

	1	$\cos 2\theta$	$\sin 2\theta$
σ_x	p	q	r
σ_y	p	$-q$	$-r$
σ_s		r	$-q$

Either table can be used. The common feature is the first column, where the influence of the angle of rotation does not exist. The constant p is called an invariant of this coordinate transformation. If we add the first two rows of the tables above, we get

$$2\bar{p} = \sigma_x + \sigma_y = 2p = \sigma_1 + \sigma_2 \tag{2.13}$$

Thus the sum of the two normal stress components remain constant, independent of the angle of rotation or ply orientation. We call this invariant the first-order invariant for stress transformation; i.e.,

$$I = \bar{p} = p \tag{2.14}$$

There is a second-order invariant that we can show as follows:

From Equation 2.12

$$\bar{q}^2 + \bar{r}^2 = \frac{1}{4}[\sigma_x - \sigma_y]^2 + \sigma_s^2$$

From Table 2.3

$$\begin{aligned} &= q^2\cos^2 2\theta + 2qr\sin 2\theta\cos 2\theta + r^2\sin^2 2\theta \\ &\quad + r^2\cos^2 2\theta - 2qr\sin 2\theta\cos 2\theta + q^2\sin^2 2\theta \\ &= q^2 + r^2 \end{aligned} \tag{2.15}$$

This is another invariant because the quantity remains the same for any value of angle or ply orientation. We label this second-order invariant as

$$R^2 = \bar{q}^2 + \bar{r}^2 = q^2 + r^2 \tag{2.16}$$

where R is the radius of the Mohr's circle for stress transformation. The geometric relationship of Equation 2.16 is shown in Figure 2.8. Also shown in this figure are the phase angle and the following trigonometric relations:

$$\begin{aligned} q &= R\cos 2\theta_o \\ r &= R\sin 2\theta_o \\ \theta_o &= \frac{1}{2}\tan^{-1}\frac{r}{q} = \frac{1}{2}\sin^{-1}\frac{r}{R} = \frac{1}{2}\cos^{-1}\frac{q}{R} \end{aligned} \tag{2.17}$$

Thus the three stress components that characterize the state of plane stress can be represented by at least three sets of variables for a given

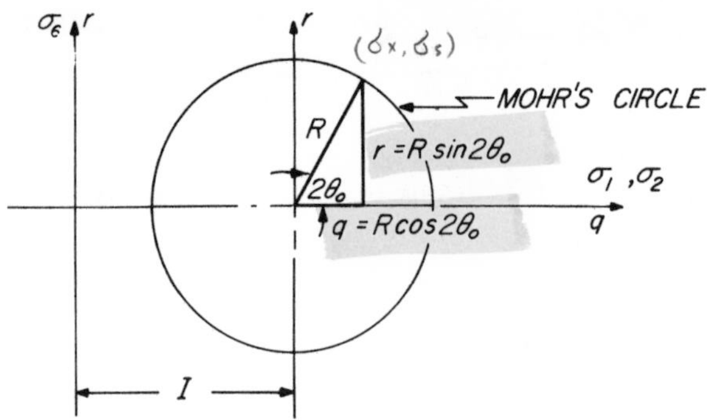

Figure 2.8 Geometric relations of second-order stress invariant R and Mohr's circle. The location of the center is specified by the first invariant I.

coordinate system; i.e., for a given angle of orientation, say, θ.

- First set: Stress components 1,2,6.
- Second set: p, q, r in accordance with Equation 2.12.
- Third set: Invariates I and R and the phase angle, defined in Equations 2.14, 2.16 and 2.17, respectively.

Similarly, the stress transformation can be formulated in terms of each of the sets above. We have done the first two; the third set can be used to derive the transformation equations by substituting Equations 2.14, 2.16 and 2.17 into the appropriate column headings in Table 2.2, the transformation in terms of p, q, r.

$$
\begin{aligned}
\sigma_x &= I + R\cos 2\theta_o \cos 2\theta + R\sin 2\theta_o \sin 2\theta \\
&= I + R\cos 2\,[\theta - \theta_o] \\
\sigma_y &= I - R\cos 2\,[\theta - \theta_o] \\
\sigma_s &= -R\cos 2\theta_o \sin 2\theta + R\sin 2\theta_o \cos 2\theta \\
&= -R\sin 2\,[\theta - \theta_o]
\end{aligned}
\tag{2.18}
$$

$$\text{where } I = \frac{1}{2}[\sigma_1 + \sigma_2]$$

$$R = \sqrt{q^2 + r^2} = \sqrt{\frac{1}{4}[\sigma_1 - \sigma_2]^2 + \sigma_6{}^2} \quad (2.19)$$

$$\theta_o = \frac{1}{2}\cos^{-1}\frac{q}{R} = \frac{1}{2}\sin^{-1}\frac{r}{R} = \frac{1}{2}\tan^{-1}\frac{r}{q}$$

This invariant formulation of the transformation equations can be shown in a matrix multiplication table as follows:

table 2.4
stress transformation in invariant functions

	I	R
σ_x	1	$\cos 2(\theta-\theta_o)$
σ_y	1	$-\cos 2(\theta-\theta_o)$
σ_s		$-\sin 2(\theta-\theta_o)$

We have seen that transformation equations can be written in different sets of functions. There are advantages and disadvantages associated with each set. From the standpoint of numerical calculation, the invariant functions in Table 2.4 may be the easiest because there are only two columns in this table, instead of three columns as in Tables 2.1–2.3. The Mohr's circle representation is also based on the invariant functions. But the direction of rotation, the magnitude and the sign of the phase angle can be troublesome. Care must be exercised in applying the last line of Equation 2.17 or 2.19 to avoid a 180 degree out of phase mistake.* Inverse trigonometric functions are not single-valued; they repeat themselves at fixed intervals. The double angle functions of the stress transformation in Tables 2.2 and 2.3 are better in the sense that the signs are correctly built in. The classical power functions formulation in Table 2.1 appears most frequently in current textbooks. This formulation is most convenient to use when one or two of the stress components are zero.

*The same care is required for the conversion of rectangular to polar coordinates.

3. numerical examples of stress transformation

Problem: Given stress in the 1-2 system

$$\sigma_i = (9,3,4) \tag{2.20}$$

Find stress components in the x-y system for $\theta = 45$ degrees. The two reference coordinates are shown in Figure 2.9 (same as Figure 2.5).

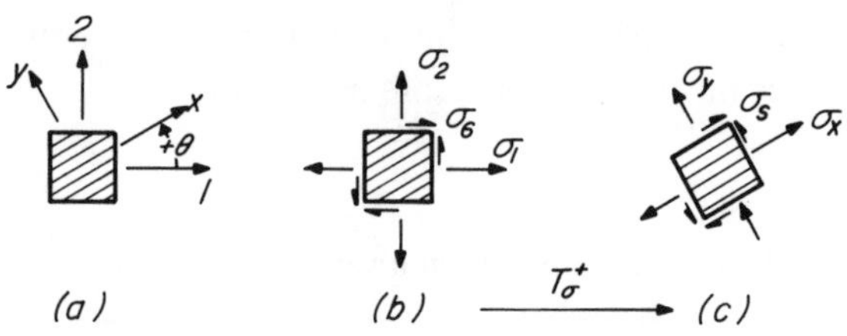

Figure 2.9 Stress transformation: changes in stress components due to coordinate transformation.

Solution:

(1) From the power function transformation in Table 2.1.

$$\cos\theta = \sin\theta = \frac{1}{\sqrt{2}} \tag{2.21}$$

$$\sigma_x = \frac{1}{2}(9 + 3 + 2 \times 4) = 10$$

$$\sigma_y = \frac{1}{2}(9 + 3 - 2 \times 4) = 2$$

$$\sigma_s = \frac{1}{2}(-9 + 3 + 0) = -3 \tag{2.22}$$

(2) From the double angle function transformation in Table 2.3:

$$\cos 2\theta = 0, \sin 2\theta = 1 \tag{2.23}$$

$$p = \frac{1}{2}(9+3) = 6$$

$$q = \frac{1}{2}(9-3) = 3 \tag{2.24}$$

$$r = 4$$

Then,

$$\sigma_x = 6 + 4 = 10$$

$$\sigma_y = 6 - 4 = 2 \tag{2.25}$$

$$\sigma_s = -3$$

(3) From the definition of invariants in Equation 2.19:

$$I = p = 6 \tag{2.26}$$

$$R = \sqrt{q^2 + r^2} = \sqrt{3^2 + 4^2} = 5 \tag{2.27}$$

$$\theta_o = \frac{1}{2}\tan^{-1}\frac{4}{3} = 26.56 \text{ degree}$$

or

$$= \frac{1}{2}\sin^{-1}\frac{4}{5} = 26.56 \text{ degree} \tag{2.28}$$

or

$$= \frac{1}{2}\cos^{-1}\frac{3}{5} = 26.56 \text{ degree}$$

From the invariant function transformation in Table 2.4 for $\theta = 45$ degrees, we have

$$\begin{aligned} \sigma_x &= 6 + 5\cos 2(45 - 26.56) = 10 \\ \sigma_y &= 6 - 5\cos 2(45 - 26.56) = 2 \\ \sigma_s &= -5\sin 2(45 - 26.56) = -3 \end{aligned} \tag{2.29}$$

As expected all three formulations yield the same answer.

By virtue of symmetry, four combinations of stress components are closely related, each corresponding to a phase angle. This is shown in the Mohr's circle in Figure 2.10.

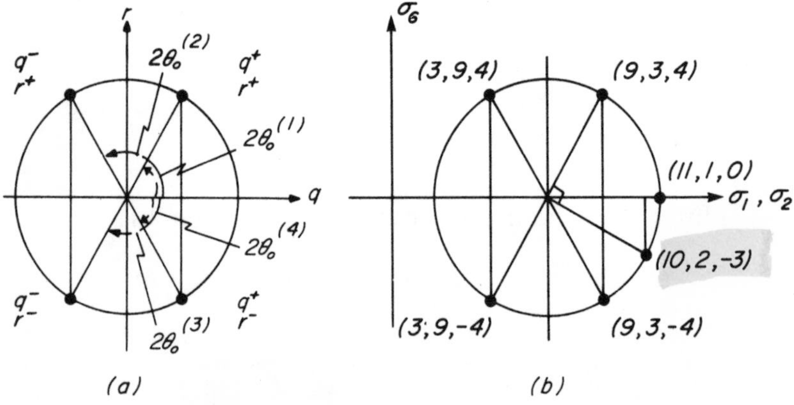

Figure 2.10 Four possible combinations of components of stress. For a given Mohr's circle (or a state of stress) the magnitude and sign of the stress components depend on the signs of q and r in Equation 2.24. Each combination is associated with a phase angle. The relations between the four phase angles are shown in this figure. The components for each phase angle are also shown.

Care must be exercised to distinguish among these combinations. They are repeated here for emphasis and shown in Figure 2.10(b).

$$\begin{aligned} \theta_o^{(1)} &= 26.56, \qquad \sigma_i = (9,3,4) \\ \theta_o^{(2)} &= 63.44, \qquad \sigma_i = (3,9,4) \\ \theta_o^{(3)} &= -63.44, \qquad \sigma_i = (3,9,-4) \\ \theta_o^{(4)} &= -26.56, \qquad \sigma_i = (9,3,-4) \end{aligned} \tag{2.30}$$

Note that the first phase angle is the given orientation for this present state of stress. Note relationships between phase angles; e.g.,

$$
\begin{aligned}
\theta_o^{(1)} + \theta_o^{(2)} &= 90 \\
\theta_o^{(3)} + \theta_o^{(4)} &= -90 \\
\theta_o^{(1)} &= -\theta_o^{(4)} \\
\theta_o^{(2)} &= -\theta_o^{(3)}
\end{aligned}
\qquad (2.31)
$$

In spite of the multivalued phase angles for the invariant formulation of the transformation, the phase angle has one important feature. When the angle of rotation θ is equal to a phase angle, say 26.56 degrees from Equation 2.30, we have from Equation 2.29:

$$
\begin{aligned}
\sigma_1 &= 6 + 5 = 11 \\
\sigma_2 &= 6 - 5 = 1 \\
\sigma_6 &= 0
\end{aligned}
\qquad (2.32)
$$

This combination of stress components are also shown in Figure 2.10(*b*). This orientation is called the principal direction. In this orientation, the shear stress is zero, and the normal stress components reach maximum and minimum values. They can be determined immediately from the two invariants:

$$
\begin{aligned}
\sigma_{\mathrm{I}} &= \sigma_{max} = I + R \\
\sigma_{\mathrm{II}} &= \sigma_{min} = I - R
\end{aligned}
\qquad (2.33)
$$

where stress components are the principal stress components. See Figure 2.11 for graphical illustration of various invariant quantities. Thus, the principal direction is derived from

$$\theta - \theta_o = 0 \qquad (2.34)$$

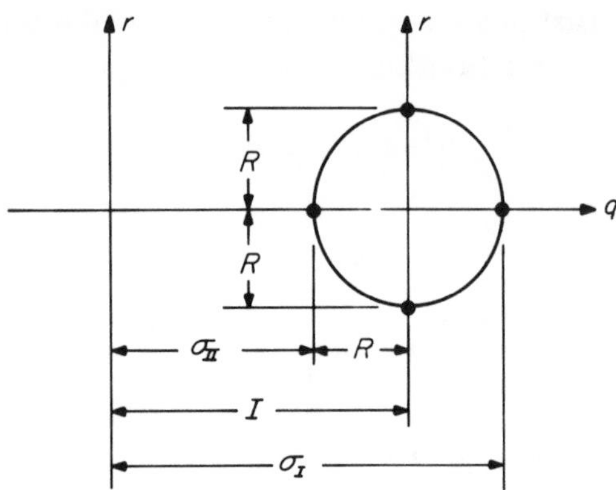

Figure 2.11 Principal stress components. They are the maximum or minimum values in the Mohr's circle.

There is another important orientation; i.e., 45 degrees from the principal direction, or when

$$\theta - \theta_o = 45 \tag{2.35}$$

At this angle, we have

$$\begin{aligned} \sigma_1 &= \ \ I = \ \ 6 \\ \sigma_2 &= \ \ I = \ \ 6 \\ \sigma_6 &= -R = -5 \end{aligned} \tag{2.36}$$

Here, both normal stress components are equal to the first invariant; the shear stress component reaches its minimum value. The latter would have been the maximum shear stress if −45 degree is used in Equation 2.35.

As a final emphasis on the importance of the sign of angles, Figure 2.12 shows the consequence of a sign error. A positive transformation from the 1-2 axes will result in the material symmetry axes, designated 1^+-2^+ axes. A sign error will result in the 1^--2^- axes which are 2θ orientation away from the correct answer.

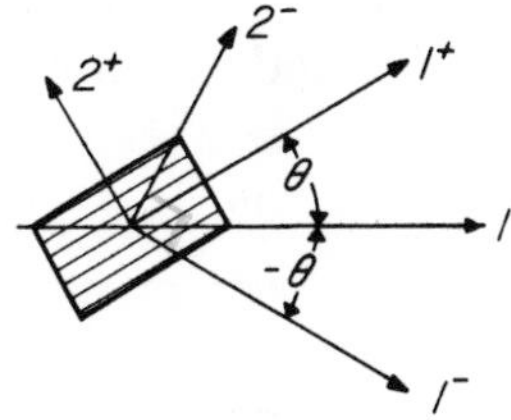

Figure 2.12 Positive and negative angles of rotation. Guesswork is not good enough for composites. Keep track of the signs.

4. transformation of strain

Strain transformation is as important as stress transformation. An identical figure to Figure 2.5 can be drawn for the strain components. This is done in Figure 2.13.

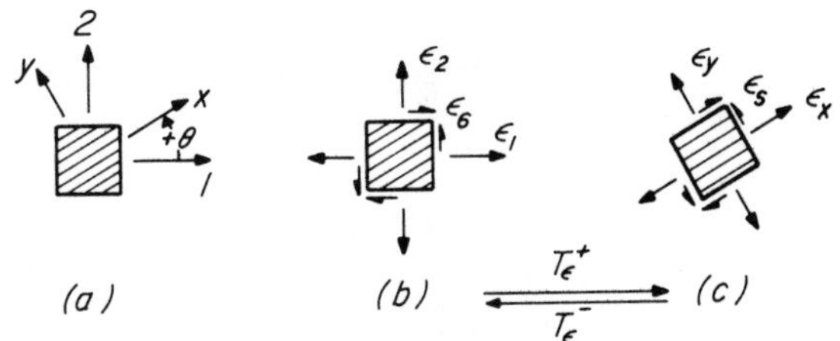

Figure 2.13 Strain transformation. Changes in strain components due to coordinate rotation or transformation.

(*a*) Relation between the 1-2 and *x*-*y* systems. Counterclockwise rotation is positive.

(*b*) The off-axis strain components, with numerical subscripts.

(*c*) The on-axis strain components, with letter subscripts.

All arrows for the components are pointing in a positive direction.

Like the definition of strain itself, strain transformation is purely geometric and involves no material property or balance of forces. Using the notation shown in Figure 2.13, the off-axes orientation is the 1-2 system, and the on-axis, the *x*-*y* system. We will now derive the strain transformation relations from the strain-displacement relations shown in Equation 1.1 and 1.4 and repeated as follows:

$$\begin{aligned} \epsilon_x &= \frac{\partial u}{\partial x} \\ \epsilon_y &= \frac{\partial v}{\partial y} \\ \epsilon_s &= \frac{\partial v}{\partial x} + \frac{\partial u}{\partial y} \end{aligned} \tag{2.37}$$

Since both displacements u and v and coordinates x and y are vectors, and are directionally dependent quantities, we only need to find the relationship between the primed and the unprimed components of a vector, shown in Figure 2.14(a) and (b), respectively,

$$\begin{aligned} x &= mx' + ny' \\ y &= -nx' + my' \end{aligned} \tag{2.38}$$

conversely,

$$\begin{aligned} x' &= mx - ny \\ y' &= nx + my \end{aligned} \tag{2.39}$$

where, as before,

$m = \cos\theta, \quad n = \sin\theta$

From Equation 2.39, we can get the following by partial differentiation:

$$\frac{\partial x'}{\partial x} = m, \quad \frac{\partial x'}{\partial y} = -n, \quad \frac{\partial y'}{\partial x} = n, \quad \frac{\partial y'}{\partial y} = m \tag{2.40}$$

The relations between displacements in the primed and unprimed coordinates are identical to those in Equations 2.38 and 2.39 because all quantities are vectors. We can simply write the following by re-

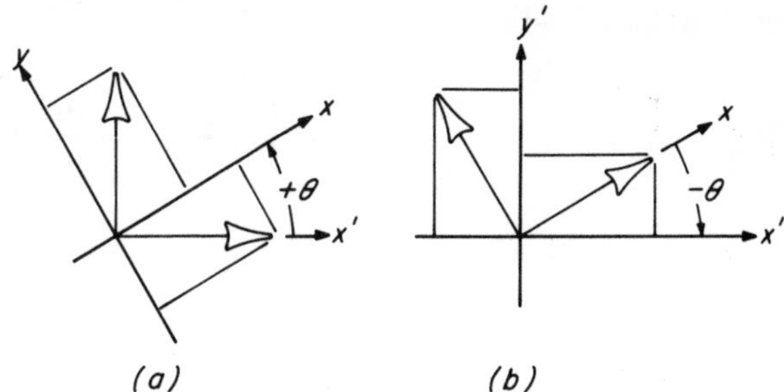

Figure 2.14 Coordinate systems between the primed or numbered and unprimed axes.
(*a*) To go from primed to unprimed, θ is positive.
(*b*) To go from unprimed to primed, θ is negative.

placing x, y, x', y', by u, v, u', v', respectively:

$$\begin{aligned} u &= mu' + nv' \\ v &= -nu' + mv' \end{aligned} \tag{2.41}$$

Conversely,

$$\begin{aligned} u' &= mu - nv \\ v' &= nu + mv \end{aligned} \tag{2.42}$$

Now we are ready to derive the strain transformation equations. From Equation 2.37

$$\epsilon_x = \frac{\partial u}{\partial x} \tag{2.43}$$

By chain differentiation

$$\epsilon_x = \frac{\partial u}{\partial x'}\frac{\partial x'}{\partial x} + \frac{\partial u}{\partial y'}\frac{\partial y'}{\partial x} \tag{2.44}$$

From Equation 2.40 and 2.41

$$\epsilon_x = \left[m \frac{\partial u'}{\partial x'} + n \frac{\partial v'}{\partial x'}\right] m + \left[m \frac{\partial u'}{\partial y'} + n \frac{\partial v'}{\partial y'}\right] n$$

$$= m^2 \epsilon_1 + n^2 \epsilon_2 + mn\epsilon_6 \tag{2.45}$$

where

$$\epsilon_1 = \epsilon_x{}' = \frac{\partial u'}{\partial x'}$$

$$\epsilon_2 = \epsilon_y{}' = \frac{\partial v'}{\partial y'} \tag{2.46}$$

$$\epsilon_6 = \epsilon_s{}' = \frac{\partial u'}{\partial y'} + \frac{\partial v'}{\partial x'}$$

Here primes are added to all the variables in the definitions of strains in Equation 2.37. We can do this because the relationship is invariant; i.e., the relationship does not change from coordinates to coordinates, and is valid for all coordinate systems. Note that the strain transformation in Equation 2.45 is very similar to the stress transformation in Equation 2.4 except the factor 2 is missing in the shear term. This difference comes about from the use of engineering shear strain as shown in Table 1.3 and Equation 1.4.

By an identical process as that used in the derivation of Equation 2.45, we can show

$$\epsilon_y = n^2 \epsilon_1 + m^2 \epsilon_2 - mn\epsilon_6 \tag{2.47}$$

$$\epsilon_s = -2mn\epsilon_1 + 2mn\epsilon_2 + [m^2 - n^2]\epsilon_6 \tag{2.48}$$

This is summarized in a matrix multiplication table as follows:

table 2.5
strain transformation equations in power functions

	ϵ_1	ϵ_2	ϵ_6
ϵ_x	m^2	n^2	mn
ϵ_y	n^2	m^2	$-mn$
ϵ_s	$-2mn$	$2mn$	m^2-n^2

We can express the transformation relations in terms of double angle and invariant functions as we did for the stress transformation. Comparable to Tables 2.2 and 2.3 for stress transformation, we can show strain transformation in double angle functions in Tables 2.6 and 2.7, respectively.

table 2.6
strain transformation in double angle function – I

	p	q	r
ϵ_x	1	$\cos 2\theta$	$\sin 2\theta$
ϵ_y	1	$-\cos 2\theta$	$-\sin 2\theta$
ϵ_s		$-2\sin 2\theta$	$2\cos 2\theta$

table 2.7
strain transformation in double angle functions – II

	1	$\cos 2\theta$	$\sin 2\theta$
ϵ_x	p	q	r
ϵ_y	p	$-q$	$-r$
ϵ_s		$2r$	$-2q$

where $$p = \frac{1}{2}[\epsilon_1 + \epsilon_2], \; q = \frac{1}{2}[\epsilon_1 - \epsilon_2], \; \boxed{r = \frac{1}{2}\epsilon_6} \qquad (2.49)$$

Note that the definition of r is different from that for the stress transformation in Equation 2.12. The use of engineering shear strain is responsible for the difference.

The invariant function comparable to Table 2.4 for the stress transformation can be derived in a similar fashion and the results are listed in a matrix multiplication table as follows:

table 2.8

strain transformation in invariant functions

	I	R
ϵ_x	I	$\cos 2(\theta - \theta_o)$
ϵ_y	I	$-\cos 2(\theta - \theta_o)$
ϵ_s		$-2\sin 2(\theta - \theta_o)$

where:

$$I = I_\epsilon = \frac{1}{2}[\epsilon_1 + \epsilon_2]$$

$$R = R_\epsilon = \sqrt{q^2 + r^2} = \sqrt{\frac{1}{4}[\epsilon_1 - \epsilon_2]^2 + \frac{1}{4}\epsilon_6{}^2} \qquad (2.50)$$

$$\theta_o = \frac{1}{2}\cos^{-1}\frac{q}{R} = \frac{1}{2}\sin^{-1}\frac{r}{R} = \frac{1}{2}\tan^{-1}\frac{r}{q}$$

The advantages and disadvantages of each formulation for the strain transformation are similar to those for the stress transformation. The double angle formulation appears to provide the best compromise and is recommended for general usage. This will be our choice for the balance of this book.

5. numerical examples of strain transformation

a. Problem: Given a state of strain in the 1-2 system

$$\epsilon_i = (9,3,4) \times 10^{-3} \qquad (2.51)$$

Find a transformed strain for θ equal to 45 degrees. See Figure 2.15.

Solution: From Table 2.7, $\cos 2\theta = 0$, $\sin 2\theta = 1$

$$p = \frac{1}{2}(9 + 3) = 6 \times 10^{-3}$$

$$q = \frac{1}{2}(9 - 3) = 3 \times 10^{-3} \qquad (2.52)$$

$$r = \frac{4}{2} = 2 \times 10^{-3}$$

Then

$$\begin{aligned} \epsilon_x &= 6 + 2 = 8 \times 10^{-3} \\ \epsilon_y &= 6 - 2 = 4 \times 10^{-3} \\ \epsilon_s &= -2 \times 3 = -6 \times 10^{-3} \end{aligned} \quad (2.53)$$

Note that the transformed strain is quite different from the transformed stress of (10,2,–3) from Equation 2.25. The factor of 2 in the engineering shear strain is responsible for this dramatic difference.

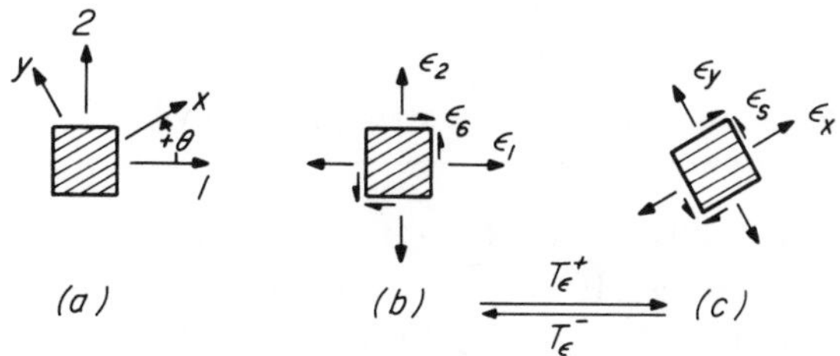

Figure 2.15 Strain transformation. To go from (*b*) to (*c*) is a positive transformation when the angle of rotation is positive, as in Problem *a*. The negative or inverse transformation in Problem *b* is in effect when the angle is negative. (This figure is the same as Figure 2.13.)

b. Problem: Try inverse transformation for strain in Equation 2.53

$$\epsilon_i = (8,4,-6) \times 10^{-3} \quad (2.54)$$

Find strain at –45 degree rotation.

Solution: From Table 2.7, $\cos 2\theta = 0$, $\sin 2\theta = -1$

$$\begin{aligned} \bar{p} &= \frac{1}{2}(8 + 4) = 6 \times 10^{-3} \\ \bar{q} &= \frac{1}{2}(8 - 4) = 2 \times 10^{-3} \\ \bar{r} &= -\frac{6}{2} = -3 \times 10^{-3} \end{aligned} \quad (2.55)$$

Then

$$\begin{aligned} \epsilon_1 &= 6+3 = 9 \times 10^{-3} \\ \epsilon_2 &= 6-3 = 3 \times 10^{-3} \\ \epsilon_6 &= 2 \times 2 = 4 \times 10^{-3} \end{aligned} \tag{2.56}$$

Note that the original strain in Equation 2.51 is recovered.

6. graphic interpretations of stress-strain relations

The stress-strain relations in Chapter 1 used components of stress and strain as the variables. If we use their linear combinations p,q,r, as defined by Equations 2.12 and 2.49, we can write the equivalent stress-strain relations for an on-axis, orthotropic material as follows:

table 2.9
equivalent on-axis stress-strain relation in terms of compliance

	$\bar{p}_\sigma$	$\bar{q}_\sigma$	$\bar{r}_\sigma$
$\bar{p}_\epsilon$	$\frac{1}{2}(S_{xx}+S_{yy}+2S_{xy})$	$\frac{1}{2}(S_{xx}-S_{yy})$	
$\bar{q}_\epsilon$	$\frac{1}{2}(S_{xx}-S_{yy})$	$\frac{1}{2}(S_{xx}+S_{yy}-2S_{xy})$	
$\bar{r}_\epsilon$			$\frac{1}{2}S_{ss}$

table 2.10
equivalent on-axis stress-strain relation in terms of modulus

	$\bar{p}_\epsilon$	$\bar{q}_\epsilon$	$\bar{r}_\epsilon$
$\bar{p}_\sigma$	$\frac{1}{2}(Q_{xx}+Q_{yy}+2Q_{xy})$	$\frac{1}{2}(Q_{xx}-Q_{yy})$	
$\bar{q}_\sigma$	$\frac{1}{2}(Q_{xx}-Q_{xy})$	$\frac{1}{2}(Q_{xx}+Q_{yy}-2Q_{xy})$	
$\bar{r}_\sigma$			$2Q_{ss}$

If the material is isotropic, we have additional relations among the components of compliance and modulus, such as those in Equation 1.23, Tables 2.9 and 2.10 become:

table 2.11
equivalent stress-strain relation of isotropic materials

	p_σ	q_σ	r_σ
p_ϵ	$S_{xx}+S_{xy}$		
q_ϵ		$S_{xx}-S_{xy}$	
r_ϵ			$S_{xx}-S_{xy}$

	p_ϵ	q_ϵ	r_ϵ
p_σ	$Q_{xx}+Q_{xy}$		
q_σ		$Q_{xx}-Q_{xy}$	
r_σ			$Q_{xx}-Q_{xy}$

All off-diagonal terms vanish; all the stress-strain combinations are uncoupled. The Mohr's circle for stress and strain are related as follows:

$$\begin{aligned} p_\epsilon &= I_\epsilon = (S_{xx}+S_{xy})I_\sigma \\ R_\epsilon &= \sqrt{q_\epsilon^2+r_\epsilon^2} = (S_{xx}-S_{xy})R_\sigma \\ \theta_\epsilon &= \theta_\sigma \end{aligned} \tag{2.57}$$

Similarly

$$\begin{aligned} p_\sigma &= I_\sigma = (Q_{xx}+Q_{xy})I_\epsilon \\ R_\sigma &= (Q_{xx}-Q_{xy})R_\epsilon \\ \theta_\sigma &= \theta_\epsilon \end{aligned} \tag{2.58}$$

The last relation is the phase angle which remains the same as we go from the stress to strain space or vice versa. The principal axes of stress and strain are coincident. The two Mohr's circles, shown in Figure 2.16, are related by two independent constants or scale factors. We expect this in isotropic materials.

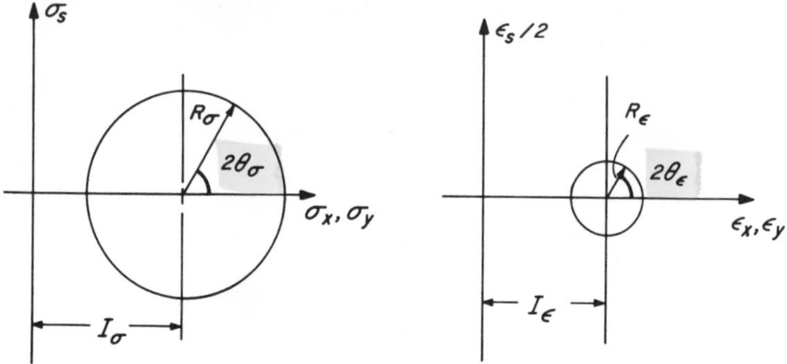

Figure 2.16 Graphic representation of stress-strain relation of an isotropic material. The phase angle remains the same. The location and size of the Mohr's circles are different.

If our material is square-symmetric, the equivalent stress-strain relations must be modified because the relation of Equation 1.23 is no longer valid. The new stress-strain relations are shown as follows:

table 2.12
equivalent on-axis stress-strain relations for square-symmetric materials

	$\bar{p}_\sigma$	$\bar{q}_\sigma$	$\bar{r}_\sigma$
$\bar{p}_\epsilon$	$S_{xx}+S_{xy}$		
$\bar{q}_\epsilon$		$S_{xx}-S_{xy}$	
$\bar{r}_\epsilon$			$\frac{1}{2}S_{ss}$

	$\bar{p}_\epsilon$	$\bar{q}_\epsilon$	$\bar{r}_\epsilon$
$\bar{p}_\sigma$	$Q_{xx}+Q_{xy}$		
$\bar{q}_\sigma$		$Q_{xx}-Q_{xy}$	
$\bar{r}_\sigma$			$2Q_{ss}$

In place of the relations in Equations 2.57 and 2.58, we have:

$$
\begin{aligned}
I_\epsilon &= [S_{xx} + S_{xy}] I_\sigma \\
R_\epsilon &= [S_{xx} - S_{xy}] \sqrt{\overline{q}_\sigma^2 + \left[\frac{1}{2}\frac{S_{ss}}{S_{xx} - S_{xy}} \overline{r}_\sigma\right]^2} \\
\theta_\epsilon &= \frac{1}{2} \tan^{-1} \left[\frac{1}{2}\frac{S_{ss}}{S_{xx} - S_{xy}} \frac{\overline{r}_\sigma}{\overline{q}_\sigma}\right]
\end{aligned}
\tag{2.59}
$$

Similarly

$$
\begin{aligned}
I_\sigma &= [Q_{xx} + Q_{xy}] I_\epsilon \\
R_\sigma &= [Q_{xx} - Q_{xy}] \sqrt{\overline{q}_\epsilon^2 + \left[\frac{2Q_{ss}}{Q_{xx} - Q_{xy}} \overline{r}_\epsilon\right]^2} \\
\theta_\sigma &= \frac{1}{2} \tan^{-1} \left[\frac{2Q_{ss}}{Q_{xx} - Q_{xy}} \frac{\overline{r}_\epsilon}{\overline{q}_\epsilon}\right]
\end{aligned}
\tag{2.60}
$$

Thus, the simple scaling applies to the location of the Mohr's circle only. The radii of the Mohr's circle depend on the specific material constants. All three independent constants are involved. Explicit relation between the two radii of the Mohr's circles does not exist. We can recover the isotropic relations in Equations 2.57 and 2.58 if the relation in Equation 1.23 is invoked. The phase angle also changes as we go from one Mohr's circle to another. The principal axes of stress will not be coincident with those of strain.

If we consider an orthotropic material, the relations in Tables 2.9 and 2.10 show that the location of the Mohr's circle as specified by the first invariant is now coupled with other components of stress (q) and other components of the compliance. The location of the circle in the Mohr's strain space, for example, depends on the specific material and the stress combinations of p and q. There is one special case where the principal axes of stress and strain coincide for orthotropic or square symmetric material. This occurs when the material symmetry axes or the on-axis orientations coincide with the principal stress or strain axes.

If we start with zero stress and gradually increase all three components of stress by the same proportion, this is called proportional loading. The unit vector of stress remains constant. The lines of loading can be shown as a straight line in the Mohr's circle space. These loading paths are added to the circles in Figure 2.16 and shown in Figure 2.17. For nonisotropic material, the phase angles will be different. The principal axes will therefore be different between the stress and strain. The concept of loading path is important to the design and sizing of composite materials. For structures, multiple loading conditions often exist. Multiple loading paths and paths other than proportional loading are all possible. Graphic illustrations in Figure 2.17 can increase understanding of the basic concept in design.

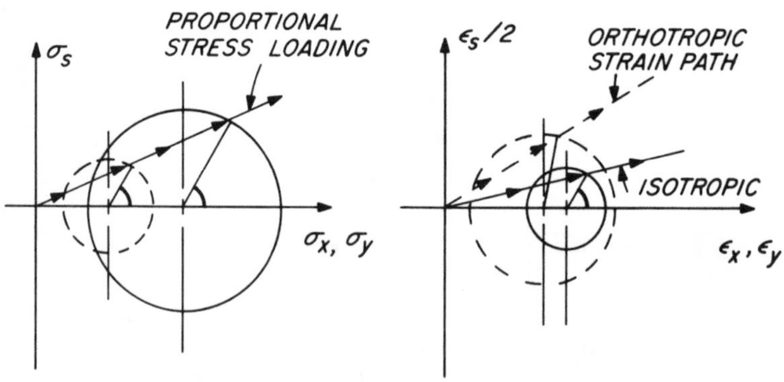

Figure 2.17 Loading paths of stress and strains in Mohr's circle space. The strain paths are dependent on elastic moduli of the material.

7. conclusions

Stresses, strains and their transformation properties are familiar concepts. For composite materials added considerations must be given. The direction of rotation or the sign of the ply orientation must be observed faithfully. While guesswork is often harmless for the conventional material because material behavior is often insensitive to the direction of rotation, we must develop strict rules in keeping track of the sign of the angles in composite materials.

The notations which we have used in this chapter is arbitrary to the extent that a number of systems could be followed. It is important to distinguish the reference coordinate axes from the material symmetry axes, the new from the old coordinate axes, etc. There is no one universally acceptable system of notation.

The transformation of stresses is independent of materials. The relations we used in this chapter are from the off-axis orientation to the on-axis orientation. They are most frequently encountered working with composite materials. The transformation of strain is purely geometric. No material properties are involved. This should not be confused by the fact that the resulting strain from an applied stress does depend on the type of material. A stiffer material will result in less strain for the same applied stress. The strain so produced is dictated by the stress-strain relation and the magnitude of the elastic constants. But the transformation of strain is purely geometric for a given state of strain. In the next chapter the transformation of stress and strain will provide the basis for the derivation of the transformation of the modulus and compliance of unidirectional composites.

The concepts of the principal directions of stress and strain are important to composite materials. Like the sign of ply orientation, the sign of the orientation of the principal axes must be treated with care. The notation and definition such as that in Figures 2.2, 2.10, et al. must be kept in mind in order to avoid bad errors.

8. homework problems

a. Express stress transformation from the p-q-r to the p'-q'-r', where the primed quantities are the linear combinations of the stress components in the new, transformed axes.

table 2.13
transformation of stress

	p_σ	q_σ	r_σ
p'_σ	1		
q'_σ		$\cos 2\theta$	$\sin 2\theta$
r'_σ		$-\sin 2\theta$	$\cos 2\theta$

What is the relation between this table and the coordinate transformation in Equation 2.38 or a rigid body rotation in the q-r space? Show this in a figure and compare it with Figure 2.14.

b. Express strain transformation in terms of the linear combinations of the strain components, analogous to the stress transformation in Problem *a*. Why are the coefficients identical?

table 2.14
transformation of strain

	p_ϵ	q_ϵ	r_ϵ
p'_ϵ	1		
q'_ϵ		$\cos 2\theta$	$\sin 2\theta$
r'_ϵ		$-\sin 2\theta$	$\cos 2\theta$

c. Simple states of stress are those with only one nonzero stress component such as

$$
\begin{aligned}
&1)\quad \sigma_1 = a,\ \sigma_2 = \sigma_6 = 0\\
&2)\quad \sigma_2 = b,\ \sigma_1 = \sigma_6 = 0 \qquad (2.61)\\
&3)\quad \sigma_6 = c,\ \sigma_1 = \sigma_2 = 0
\end{aligned}
$$

Show these states in solid dots in the Mohr's circle space as in Figure 2.11. If each of the applied stress above is equal to 100 MPa, find the resulting strain components and show the strain to scale in the Mohr's circle space for aluminum, T300/5208 cross ply and unidirectional (isotropic, square-symmetric and orthotropic materials, respectively, as listed in the Homework Problems in Chapter 1).

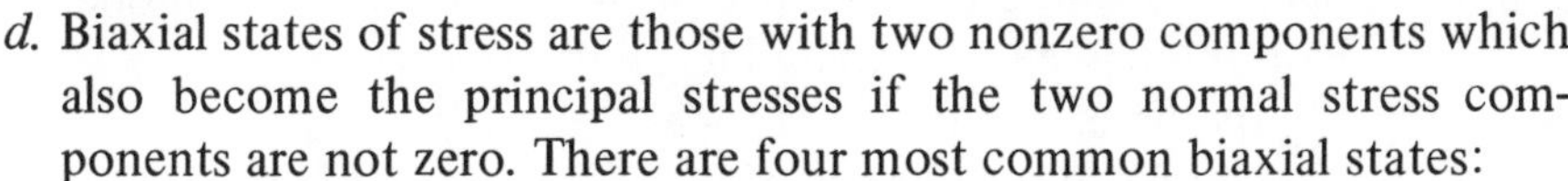

d. Biaxial states of stress are those with two nonzero components which also become the principal stresses if the two normal stress components are not zero. There are four most common biaxial states:

$$
\begin{aligned}
&1) \quad \sigma_1 = \sigma_2 = P, \sigma_6 = 0 \\
&2) \quad \sigma_1 = \sigma_2 = -P, \sigma_6 = 0 \\
&3) \quad \sigma_1 = -\sigma_2 = Q, \sigma_6 = 0 \\
&4) \quad \sigma_1 = -\sigma_2 = -Q, \sigma_6 = 0
\end{aligned}
\tag{2.62}
$$

Show these states in solid dots in the Mohr's circle space. If each of the applied stress is equal to 100 MPa, show the resulting states of strain for the same materials in Problem *c*.

e. Show the relation between the following two states of stress:

$$
\begin{aligned}
&\sigma_6 = Q, \text{ and} \\
&\sigma_1 = -\sigma_2 = Q
\end{aligned}
\tag{2.63}
$$

How can this be used to establish relations between engineering constants for isotropic material as in Equation 1.23? Can this be used for nonisotropic materials?

nomenclature

I	= First order invariant of stress or strain, depending on the subscript
m, n	= $\cos\theta$, $\sin\theta$
p, q, r	= Linear combinations of stress or strain components
$\bar{p}, \bar{q}, \bar{r}$	= Special linear combinations of stress or strain components with reference to the material symmetry axes
Q_{ij}	= Components of modulus; $i,j = x,y,s$ or 1,2,6
R	= A second-order invariant of stress or strain; it is the radius of the Mohr's circle
S_{ij}	= Components of compliance, $i,j = x,y,s$ or 1,2,6
T^+, T^-	= Positive, and negative or inverse transformation of stress or strain, depending on the subscript. The sign corresponds to that of the ply orientation
x, y	= New or transformed coordinate axes, usually refer to the on-axis orientation
1, 2	= Reference coordinate axes, usually refer to some off-axis orientation
u, v	= Displacements along the x and y axes
u', v'	= Displacements along the 1 and 2 axes
σ_i	= Stress components in the material symmetry axes, $i = x,y,s$; or in the 1-2 reference axes, $i = 1,2,6$
$\sigma_{I,\ II}$	= Principal stress components
ϵ_i	= Strain components in the material symmetry axes, $i = x,y,s$; or in the 1-2 reference axes, $i = 1,2,6$
$\epsilon_{I,\ II}$	= Principal strain components
θ	= Angle of ply orientation; counterclockwise rotation is positive
θ_o	= Phase angle for stress or strain transformation in the invariant formulation; it is the orientation of the principal axes measured in the reference axes 1-2

chapter 3
off-axis stiffness of unidirectional composites

The stiffness of unidirectional composites with off-axis ply orientation is important because composite laminates are normally made of off-axis in addition to on-axis plies. We must know how to determine the contribution to the laminate stiffness by each ply or ply group. We will need the transformation of stiffness and compliance to determine the off-axis stiffness. Like those for stress and strain, the transformation relations of stiffness and compliance can be formulated in terms of the power functions, the multiple angle functions and the invariants. Examples of a specific graphite-epoxy composite are used to illustrate the off-axis stiffness of unidirectional composites.

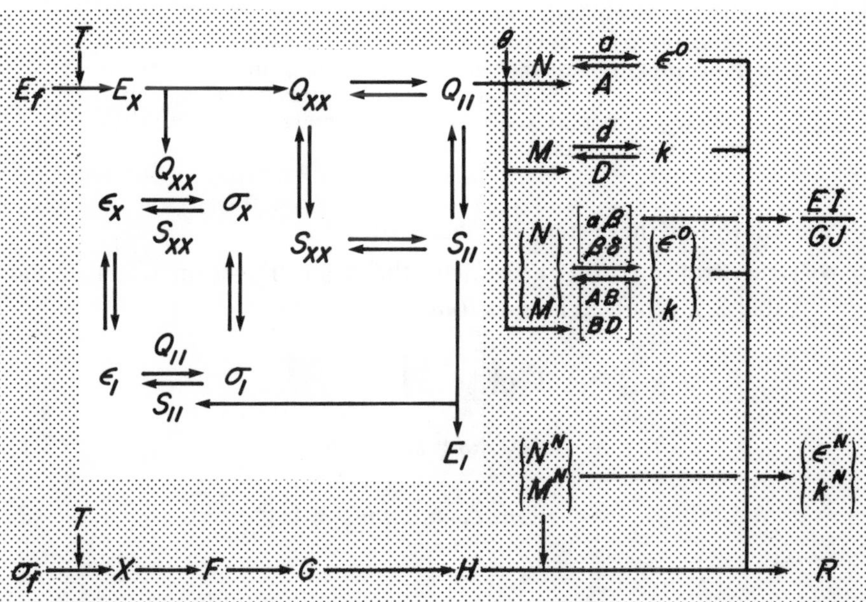

1. off-axis stiffness modulus

As we have shown in Figure 2.4 and repeated here in Figure 3.1, the off-axis stiffness can be determined in three steps: the off-axis to on-axis strain transformation, the on-axis stress-strain relations, and the inverse or the on-axis to off-axis stress transformation. This process was initiated by a given strain in Figure 3.1(*a*) and led us eventually to the induced stress in Figure 3.1(*d*). The off-axis compliance can be similarly derived in three steps, as shown in Figure 2.3. The purpose here is to derive the off-axis stiffness and the off-axis stress-strain relation for an arbitrary angle of orientation. Then we can go directly from (*a*) to (*d*) in Figure 3.1 in one step.

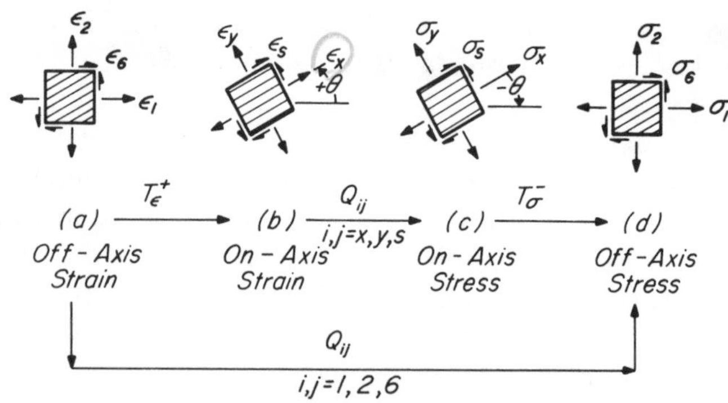

Figure 3.1 Determination of the off-axis stiffness:
From (*a*) to (*b*): use positive strain transformation.
From (*b*) to (*c*): use the on-axis stress-strain relations in stiffness.
From (*c*) to (*d*): use negative or inverse stress transformation.
We can go from (*a*) to (*d*) directly if we merge these three steps into one. This is the same as Figure 2.4.

We will follow these steps in Figure 3.1.

- To go from (*a*) to (*b*), we need the strain transformation listed in Table 2.5, repeated here as follows:

$$\begin{aligned} \epsilon_x &= m^2\epsilon_1 + n^2\epsilon_2 + mn\epsilon_6 \\ \epsilon_y &= n^2\epsilon_1 + m^2\epsilon_2 - mn\epsilon_6 \\ \epsilon_s &= -2mn\epsilon_1 + 2mn\epsilon_2 + [m^2 - n^2]\epsilon_6 \end{aligned} \tag{3.1}$$

- To go from (*b*) to (*c*) in Figure 3.1, we need the on-axis orthotropic stress-strain relation in modulus in Table 1.6 which, when combined with the results in Equation 3.1, produces:

$$\begin{aligned} \sigma_x &= Q_{xx}[m^2\epsilon_1 + n^2\epsilon_2 + mn\epsilon_6] \\ &\quad + Q_{xy}[n^2\epsilon_1 + m^2\epsilon_2 - mn\epsilon_6] \\ &= [m^2 Q_{xx} + n^2 Q_{xy}]\epsilon_1 + [n^2 Q_{xx} + m^2 Q_{xy}]\epsilon_2 \\ &\quad + [mnQ_{xx} - mnQ_{xy}]\epsilon_6 \end{aligned} \tag{3.2}$$

Similarly,

$$\begin{aligned} \sigma_y &= [m^2 Q_{xy} + n^2 Q_{yy}]\epsilon_1 + [n^2 Q_{xy} + m^2 Q_{yy}]\epsilon_2 \\ &\quad + [mnQ_{xy} - mnQ_{yy}]\epsilon_6 \end{aligned} \tag{3.3}$$

$$\sigma_s = -2mnQ_{ss}\epsilon_1 + 2mnQ_{ss}\epsilon_2 + [m^2 - n^2]Q_{ss}\epsilon_6 \tag{3.4}$$

- To go from (*c*) to (*d*) in Figure 3.1, we need to modify the stress transformation as listed in Table 2.1. The angle of rotation is now negative. The numeric and letter subscripts are interchanged. The letter subscripts now refer to the old (before transformation), and the numeric subscripts, the new (after transformation).

$$\sigma_1 = m^2\sigma_x + n^2\sigma_y - 2mn\sigma_s \tag{3.5}$$

$$\begin{aligned} &= m^2\,[(m^2 Q_{xx} + n^2 Q_{xy})\epsilon_1 + (n^2 Q_{xx} + m^2 Q_{xy})\epsilon_2 \\ &\quad + (mnQ_{xx} - mnQ_{xy})\epsilon_6] \\ &\quad + n^2\,[(m^2 Q_{xy} + n^2 Q_{yy})\epsilon_1 + (n^2 Q_{xy} + m^2 Q_{yy})\epsilon_2 \\ &\quad + (mnQ_{xy} - mnQ_{yy})\epsilon_6] \\ &\quad - 2mn\,[-2mnQ_{ss}\epsilon_1 + 2mnQ_{ss}\epsilon_2 + (m^2 - n^2)Q_{ss}\epsilon_6] \end{aligned} \tag{3.6}$$

$$= [m^4 Q_{xx} + n^4 Q_{yy} + 2m^2 n^2 Q_{xy} + 4m^2 n^2 Q_{ss}] \epsilon_1$$

$$+ [m^2 n^2 Q_{xx} + m^2 n^2 Q_{yy} + (m^4 + n^4) Q_{xy} - 4m^2 n^2 Q_{ss}] \epsilon_2 \quad (3.7)$$

$$+ [m^3 n Q_{xx} - mn^3 Q_{yy} + (mn^3 - m^3 n) Q_{xy} + 2 (mn^3 - m^3 n) Q_{ss}] \epsilon_6$$

$$\sigma_1 = Q_{11} \epsilon_1 + Q_{12} \epsilon_2 + Q_{16} \epsilon_6 \quad (3.8)$$

Similarly,

$$\sigma_2 = Q_{21} \epsilon_1 + Q_{22} \epsilon_2 + Q_{26} \epsilon_6 \quad (3.9)$$

$$\sigma_6 = Q_{61} \epsilon_1 + Q_{62} \epsilon_2 + Q_{66} \epsilon_6 \quad (3.10)$$

This is the off-axis stress-strain relation that directly relates the given strain in Figure 3.1(*a*) to the resulting stress in 3.1(*d*), redrawn in Figure 3.2. This relation can also be arranged in a matrix multiplication table as follows:

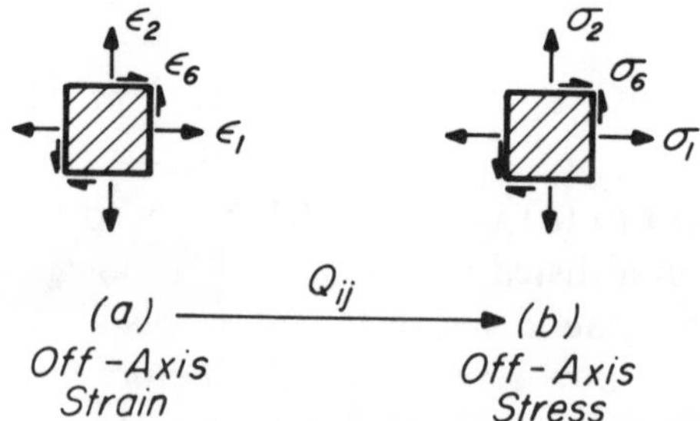

Figure 3.2 The off-axis stress-strain relations in stiffness. We have merged the three steps in Figure 3.1 into one.

table 3.1

off-axis stress-strain relation for unidirectional composites in terms of stiffness

	ϵ_1	ϵ_2	ϵ_6
σ_1	Q_{11}	Q_{12}	Q_{16}
σ_2	Q_{21}	Q_{22}	Q_{26}
σ_6	Q_{61}	Q_{62}	Q_{66}

The major difference between the on-axis stress-strain relation in Table 1.6 and the off-axis relation in Table 3.1 lies in the additional components in the stiffness. These components with subscripts 16 and 26 are shear coupling terms that relate the shear strain to normal stress. Those with 61 and 62 superscripts are normal coupling terms to relate normal strain to shear stress. Such couplings do not exist in conventional materials, or in unidirectional composites in their on-axis orientation. Geometric illustration of these coupling effects will be done when we develop the off-axis orthotropic compliance. Symmetry of these components can also be demonstrated in a manner similar to that used for the Q_{12} component in Chapter 1. The stored energy in Equation 1.16 must contain interaction terms of $\sigma_1 \sigma_6$ and $\sigma_2 \sigma_6$, or their equivalent in strain components $\epsilon_1 \epsilon_6$ and $\epsilon_2 \epsilon_6$.

The relationship between the stiffness components of the on-axis and the off-axis orientations can be summarized in Table 3.2 where matrix multiplication is implied. These relations result from the derivation of Equation 3.8 and what was omitted in Equations 3.9 and 3.10. These relations are limited to transformation from the on-axis, orthotropic orientation where shear coupling components are zero. Note that Q_{xs} and Q_{ys} do not appear as column headings in this table.*

table 3.2

transformation of stiffness from on-axis unidirectional composites in power functions

	Q_{xx}	Q_{yy}	Q_{xy}	Q_{ss}
Q_{11}	m^4	n^4	$2m^2n^2$	$4m^2n^2$
Q_{22}	n^4	m^4	$2m^2n^2$	$4m^2n^2$
Q_{12}	m^2n^2	m^2n^2	m^4+n^4	$-4m^2n^2$
Q_{66}	m^2n^2	m^2n^2	$-2m^2n^2$	$(m^2-n^2)^2$
Q_{16}	m^3n	$-mn^3$	mn^3-m^3n	$2(mn^3-m^3n)$
Q_{26}	mn^3	$-m^3n$	m^3n-mn^3	$2(m^3n-mn^3)$

$m = \cos\theta$, $n = \sin\theta$

*If the transformation is from one off-axis orientation to another, additional columns for Q_{16} and Q_{26} must be present. This unabridged transformation relation can be found in the Appendix A of this book.

This formulation appears most frequently in the literature. This is easy to use if the ply orientation is ±45 degrees, or when one or more on-axis moduli are zero. Note that all the sums of exponents of the trigonometric functions in this table are in the fourth power which are by definition characteristic of the 4th rank tensor. The stress transformation equations are governed by the 2nd power functions, as we have shown in Table 2.1, and belong to the 2nd rank tensor. The strain transformation equations in Table 2.5 are also governed by 2nd power functions, but are different from those for stress because engineering shear strain is used which is twice the tensorial shear strain.

The critical issue is again the sign convention. The angle used in this table is the ply orientation. Because of its importance, Figure 2.2 is shown here again for emphasis. For unidirectional composites, the on-axis, orthotropic, and material symmetry axes coincide. We use the x-y axes to denote this configuration. The off-axis, generally orthotropic configuration refers to ply orientations other than 0 or 90 degrees. We use the 1-2 axes for the off-axis situation. This is shown in Figure 3.3(a). But for multidirectional laminates, there can be many ply orientations. The 1-2 axes remain as the reference coordinates for the laminate. Each ply orientation θ_i can be designated by x_i-y_i axes. The angle used in Table 3.2 is that shown in Figure 3.3. This sign convention is not used universally. Some authors define ply orientation opposite to that shown in Figure 3.3. Then their transformation relations will be different from those shown in Table 3.2. In particular, any term that has the odd power of sines (n and n^3) must change its sign. The shear or normal coupling terms Q_{16} and Q_{26} are the only components affected.

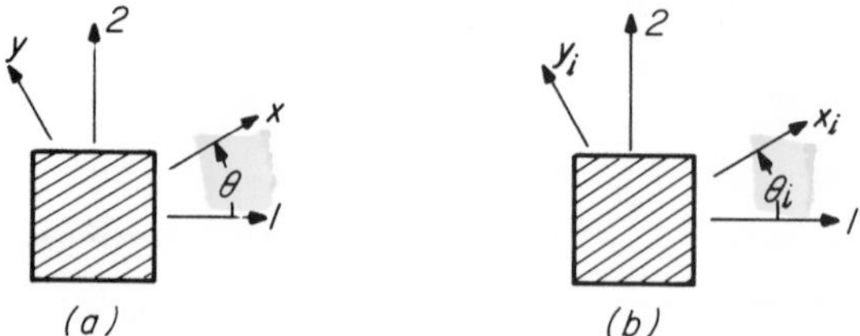

Figure 3.3 Positive ply orientation is shown. The notation for unidirectional composites normally follows that in (a); that for multidirectional composites, in (b) where θ_i is the orientation of the i-th ply or ply group.

In order to clearly define the ply orientation or the angle of transformation, Figure 3.4 shows alternative relations between the coordinates x-y and 1-2. The transformation relations of Table 3.2 are applicable to the angles so defined in Figure 3.4.

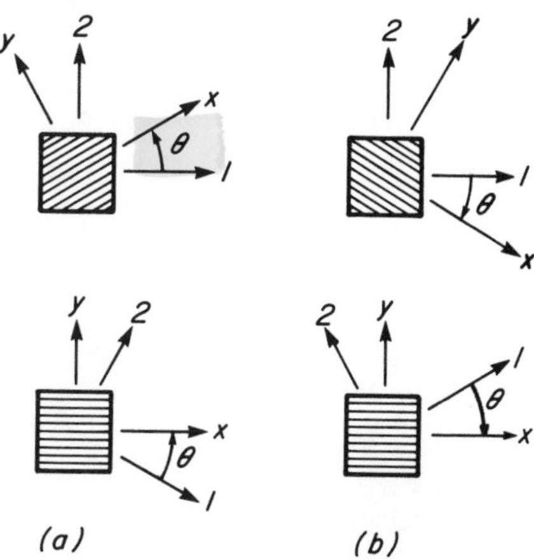

Figure 3.4 Angle of transformation for Table 3.2. Angle is positive in (*a*); and negative in (*b*). Many authors use the relation in (*b*), but call the angle positive. Then the relation in Table 3.2 for the 16 and 26 components must change sign.

There is a fundamental difference between the transformation relations for the stiffness and those for stress and strain.* For the stiffness, the relation is based on a transformation from the on-axis or material symmetry axis to an arbitrary reference axis. The column headings in Table 3.2 have subscripts x, y, s; while the row headings, 1, 2, 6. We go from x-y to 1-2 coordinate system for the stiffness. But for stress and strain, the transformation, such as those shown in Tables 2.1 and 2.5, we go from the off-axis to the on-axis, or 1-2 to x-y coordinate system. While the selection of 1-2 or x-y or any other description of the axes is arbitrary, the stress and strain transformations shown above are valid provided the angle of rotation follows the established signs convention; i.e., positive for counterclockwise rotation. The transformation of stiffness shown in Table 3.2 is valid only when we go from on-axis to off-axis. If we want to go from one off-axis orientation to another

*The difference is referred to in Appendix A as the material versus behavioral quantities.

off-axis, additional terms are required for the transformation relations. This can be readily derived by adding the contribution of the normal coupling term in Equation 3.2. The result is shown in Appendix A.

We can further develop a multiple-angle formulation for the stiffness transformation in place of the power functions in Table 3.2. This process can be done directly by substituting the following trigonometric identities into Table 3.2.*

$$\begin{aligned}
m^4 &= \frac{1}{8}(3 + 4\cos 2\theta + \cos 4\theta) \\
m^3 n &= \frac{1}{8}(2\sin 2\theta + \sin 4\theta) \\
m^2 n^2 &= \frac{1}{8}(1 - \cos 4\theta) \\
mn^3 &= \frac{1}{8}(2\sin 2\theta - \sin 4\theta) \\
n^4 &= \frac{1}{8}(3 - 4\cos 2\theta + \cos 4\theta)
\end{aligned} \tag{3.11}$$

We will now show three examples of the substitution of values from Equation 3.11 into Table 3.2:

$$\begin{aligned}
Q_{11} &= m^4 Q_{xx} + n^4 Q_{yy} + 2m^2 n^2 [Q_{xy} + 2Q_{ss}] \\
&= \frac{1}{8}(3 + 4\cos 2\theta + \cos 4\theta)Q_{xx} + \frac{1}{8}(3 - 4\cos 2\theta + \cos 4\theta)Q_{yy} \\
&\quad + \frac{1}{4}(1 - \cos 4\theta)[Q_{xy} + 2Q_{ss}] \\
&= \frac{1}{8}[3Q_{xx} + 3Q_{yy} + 2Q_{xy} + 4Q_{ss}] + \frac{1}{2}[Q_{xx} - Q_{yy}]\cos 2\theta \\
&\quad + \frac{1}{8}[Q_{xx} + Q_{yy} - 2Q_{xy} - 4Q_{ss}]\cos 4\theta \\
&= U_1 + U_2\cos 2\theta + U_3\cos 4\theta
\end{aligned} \tag{3.12}$$

*This was suggested to us by P. W. Mast, U.S. Naval Research Laboratory, Washington, D.C.

$$Q_{12} = m^2n^2\,[Q_{xx} + Q_{yy} - 4Q_{ss}] + [m^4 + n^4]\,Q_{xy}$$

$$= \frac{1}{8}(1 - \cos 4\theta)[Q_{xx} + Q_{yy} - 4Q_{ss}] + \frac{1}{8}(6 + 2\cos 4\theta)Q_{xy}$$

$$= \frac{1}{8}[Q_{xx} + Q_{yy} + 6Q_{xy} - 4Q_{ss}]$$

$$-\frac{1}{8}[Q_{xx} + Q_{yy} - 2Q_{xy} - 4Q_{ss}]\cos 4\theta$$

$$= U_4 - U_3\cos 4\theta \qquad (3.13)$$

$$Q_{16} = m^3nQ_{xx} - mn^3Q_{yy} + [mn^3 - m^3n][Q_{xy} + 2Q_{ss}]$$

$$= \frac{1}{8}(2\sin 2\theta + \sin 4\theta)Q_{xx} - \frac{1}{8}[2\sin 2\theta - \sin 4\theta]\,Q_{yy}$$

$$-\frac{1}{4}\sin 4\theta\,[Q_{xy} + 2Q_{ss}]$$

$$= \frac{1}{8}[2Q_{xx} - 2Q_{yy}]\sin 2\theta + \frac{1}{8}[Q_{xx} + Q_{yy} - 2Q_{xy} - 4Q_{ss}]\sin 4\theta$$

$$= \frac{1}{2}U_2\sin 2\theta + U_3\sin 4\theta \qquad (3.14)$$

We can repeat the process for the other three components of the off-axis modulus and list the results in Table 3.3 in matrix multiplication format and using the following definitions of the linear combinations of modulus:

$$U_1 = \frac{1}{8}[3Q_{xx} + 3Q_{yy} + 2Q_{xy} + 4Q_{ss}]$$

$$U_2 = \frac{1}{2}[Q_{xx} - Q_{yy}]$$

$$U_3 = \frac{1}{8}[Q_{xx} + Q_{yy} - 2Q_{xy} - 4Q_{ss}] \qquad (3.15)$$

(continues)

$$U_4 = \frac{1}{8}[Q_{xx} + Q_{yy} + 6Q_{xy} - 4Q_{ss}]$$

$$U_5 = \frac{1}{8}[Q_{xx} + Q_{yy} - 2Q_{xy} + 4Q_{ss}] \tag{3.15}$$

(concluded)

table 3.3
transformed stiffness from on-axis unidirectional composites in multiple angle functions

	I	U_2	U_3
Q_{11}	U_1	$\cos 2\theta$	$\cos 4\theta$
Q_{22}	U_1	$-\cos 2\theta$	$\cos 4\theta$
Q_{12}	U_4		$-\cos 4\theta$
Q_{66}	U_5		$-\cos 4\theta$
Q_{16}		$\frac{1}{2}\sin 2\theta$	$\sin 4\theta$
Q_{26}		$\frac{1}{2}\sin 2\theta$	$-\sin 4\theta$

This formulation has two distinct advantages over the power function formulation. First, the invariants are explicit. Secondly, the integration and differentiation of multiple angle functions are easier than those of power functions.

From the transformation equations in Table 3.3, we can show the off-axis combinations listed in Equation 3.15 are:

$$U_1' = \frac{1}{8}[3Q_{11} + 3Q_{22} + 2Q_{12} + 4Q_{66}] \tag{3.16}$$

From Table 3.3:

$$= \frac{1}{8}[6U_1 + 2U_4 + 4U_5] \tag{3.17}$$

From Equation 3.15:

$$= \frac{1}{64}\{[18Q_{xx} + 18Q_{yy} + 12Q_{xy} + 24Q_{ss}] + [2Q_{xx} + 2Q_{yy} + 12Q_{xy} - 8Q_{ss}] + [4Q_{xx} + 4Q_{yy} - 8Q_{xy} + 16Q_{ss}]\}$$

$$= \frac{1}{64}[24Q_{xx} + 24Q_{yy} + 16Q_{xy} + 32Q_{ss}] \tag{3.18}$$

$$= \frac{1}{8}[3Q_{xx} + 3Q_{yy} + 2Q_{xy} + 4Q_{ss}]$$

$$= U_1 = \text{an invariant}$$

Similarly

$$\begin{aligned} U_2' &= U_2\cos 2\theta &&= \text{not invariant} \\ U_3' &= U_3\cos 4\theta &&= \text{not invariant} \\ U_4' &= U_4 &&= \text{an invariant} \\ U_5' &= U_5 &&= \text{an invariant} \end{aligned} \tag{3.19}$$

When the off-axis(primed) and on-axis(un-primed) combinations are equal, they are by definition invariant. This is analogous to the stress invariants in Equation 2.14. Note that U_1, U_4, and U_5 are first-order or linear invariants, of which two are independent because we can show from Equation 3.15 that

$$U_5 = \frac{1}{2}[U_1 - U_4] \tag{3.20}$$

This relation between the invariants is analogous to that between the modulus of isotropic materials shown in Equation 1.23. We have shown that stress and strain possess a second-order or quadratic invariant each; i.e., the radius of Mohr's circle. The modulus also possesses second-order invariants. They can be derived as follows. We will first define two additional linear combinations for the modulus.

$$\begin{cases} U_6' = \frac{1}{2}[Q_{16} + Q_{26}] = \frac{1}{2}U_2 \sin 2\theta \\ U_7' = \frac{1}{2}[Q_{16} - Q_{26}] = U_3 \sin 4\theta \end{cases} \tag{3.21}$$

We can now derive two second-order invariants as follows:

$$\begin{aligned} R_1^2 &= U_2'^2 + 4U_6'^2 \\ &= U_2^2 [\cos^2 2\theta + \sin^2 2\theta] \\ &= U_2^2 \end{aligned} \tag{3.22}$$

or

$$R_1 = \pm U_2 \tag{3.23}$$

Similarly,

$$\begin{aligned} R_2^2 &= U_3'^2 + U_7'^2 \\ &= U_3^2 [\cos^2 4\theta + \sin^2 4\theta] \\ &= U_3^2 \end{aligned} \tag{3.24}$$

or

$$R_2 = \pm U_3 \tag{3.25}$$

where R_1 and R_2 are invariants and U_2 and U_3 are not. The values of U_2' and U_3' stated in Equation 3.19 are based on the on-axis orientation for which the shear/normal coupling terms are zero. For off-axis orientations, terms containing U_6' and U_7' must be added to the relations in Equations 3.19 and 3.21. This is shown in Appendix A. The R's are radii of the equivalent Mohr's circles for the modulus. (See Figure 3.9.) They must always be positive. The U's can be positive or negative. In fact, U_2 would be negative if the longitudinal and transverse directions are interchanged. The correct sign in Equations 3.23 and 3.24 must be picked to make R_1 and R_2 positive.

We can derive the transformation equations for stiffness in terms of invariants as we did for stress and strain transformations and shown in Tables 2.4 and 2.8. If we limit ourselves to transformed stiffness from the material symmetry or orthotropic axes, the results will be identical to those shown in Table 3.3; in which case, positive values of U_2 and U_3 are assigned to R_1 and R_2, respectively. This is done in Table 3.4, where matrix multiplication is implied.

table 3.4
transformed stiffness from on-axis unidirectional composites in invariant functions

	I	R_1	R_2
Q_{11}	U_1	$\cos 2\theta$	$\cos 4\theta$
Q_{22}	U_1	$-\cos 2\theta$	$\cos 4\theta$
Q_{12}	U_4		$-\cos 4\theta$
Q_{66}	U_5		$-\cos 4\theta$
Q_{16}		$\frac{1}{2}\sin 2\theta$	$\sin 4\theta$
Q_{26}		$\frac{1}{2}\sin 2\theta$	$-\sin 4\theta$

This table is valid for the x-axis to be pointed along the fiber orientation, like that in Figures 3.3 and 3.4. If a material is not orthotropic but anisotropic, the table must be modified. The definitions of R's will remain the same as Equations 3.22 and 3.24. But the off-axis linear combinations of stiffness U_2', U_3', U_6' and U_7' will have additional terms. There will also be two phase angles, analogous to that for the stress transformation in Equation 2.19. The transformation of anisotropic modulus will be listed in Appendix A.

2. examples of off-axis stiffness

We will show in this section the transformed stiffness for a graphite-epoxy composite. The particular material system for our example is the Union Carbide and Toray T300 filament and Narmco 5208 resin, or T300/5208 for short. The stiffness data for this material were listed in Table 1.9. We can immediately calculate the transformed stiffness by substituting the data into the transformation equations in Table 3.2. Numerical data for the transformed stiffness are listed in Table 3.5 and plotted in Figure 3.5. All six transformed components of the stiffness are shown. The angle of ply orientation is also shown where counter-clockwise direction is positive.

table 3.5
transformed stiffness of T300/5208 unidirectional composites (GPa)

θ	Q_{11}	Q_{22}	Q_{12}	Q_{66}	Q_{16}	Q_{26}
0	181.8	10.3	2.90	7.17	0	0
15	160.4	11.9	12.75	17.05	38.50	4.36
30	109.3	23.6	32.46	36.78	54.19	20.05
45	56.6	56.6	42.32	46.59	42.87	42.87
60	23.6	109.3	32.46	36.78	20.05	54.19
75	11.9	160.4	12.75	17.05	4.36	38.50
90	10.3	181.8	2.90	7.17	0	0

An alternative method of arriving at the same transformed stiffness is the use of the multiple-angle or the invariant formulation in Table 3.3 or Table 3.4, respectively. We must first determine the values of the U's using Equation 3.15 and data in Table 1.9. The results of several unidirectional composites including T300/5208 are listed in Table 3.6. A typical calculation is listed as follows:

From Equation 3.15

$$U_1 = \frac{1}{8}[3Q_{xx} + 3Q_{yy} + 2Q_{xy} + 4Q_{ss}]$$

From Table 1.9 for T300/5208

$$U_1 = \frac{1}{8}(3 \times 181.8 + 3 \times 10.34 + 2 \times 2.897 + 4 \times 7.17)$$

$$= 76.37 \text{ GPa} \tag{3.26}$$

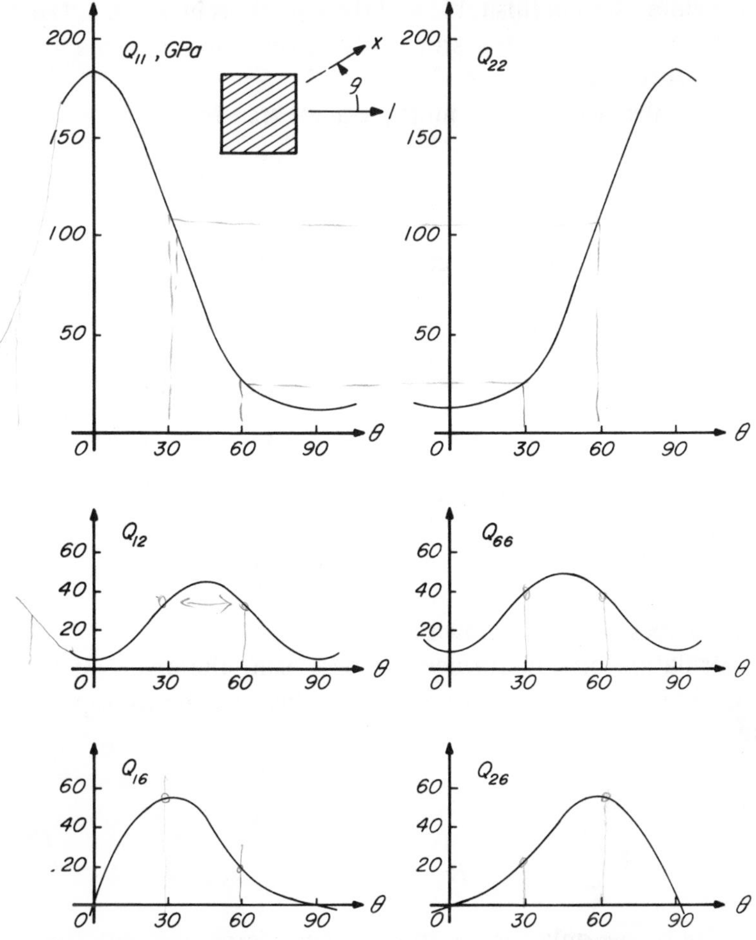

Figure 3.5 Transformed, off-axis stiffness of T300/5208. The angle is the ply orientation, and is positive for counterclockwise rotation.

obtained by on-axis stiffness

table 3.6

linear combinations of stiffness for transformation of modulus (GPa)

	U_1	$U_2 = R_1$	$U_3 = R_2$	U_4	U_5
T300/5208	76.37	85.73	19.71	22.61	26.88
B(4)/5505	87.80	93.21	23.98	28.26	29.77
AS/3501	59.65	64.89	14.25	16.95	21.35
Scotchply 1002	20.45	15.38	3.32	5.51	7.46
Kevlar 49/Epoxy	32.44	35.54	8.65	10.53	10.95

Question Al ?

With the values listed in this table and the equations in Table 3.3 or 3.4, we can arrive at the same transformed stiffness listed in Table 3.5 and shown in Figure 3.5.

A typical calculation by the multiple-angle transformation is listed as follows:

From Table 3.4:

$$Q_{11} = U_1 + U_2\cos 2\theta + U_3\cos 4\theta \tag{3.27}$$

From Table 3.6 for T300/5208:

$$Q_{11} = 76.37 + 85.73\cos 2\theta + 19.71\cos 4\theta \tag{3.28}$$

when

$$\theta = 45 \text{ degrees}, \cos 2\theta = 0, \cos 4\theta = -1$$

$$Q_{11} = 76.37 - 19.71 = 56.66 \text{ GPa} \tag{3.29}$$

This result agrees with that shown in Table 3.5.

The calculation of transformed stiffness using the invariant formulation as shown in Table 3.4 will be identical to the multiple angle because

$$\begin{aligned} R_1 &= +U_2 \\ R_2 &= +U_3 \end{aligned} \tag{3.30}$$

This identity is true only for the stiffness transformation with the fiber axis placed along the l-axis; i.e.,

$$Q_{11} > Q_{22}$$

As we shall see later, the transformed compliance calls for negative signs in Equation 3.23. Then the multiple-angle formulation is not identical to the invariant formulation because of this sign change.

A number of general remarks can be made about the transformed stiffness, applicable to orthotropic composites listed in Table 3.6.

Mirror image exists between Q_{11} and Q_{22}, and Q_{16} and Q_{26}. This can be shown by substituting $\theta+90$ into appropriate equations in Tables 3.3 or 3.4, and seen in Figure 3.5.

$$
\begin{aligned}
Q_{11}(\theta+90) &= U_1 + U_2\cos 2(\theta+90) + U_3\cos 4(\theta+90) \\
&= U_1 - U_2\cos 2\theta + U_3\cos 4\theta \qquad (3.31) \\
&= Q_{22}(\theta)
\end{aligned}
$$

These two components can be superimposed by a displacement of 90 degrees along the θ-axis. This is expected because cosine functions are even. We can also show

$$
Q_{26}(\theta) = -Q_{26}(\theta) = -Q_{16}(\theta-90) = +Q_{16}(90-\theta) \qquad (3.32)
$$

These two components can be superimposed by a displacement and a rotation. This is also expected because these components are dependent on sine functions (see Table 3.4) which are odd.

The angular dependency and amplitude of Q_{12} and Q_{66} are the same; i.e., 4θ and U_3, respectively. (See Figure 3.6.) The two transformed components are vertically displaced by the amount of $U_5 - U_4$. The Poisson and shear components are dependent on cosine functions, and therefore symmetric about the $\theta = 0$, 90, and ±45 degrees. We can readily derive from Equation 3.15 that

$$
U_5 - U_4 = Q_{ss} - Q_{xy} \qquad (3.33)
$$

This is another invariant. This is not an independent invariant. We can say that any linear combination of invariants is an invariant.

The first four transformed components in Table 3.3 (i.e. the normal, Poisson, and shear components) are governed by cosine functions. They are even functions and not sensitive to the sign of ply orientation. So an error in the sign will not affect these components. But the last two components in Table 3.3 are the shear or normal coupling components and are governed by sine or odd functions. A sign error will lead to a real error.

Because of the symmetry relations between these transformed components, we only need to show three curves instead of six in Figure 3.5; either the three curves on the left of this figure or the three on the right. In Table 3.5, Q_{22} and Q_{26} can be deleted, if we know their symmetry relation with Q_{11} and Q_{16}, respectively.

The Q_{11} or Q_{22} component is made up of three terms; one constant

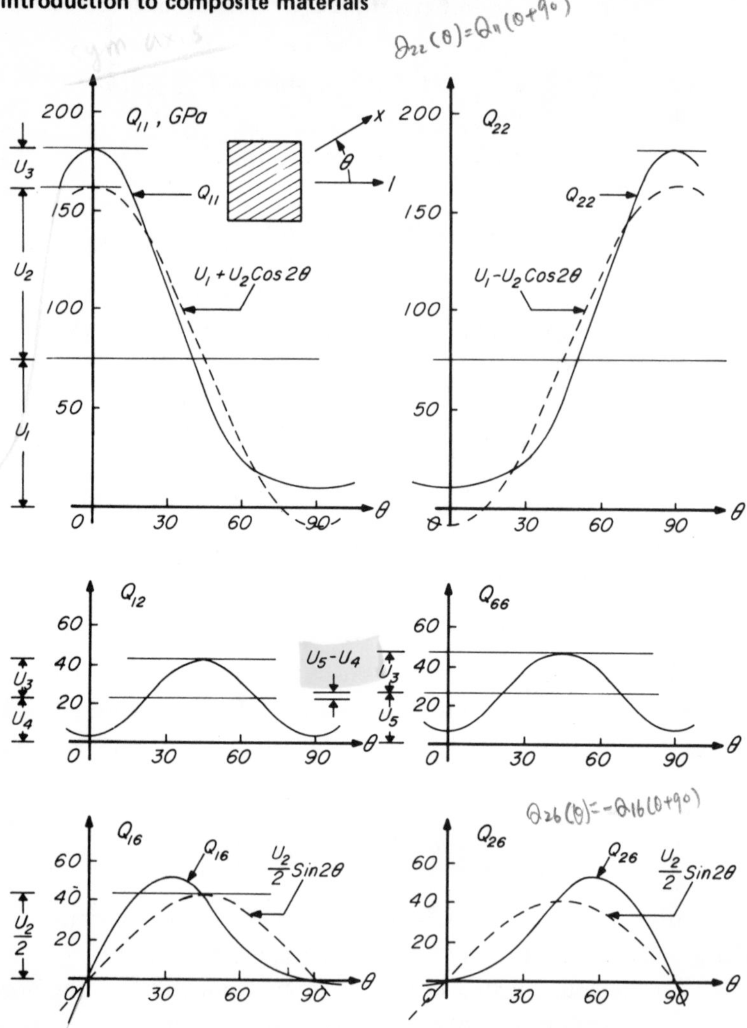

Figure 3.6 Transformed stiffness as functions of U's. The relations in Table 3.3 are shown graphically. Each transformed component contains an invariant and/or cyclic terms.

or invariant term and two cyclic terms with angular rotations 2 and 4 times that of the coordinate rotation; see Equation 3.27. Since the cyclic terms do not contribute to the total area under this curve, this area is simply proportional to U_1. Thus, this invariant represents the Young's modulus or normal (as opposed to shear) stiffness potential of this unidirectional composite. The cyclic terms contribute to the directional changes. The increase in the stiffness in one direction must be made up by a decrease in some other direction while the total area under the transformed modulus remains constant and invariant. This is like an incompressible material that can undergo a shape change without any volume change. Figure 3.6 shows the contribution of the U's to the stiffness components.

The shear and normal coupling terms Q_{16} and Q_{26} have no invariant associated with the transformation. They are not independent in the sense that they are derivable from Q_{11} and Q_{22} by differentiation, respectively. From Table 3.3 or 3.4,

$$\frac{\partial Q_{11}}{\partial \theta} = -2U_2 \sin 2\theta - 4U_3 \sin 4\theta = -4Q_{16} \tag{3.34}$$

$$\frac{\partial Q_{22}}{\partial \theta} = 2U_2 \sin 2\theta - 4U_3 \sin 4\theta = 4Q_{26} \tag{3.35}$$

From Equation 3.34

$$Q_{16} = 0, \text{ when } \theta = 0 \text{ and } 90 \text{ degrees, or} \tag{3.36}$$

$$\text{when } U_2 + 4U_3 \cos 2\theta = 0, \text{ or}$$

$$\cos 2\theta = -\frac{U_2}{4U_3} \tag{3.37}$$

For T300/5208 from Table 3.6

$$\frac{U_2}{4U_3} = 85.73/4 \times 19.71 = 1.08 > 1 \tag{3.38}$$

There is, therefore, no solution for θ from Equation 3.37. The shear coupling goes to zero only at 0 and 90 degrees. The same holds true for Q_{26}, except the sign in Equation 3.37 is positive. Because of these relations, the tangents, maxima and points of inflection between Q_{11} and Q_{16} can be obtained from

$$\frac{\partial^2 Q_{11}}{\partial \theta^2} = -4U_2 \cos 2\theta - 16U_3 \cos 4\theta = 0 \tag{3.39}$$

Substituting $\cos 4\theta = 2\cos^2 2\theta - 1$, and rearranging, we get

$$\cos^2 2\theta + \frac{U_2}{8U_3}\cos 2\theta - \frac{1}{2} = 0 \tag{3.40}$$

The solutions for θ are

$$\cos 2\theta = -\frac{U_2}{16U_3} \pm \sqrt{\left(\frac{U_2}{16U_3}\right)^2 + \frac{1}{2}} \tag{3.41}$$

For T300/5208,

$$\begin{aligned} \cos 2\theta &= 0.485; -1.029 \\ \theta &= 30.4 \text{ degrees; no solution} \end{aligned} \tag{3.42}$$

At this angle, it is the point of inflection in the Q_{11} curve and a maximum in the Q_{16}. Similarly, at 59.6 degrees, Q_{22} has the inflection, and Q_{26}, the maximum. These relations are shown in Figure 3.7. Points of inflection of Q_{16} can be found from letting

$$\frac{\partial^3 Q_{11}}{\partial \theta^3} = 8U_2 \sin 2\theta + 64U_3 \sin 4\theta = 0 \tag{3.43}$$

$$8\sin 2\theta[U_2 + 16U_3 \cos 2\theta] = 0 \tag{3.44}$$

$$\begin{aligned} \sin 2\theta &= 0 \text{ or } \theta = 0, 90 \text{ for all components} \\ \cos 2\theta &= -\frac{U_2}{16U_3} \end{aligned} \tag{3.45}$$

$$\theta = 52.88 \text{ degrees for T300/5208}$$

The power functions formulation for the stiffness transformation can be used to demonstrate the dominance of the longitudinal properties of unidirectional composites. Since the first column of Table 3.2 is

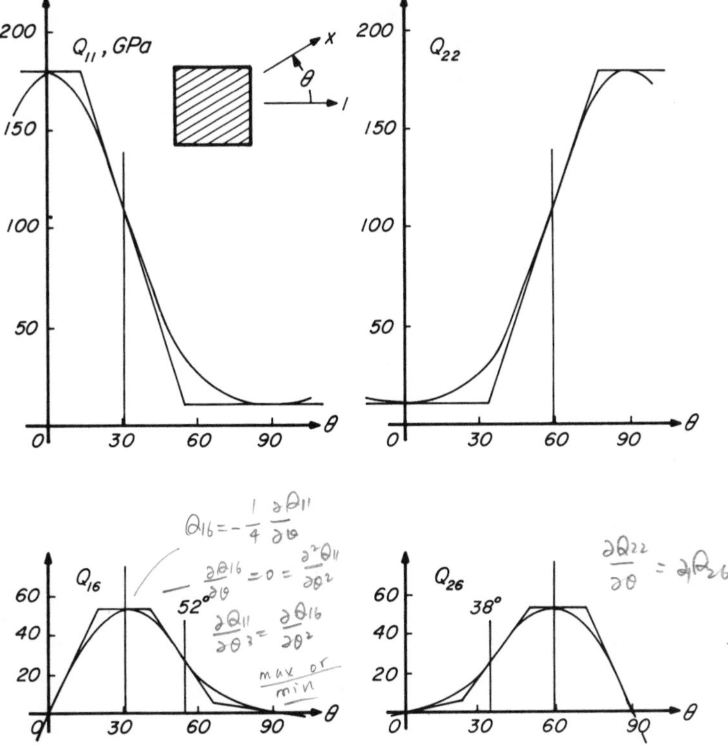

Figure 3.7 Relationships between transformed stiffness. Special relationships are expressed in Equations 3.34 to 3.45. Note the point of inflection in Q_{11} is the point of maximum value in Q_{16}. At 0 and 90 degrees, Q_{16} are zero, and the slopes of Q_{11} are also zero. The points of inflection for Q_{16} are also shown.

many times higher than the other components; i.e., 181 GPa versus 10, 3 and 7 for T300/5208 based on the data in Table 1.9, we can show the contribution of the first column or that of Q_{xx} in Figure 3.8. The dashed lines are those transformed stiffness based on the first column of Table 3.2; the solid lines are the complete solutions, as those in Figure 3.5. We can see that for a highly anisotropic unidirectional composite such as T300/5208, this approximation is fairly close to the exact. By the same token, we can approximate the values of U's in Equation 3.15 as:

$$
\begin{aligned}
U_1 &= \frac{3Q_{xx}}{8} \\
U_2 &= \frac{Q_{xx}}{2} \\
U_3 = U_4 = U_5 &= \frac{Q_{xx}}{8}
\end{aligned}
\tag{3.46}
$$

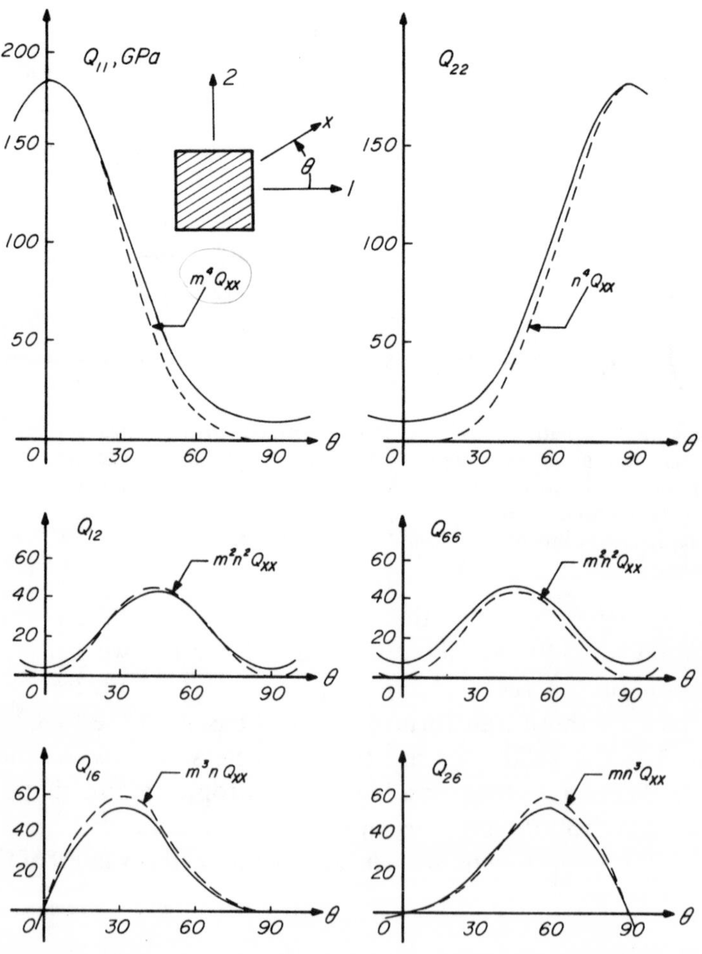

Figure 3.8 Approximation of transformed stiffness. Only the Q_{xx} term for T300/5208 is used. The dashed lines are approximate; the solid lines, exact.

The approximate transformation equations can be simplified from Table 3.3 as follows:

$$Q_{11} = \frac{1}{8}(3 + 4\cos 2\theta + \cos 4\theta)Q_{xx} = m^4 Q_{xx} \tag{3.47}$$

This and the other components of transformed stiffness will be the same as the first column of Table 3.2. Thus, both power functions and multiple-angle functions for the stiffness transformation reduce to the same limiting case when only the Q_{xx} is present.

The transformation relations in Tables 3.3 and 3.4 can be illustrated by two generalized Mohr's circles similar to the Mohr's circle for the transformation of stress or strain. From the first equation in Table 2.4 for the stress transformation:

$$\sigma_x = I + R\cos 2[\theta - \theta_o] \tag{3.48}$$

A Mohr's circle can be constructed as shown in Figure 2.8. The location of the center is I, the radius of the circle is R, and the phase angle determines the specific stress components for a given state of stress.

The transformation of modulus is governed by similar equations such as the first one from Table 3.4:

$$Q_{11} = U_1 + R_1\cos 2\theta + R_2\cos 4\theta \tag{3.49}$$

This relation can also be shown as Mohr's circles. Now we have two circles with radius R_1 and R_2, and angular rotations two and four times that of the coordinate axes. The distance between the circle is the invariant U_1. The generalized Mohr's circles for the stiffness of an orthotropic material such as a unidirectional composite T300/5208 are shown in Figure 3.9 using the data in Table 3.6. The radii of the circles dictate the degree of orthotropy. The direction of rotation corresponding to a positive ply orientation θ is also shown in this figure. As we will see in the next chapter, the effective radii of the generalized Mohr's circles reduce as unidirectional plies are made into multidirectional laminates. The distance between the centers, however, remains invariant and fixed. In the limit, the radii can go to zero and we are left with only the isotropic constant U_1. The process of lamination always reduces the radii of the generalized Mohr's circle.

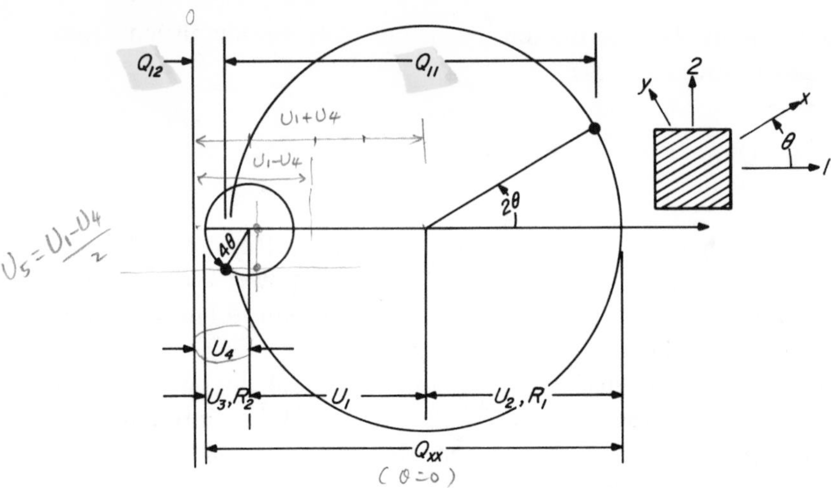

Figure 3.9 Generalized Mohr's circles for the stiffness of T300/5208. The distance between the two centers is the first invariant. The radii for a given material are also fixed. The angle of rotation of the coordinate axes is magnified to two and four times in the Mohr's circles.

3. off-axis compliance

Analogous to the approach for the off-axis modulus, we can derive the off-axis compliance following the sequence of Figure 3.10, which is a repeat of Figure 2.3.

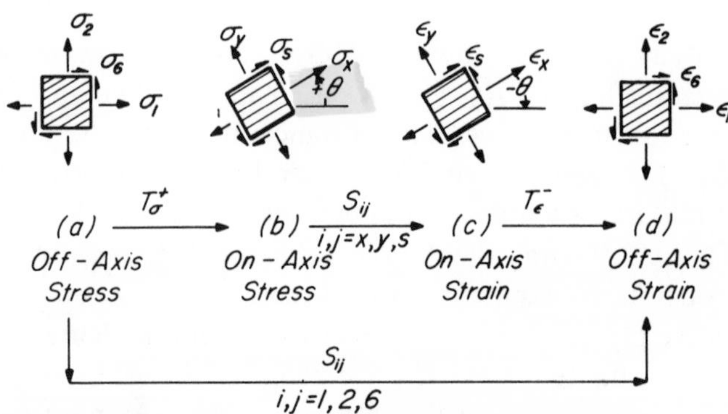

Figure 3.10 Derivation of off-axis compliance.
From (*a*) to (*b*): Positive stress transformation.
From (*b*) to (*c*): Stress-strain relations in compliance.
From (*c*) to (*d*): Negative strain transformation.
We can go directly from (*a*) to (*d*) by merging the three steps into one.

Since the derivation of the compliance transformation is analogous to that of the stiffness transformation from Equations 3.1 to 3.10, we will write only the first line of each equation for this derivation.

- To go from (*a*) to (*b*) in Figure 3.10, we use the equations for stress transformation from Table 2.1:

$$\sigma_x = m^2\sigma_1 + n^2\sigma_2 + 2mn\sigma_6 \tag{3.50}$$

- To go from (*b*) to (*c*) in Figure 3.10, we need the on-axis stress strain relation in compliance from Table 1.5.

$$\begin{aligned} \epsilon_x = \; & S_{xx}\,[m^2\sigma_1 + n^2\sigma_2 + 2mn\sigma_6] \\ & + S_{xy}\,[n^2\sigma_1 + m^2\sigma_2 - 2mn\sigma_6] \end{aligned} \tag{3.51}$$

- To go from (*c*) to (*d*) in Figure 3.10, we need the negative strain transformation in Table 2.5, where the sine functions now have negative signs.

$$\epsilon_1 = m^2\epsilon_x + n^2\epsilon_y - mn\epsilon_s \tag{3.52}$$

$$\begin{aligned} = \; & \{m^4S_{xx} + n^4S_{yy} + 2m^2n^2S_{xy} + m^2n^2S_{ss}\}\sigma_1 \\ & + \{m^2n^2[S_{xx} + S_{yy}] + [m^4 + n^4]S_{xy} - m^2n^2S_{ss}\}\sigma_2 \\ & + \{2m^3nS_{xx} - 2mn^3S_{yy} + [mn^3 - m^3n]\,[2S_{xy} + S_{ss}]\}\sigma_6 \end{aligned} \tag{3.53}$$

$$\epsilon_1 = S_{11}\sigma_1 + S_{12}\sigma_2 + S_{16}\sigma_6 \tag{3.54}$$

Note that shear coupling terms appear in this off-axis unidirectional composite, in an analogous fashion as the off-axis stiffness equations 3.8 to 3.10. The off-axis stress-strain relation in terms of compliance is presented in a matrix multiplication table in Table 3.7. This is similar to the off-axis stress-strain relation in terms of the stiffness in Table 3.1. The symmetry relations for the transformed compliance can be shown by including interaction terms $\sigma_1\sigma_6$ and $\sigma_2\sigma_6$ in addition to $\sigma_1\sigma_2$ in the stored energy expression in Equation 1.16. We can then show that

$$S_{16} = S_{61},\; S_{26} = S_{62} \tag{3.55}$$

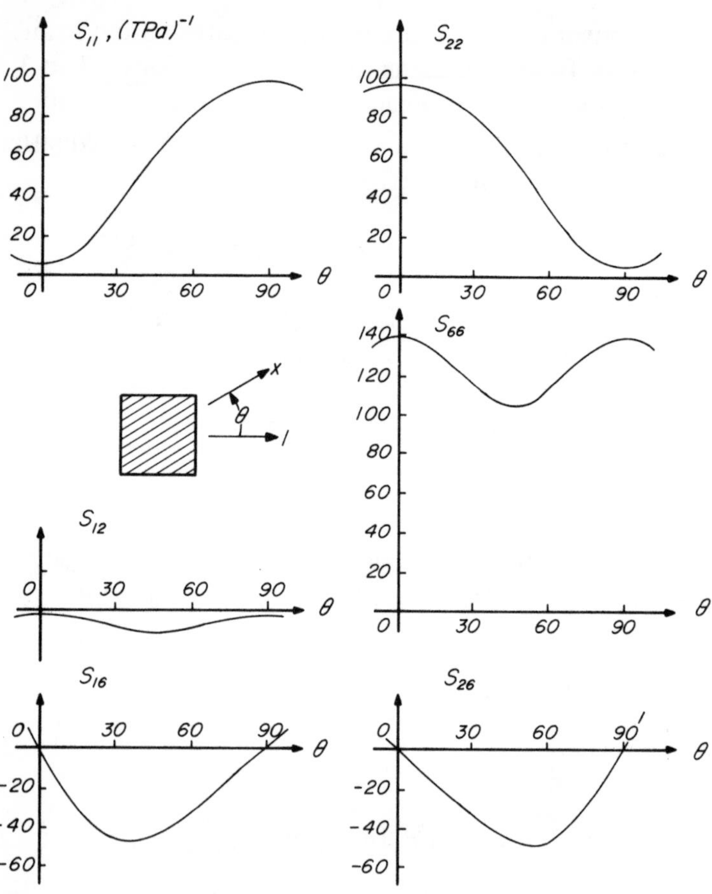

Figure 3.11 Transformed, off-axis compliance of T300/5208. The angle of rotation is positive when it is counterclockwise.

table 3.7

off-axis stress-strain relation for unidirectional composites in terms of compliance

	σ_1	σ_2	σ_6
ϵ_1	S_{11}	S_{12}	S_{16}
ϵ_2	S_{21}	S_{22}	S_{26}
ϵ_6	S_{61}	S_{62}	S_{66}

The transformation equations for the compliance are taken from matching like terms in Equations 3.53 and 3.54, and will be shown in Table 3.8. This table is analogous to the transformed stiffness in Table 3.2.

table 3.8

transformation of compliance of on-axis unidirectional composites in power functions

	S_{xx}	S_{yy}	S_{xy}	S_{ss}
S_{11}	m^4	n^4	$2m^2n^2$	m^2n^2
S_{22}	n^4	m^4	$2m^2n^2$	m^2n^2
S_{12}	m^2n^2	m^2n^2	m^4+n^4	$-m^2n^2$
S_{66}	$4m^2n^2$	$4m^2n^2$	$-8m^2n^2$	$(m^2-n^2)^2$
S_{16}	$2m^3n$	$-2mn^3$	$2(mn^3-m^3n)$	mn^3-m^3n
S_{26}	$2mn^3$	$-2m^3n$	$2(m^3n-mn^3)$	m^3n-mn^3

$m = \cos\theta,\ n = \sin\theta$

Note that the difference between this table and Table 3.2 for the transformation of stiffness can be traced to the use of engineering shear strain. For each component with single subscript 6, the coefficients on each row shall be multiplied by 2. For components with double subscript 6, such as S_{66}, the coefficients shall be multiplied by 4. In the last column of Table 3.8, the effect of double subscript s is to divide each coefficient by 4. All the differences between Tables 3.8 and 3.2 can be accounted for with these corrections.

The multiple-angle formulation of the transformation of compliance follows precisely the same pattern as that for the transformed stiffness. The multiple-angle trigonometric identities in Equation 3.11 can be substituted into the coefficients in Table 3.8. By following the same process as in Equation 3.12 to 3.14, we can derive the multiple-angle representation of the transformed compliance in Table 3.9.

table 3.9
transformed compliance for on-axis unidirectional composites in multiple-angle functions

	I	U_2	U_3
S_{11}	U_1	$\cos 2\theta$	$\cos 4\theta$
S_{22}	U_1	$-\cos 2\theta$	$\cos 4\theta$
S_{12}	U_4		$-\cos 4\theta$
S_{66}	U_5		$-4\cos 4\theta$
S_{16}		$\sin 2\theta$	$2\sin 4\theta$
S_{26}		$\sin 2\theta$	$-2\sin 4\theta$

The definitions of the U's are:

$$U_1 = \frac{1}{8}[3S_{xx} + 3S_{yy} + 2S_{xy} + S_{ss}]$$

$$U_2 = \frac{1}{2}[S_{xx} - S_{yy}]$$

$$U_3 = \frac{1}{8}[S_{xx} + S_{yy} - 2S_{xy} - S_{ss}] \qquad (3.56)$$

$$U_4 = \frac{1}{8}[S_{xx} + S_{yy} + 6S_{xy} - S_{ss}]$$

$$U_5 = \frac{1}{2}[S_{xx} + S_{yy} - 2S_{xy} + S_{ss}]$$

The difference between the U's of this equation and those for the stiffness in Equation 3.15 can again be traced to the use of engineering shear strain. The shear invariant U_5 and the S_{ss} component must be multiplied and divided by four, respectively, in order to match Equation 3.56 with 3.15.

Of the three linear or first-order invariants in Equation 3.56 only two are independent. The following relationship shows that the third invariant is dependent on the other two.

$$U_5 = 2[U_1 - U_4] \tag{3.57}$$

There are also two quadratic or second-order invariants which can be derived from the second and third columns in Table 3.9.

$$R_1^2 = \frac{1}{4}[S_{11} - S_{22}]^2 + \frac{1}{4}[S_{16} + S_{26}]^2 = U_2^2 \tag{3.58}$$

or

$$R_1 = \pm U_2 \tag{3.59}$$

$$R_2^2 = \frac{1}{64}[S_{11} + S_{22} - 2S_{12} - S_{66}]^2 + \frac{1}{16}[S_{16} - S_{26}]^2 = U_3^2 \tag{3.60}$$

or

$$R_2 = \pm U_3 \tag{3.61}$$

From the relationship above, we can derive the transformation equations in terms of the invariants. For the compliance of an on-axis unidirectional composite, U_2 and U_3 are negative if the longitudinal(the x-axis) stiffness is higher than the transverse(the y-axis) stiffness. (See data in Table 3.11.) The transformation of compliance using the invariant functions is listed in a matrix multiplication table as follows:

table 3.10
transformed compliance of on-axis unidirectional composites in invariant functions

	I	R_1	R_2
S_{11}	U_1	$-\cos 2\theta$	$-\cos 4\theta$
S_{22}	U_1	$\cos 2\theta$	$-\cos 4\theta$
S_{12}	U_4		$\cos 4\theta$
S_{66}	U_5		$4\cos 4\theta$
S_{16}		$-\sin 2\theta$	$-2\sin 4\theta$
S_{26}		$-\sin 2\theta$	$2\sin 4\theta$

Note that the signs of the trigonometric functions are changed from those in Table 3.9 because U_2 and U_3 have negative values. So negative signs must be used in Equations 3.59 and 3.61, or

$$R_1 = -U_2$$
$$R_2 = -U_3 \tag{3.62}$$

This choice of signs is different from the invariant functions of the modulus transformation in Table 3.4 where the positive values of U_2 and U_3 were picked. This was so because Q_{xx} is greater than Q_{yy} for most unidirectional composites. Table 3.10 as well as Tables 3.8 and 3.9 are limited to orthotropic compliance. For anisotropic compliance, comparable transformation tables are shown in Appendix A.

4. examples of off-axis compliance

We will show in this section the transformed compliance for T300/5208. The orthotropic components of the compliance are listed in Table 1.8 for this composite. When these values are substituted into the transformation equations in Table 3.8, we will get the transformed compliance.

Alternatively, we can arrive at the same transformed components if we use the U's computed from the relations given in Equation 3.56, and the values for typical composites are listed in Table 3.11. The transformed compliance can then be computed from the relations in Table 3.9. The numerical results are listed in Table 3.12, and curves plotted in Figure 3.11.

table 3.11

typical values of linear combinations of compliance for on-axis unidirectional composites (TPa)$^{-1}$

	U_1	$U_2 = -R_1$	$U_3 = -R_2$	U_4	U_5
T300/5208	55.53	−45.78	− 4.22	− 5.77	122.6
B(4)/5505	43.42	−24.55	−13.94	−15.06	117.0
AS/3501	61.62	−52.18	− 2.20	− 4.38	132.0
Scotchply 1002	83.50	−47.50	−10.20	−16.90	200.8
Kevlar 49/Epoxy	126.40	−84.33	−28.86	−33.33	319.4

table 3.12
transformed compliance for T300/5208 unidirectional composites (TPa)$^{-1}$

θ	S_{11}	S_{22}	S_{12}	S_{66}	S_{16}	S_{26}
0	5.52	97.09	−1.55	139.4	0	0
15	13.77	93.06	−3.66	131.0	−30.20	−15.58
30	34.75	80.53	−7.88	114.1	−46.96	−32.34
45	59.75	59.75	−9.99	105.7	−45.78	−45.78
60	80.53	34.75	−7.88	114.1	−32.34	−46.96
75	93.06	13.77	−3.66	131.0	−15.58	−30.20
90	97.09	5.52	−1.54	139.4	0	0

The general remarks on the transformed compliance are very similar to those on the stiffness. We will simply repeat the relevant features without further detailed discussions.

Mirror image exists between S_{11} and S_{22}, and S_{16} and S_{26}. This can be seen from Table 3.12 and Figure 3.11.

The amplitude of S_{12} is now one quarter that of S_{66}. The angle 4θ remains the same for both transformed components.

Again, only the shear and normal coupling components are affected by the sign of the angle of rotation.

Because of the symmetry, only three transformed components S_{11}, S_{12}(or S_{66}) and S_{16} need to be drawn.

In Figure 3.12 the dashed lines show the approximation of the transformed components using only the first one or two columns of the complete transformation equations in Table 3.9. Because of the particular values for T300/5208, the approximations give excellent results. This is analogous to the transformed stiffness shown in Figure 3.8.

5. inverse relationship between modulus and compliance

The off-axis stress-strain relations as listed in Tables 3.1 and 3.7 are based on stiffness and compliance, respectively, and are repeated here as Tables 3.13 and 3.14. The difference between these stress-strain relations is that the role of stress and strain are the inverse of each other. In Table 3.13, the strain is the independent variable; in Table 3.14, the stress is the independent. We inverted the on-axis stress-strain relations in Chapter 1. We went from Equations 1.8 to 1.11 by simply solving the simultaneous equations. We need only to repeat the same process for the off-axis case, where shear and normal coupling terms are no

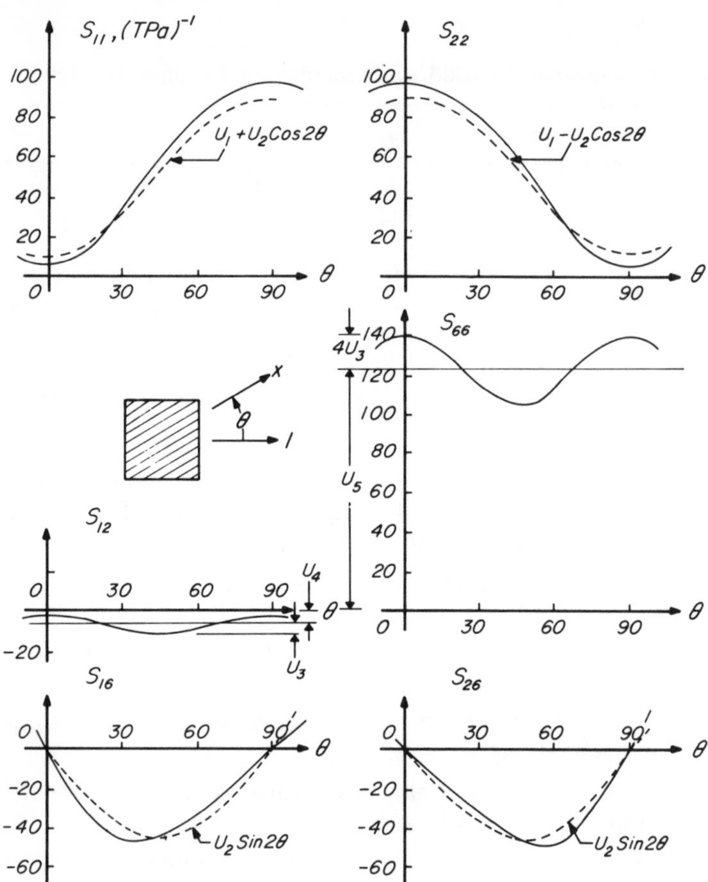

Figure 3.12 Comparison of exact and approximate transformed compliance in terms of the multiple-angle functions for T300/5208. The dashed lines are approximations without the last column in Table 3.9 or $U_3 = 0$.

table 3.13

off-axis stress-strain relation for unidirectional composites in terms of modulus

	ϵ_1	ϵ_2	ϵ_6
σ_1	Q_{11}	Q_{12}	Q_{16}
σ_2	Q_{21}	Q_{22}	Q_{26}
σ_6	Q_{61}	Q_{62}	Q_{66}

table 3.14
off-axis stress-strain relation for unidirectional composites in terms of compliance

	σ_1	σ_2	σ_6
ϵ_1	S_{11}	S_{12}	S_{16}
ϵ_2	S_{21}	S_{22}	S_{26}
ϵ_6	S_{61}	S_{62}	S_{66}

longer zero. We can proceed with the inversion or solution of these simultaneous equations by the method of determinant as follows:

We will assume that we are given the equations in Table 3.13. We will first obtain the determinant of the stiffness components:

$$\text{Determinant of Stiffness} = \det Q_{ij} = \Delta \tag{3.63}$$

$$= Q_{11}Q_{22}Q_{66} + 2Q_{12}Q_{26}Q_{61} - Q_{22}Q_{16}^2 - Q_{66}Q_{12}^2 - Q_{11}Q_{62}^2 \tag{3.64}$$

$$\begin{aligned}
S_{11} &= (Q_{22}Q_{66} - Q_{26}^2)/\Delta \\
S_{22} &= (Q_{11}Q_{66} - Q_{16}^2)/\Delta \\
S_{12} &= (Q_{16}Q_{26} - Q_{12}Q_{66})/\Delta \\
S_{66} &= (Q_{11}Q_{22} - Q_{12}^2)/\Delta \\
S_{16} &= (Q_{12}Q_{26} - Q_{22}Q_{16})/\Delta \\
S_{26} &= (Q_{12}Q_{16} - Q_{11}Q_{26})/\Delta
\end{aligned} \tag{3.65}$$

We have obtained the components of compliance from those of stiffness. If we are given the compliance and want to know the stiffness, we simply interchange the Q's and S's in these equations.

Thus there are many ways that we can compute the off-axis stiffness or compliance. This is diagrammed in Figure 3.13. The following operations are involved.

- From engineering constants, compute the on-axis compliance and stiffness shown in Tables 1.8 and 1.9, respectively.
- Compute transformed stiffness using its transformation equations in Tables 3.2, 3.3, or 3.4; and transformed compliance using Tables 3.8, 3.9, or 3.10.
- Alternatively, we can go directly from stiffness to compliance by inversion in Equations 3.64 and 3.65; or from compliance to stiffness also by inversion.
- Off-axis engineering constants, to be shown in the next section, must be obtained from the off-axis compliance.

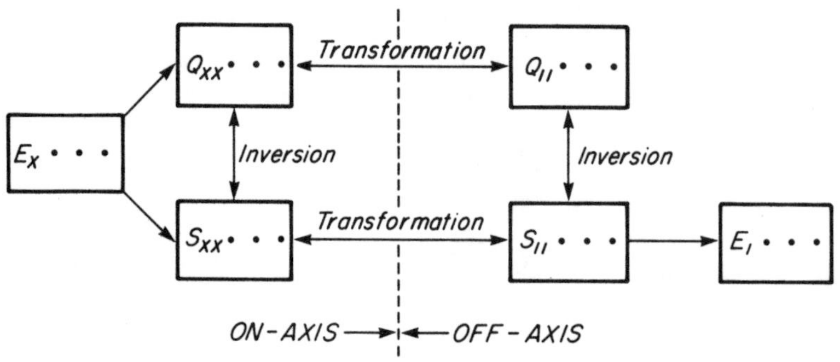

Figure 3.13 Relations among the on-axis and off-axis stiffness, compliance and engineering constants. The connecting lines indicate the paths of mathematical operations. There is no direct link between on-axis stiffness and off-axis compliance, or on-axis and off-axis engineering constants.

6. off-axis engineering constants

We first defined engineering constants in Chapter 1 for orthotropic, on-axis unidirectional composites. Equations 1.9 and 1.10 show the relations between engineering constants and the components of compliance. In the case of anisotropic, off-axis unidirectional composites, we can relate the compliance components in Table 3.7 to engineering constants by performing the following simple tests:

- Uniaxial tension test along the 1-axis.

$$\sigma_1 \neq 0 \qquad \sigma_2 = \sigma_6 = 0 \tag{3.66}$$

From Table 3.7.

$$\epsilon_1 = S_{11}\sigma_1 = \frac{\sigma_1}{E_1}$$

$$\epsilon_2 = S_{21}\sigma_1 \qquad (3.67)$$

$$\epsilon_6 = S_{61}\sigma_1$$

Now define coupling coefficients

$$\nu_{21} = -\frac{\epsilon_2}{\epsilon_1}$$

$$(3.68)$$

$$\nu_{61} = \frac{\epsilon_6}{\epsilon_1}$$

Combining with Equation 3.67, we obtain

$$\nu_{21} = -\frac{S_{21}}{S_{11}}$$

$$(3.69)$$

$$\nu_{61} = \frac{S_{61}}{S_{11}}$$

Note the terms are the longitudinal Poisson's ratio and shear coupling coefficient, respectively. The latter does not have a counterpart in conventional materials.

$$S_{21} = -\nu_{21}S_{11} = -\frac{\nu_{21}}{E_1}$$

$$(3.70)$$

$$S_{61} = \nu_{61}S_{11} = \frac{\nu_{61}}{E_1}$$

- Uniaxial tension along the 2-axis:

$$\begin{aligned} \sigma_2 &\neq 0 \\ \sigma_1 = \sigma_6 &= 0 \end{aligned} \tag{3.71}$$

From Table 3.7

$$\begin{aligned} \epsilon_2 &= S_{22}\sigma_2 = \frac{\sigma_2}{E_2} \\ \epsilon_1 &= S_{12}\sigma_2 \\ \epsilon_6 &= S_{62}\sigma_2 \end{aligned} \tag{3.72}$$

Similarly, we can define:

$$\begin{aligned} \nu_{12} &= -\frac{\epsilon_1}{\epsilon_2} = -\frac{S_{12}}{S_{22}} \\ \nu_{62} &= \frac{\epsilon_6}{\epsilon_2} = \frac{S_{62}}{S_{22}} \end{aligned} \tag{3.73}$$

These terms are the transverse Poisson's ratio and shear coupling coefficient associated with the 2-axis.

$$\begin{aligned} S_{12} &= -\nu_{12}S_{22} = -\frac{\nu_{12}}{E_2} \\ S_{62} &= \nu_{62}S_{22} = \frac{\nu_{62}}{E_2} \end{aligned} \tag{3.74}$$

- Pure shear test along the 1-2 axes

$$\begin{aligned} \sigma_6 &\neq 0 \\ \sigma_1 = \sigma_2 &= 0 \end{aligned} \tag{3.75}$$

From Table 3.7

$$\epsilon_6 = S_{66}\sigma_6 = \frac{\sigma_6}{E_6}$$

$$\epsilon_1 = S_{16}\sigma_6 \qquad (3.76)$$

$$\epsilon_2 = S_{26}\sigma_6$$

We can define:

$$\nu_{16} = \frac{\epsilon_1}{\epsilon_6} = \frac{S_{16}}{S_{66}}$$

$$(3.77)$$

$$\nu_{26} = \frac{\epsilon_2}{\epsilon_6} = \frac{S_{26}}{S_{66}}$$

These terms are the normal coupling coefficients. Conventional materials do not have such coupling. By rearrangement:

$$S_{16} = \nu_{16}S_{66} = \frac{\nu_{16}}{E_6}$$

$$(3.78)$$

$$S_{26} = \nu_{26}S_{66} = \frac{\nu_{26}}{E_6}$$

Thus in place of Table 3.7, the stress-strain relation for an off-axis unidirectional composite in terms of engineering constants can be shown in a matrix multiplication table as follows:

table 3.15

off-axis stress-strain relation for unidirectional composites in terms of engineering constants

	σ_1	σ_2	σ_6
ϵ_1	$\frac{1}{E_1}$	$-\frac{\nu_{12}}{E_2}$	$\frac{\nu_{16}}{E_6}$
ϵ_2	$-\frac{\nu_{21}}{E_1}$	$\frac{1}{E_2}$	$\frac{\nu_{26}}{E_6}$
ϵ_6	$\frac{\nu_{61}}{E_1}$	$\frac{\nu_{62}}{E_2}$	$\frac{1}{E_6}$

Since the compliance matrix is symmetric,

$$S_{12} = S_{21}$$

$$S_{16} = S_{61} \tag{3.79}$$

$$S_{26} = S_{62}$$

We can rearrange Table 3.15 and obtain an alternative arrangement shown in Table 3.16, where each row instead of each column is now normalized by a constant.

table 3.16

alternative arrangement of stress-strain relation of an off-axis unidirectional composite

	σ_1	σ_2	σ_6
ϵ_1	$\frac{1}{E_1}$	$-\frac{\nu_{21}}{E_1}$	$\frac{\nu_{61}}{E_1}$
ϵ_2	$-\frac{\nu_{12}}{E_2}$	$\frac{1}{E_2}$	$\frac{\nu_{62}}{E_2}$
ϵ_6	$\frac{\nu_{16}}{E_6}$	$\frac{\nu_{26}}{E_6}$	$\frac{1}{E_6}$

From the same symmetry property in Equation 3.79, we can immediately derive the following reciprocal relations:

$$\frac{\nu_{21}}{\nu_{12}} = \frac{E_1}{E_2} = \frac{S_{22}}{S_{11}} = a$$

$$\frac{\nu_{61}}{\nu_{16}} = \frac{E_1}{E_6} = \frac{S_{66}}{S_{11}} = b \tag{3.80}$$

$$\frac{\nu_{62}}{\nu_{26}} = \frac{E_2}{E_6} = \frac{S_{66}}{S_{22}} = c$$

where a is the ratio of the Young's moduli; and b and c, useful ratios to designers interested in the relative stiffness between bending and twisting. While the symmetry condition for the compliances holds; i.e.,

$$S_{ij} = S_{ji} \tag{3.81}$$

similar symmetry condition for the anisotropic coupling coefficients does not hold; i.e.,

$$\nu_{ij} \neq \nu_{ji} \tag{3.82}$$

Using the data of T300/5208, we can calculate the numerical values of all the engineering constants and tabulate typical results in Table 3.17.

table 3.17
off-axis engineering constants of T300/5208 unidirectional composites (GPa) or dimensionless

θ	E_1, GPa	E_6, GPa	ν_{21}	ν_{61}	ν_{16}
0	181.0	7.17	0.280	0	0
5	154.4	7.22	0.278	−1.673	−0.0782
10	107.8	7.37	0.273	−2.273	−0.155
15	72.62	7.63	0.265	−2.193	−0.230
30	28.78	8.76	0.226	−1.351	−0.411
45	16.73	9.46	0.167	−0.766	−0.433
60	12.41	8.76	0.0978	−0.401	−0.283
90	10.3	7.17	0.0159	0	0

The data in Table 3.17 are plotted in Figure 3.14. The Poisson's ratios are even functions of ply orientation; and the shear and normal coupling ratios are odd functions. When ply orientation is positive, which is the case in Figure 3.14, the shear and normal coupling ratios are negative for T300/5208 and for other composite materials listed in Table 1.7.

In Figure 3.15, the deformed shapes of squares under uniaxial tensile and compressive stresses are shown. Due to nonzero shear coupling coefficients, shear is induced. There is no counterpart of this material's response in conventional materials. In Figure 3.16, the deformed shapes of squares under pure shear are shown. Due to normal coupling, the area of the squares undergo contraction or expansion depending on the sign of the applied shear and that of the normal coupling coefficient.

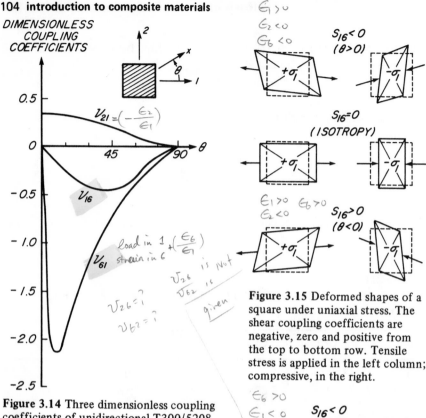

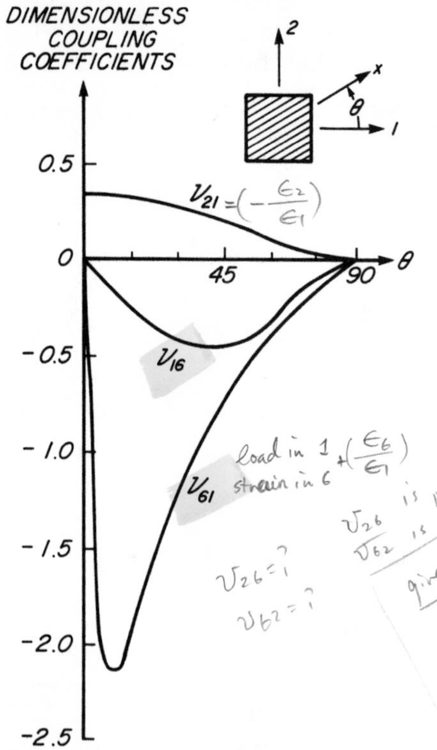

Figure 3.14 Three dimensionless coupling coefficients of unidirectional T300/5208. Data are listed in last three columns of Table 3.17.

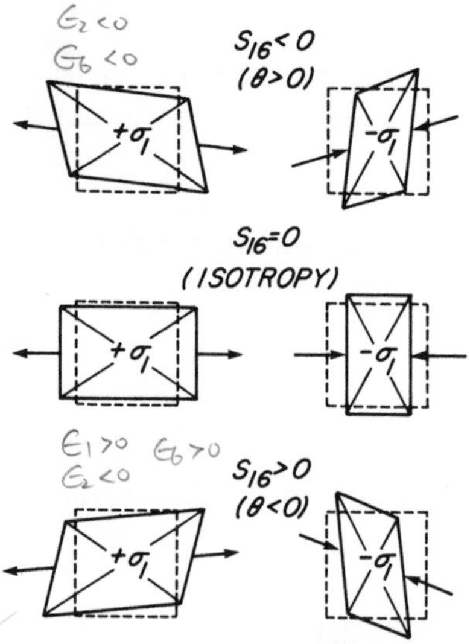

Figure 3.15 Deformed shapes of a square under uniaxial stress. The shear coupling coefficients are negative, zero and positive from the top to bottom row. Tensile stress is applied in the left column; compressive, in the right.

The relation between off-axis engineering constants and components of stiffness can best be expressed by applying simple tests such as uniaxial tension and pure shear tests like those used earlier in this chapter. By substituting the strains induced by a uniaxial tension test, shown in Equation 3.67 et al., into the stress-strain relation in terms of stiffness, we have

$$\sigma_1 = Q_{11}\epsilon_1 + Q_{12}\epsilon_2 + Q_{16}\epsilon_6$$

$$= [Q_{11} - \nu_{21}Q_{12} + \nu_{61}Q_{16}]\frac{\sigma_1}{E_1} \tag{3.83}$$

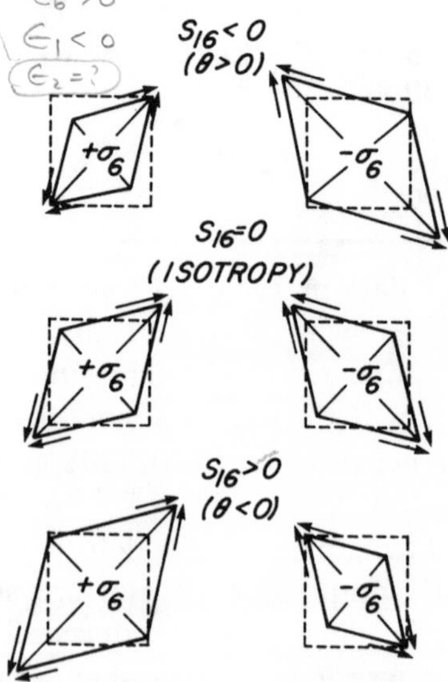

Figure 3.16 Deformed shapes of squares under pure shear. The normal coupling induces areal changes. The coupling coefficients are negative, zero and positive as we move from top to bottom.

Then

$$E_1 = Q_{11} - \nu_{21}Q_{12} + \nu_{61}Q_{16} \tag{3.84}$$

If the ply orientation is zero, we recover the on-axis relation of Equation 1.13 because at this orientation

$$\begin{aligned} \nu_{21} &= \nu_x = \frac{Q_{yx}}{Q_{yy}} \\ \nu_{61} &= 0 \end{aligned} \tag{3.85}$$

Then

$$E_x = Q_{xx}/m \tag{3.86}$$

where m is defined in Equation 1.13. Note that from Equation 3.84 we can say for anisotropic material

$$E_1 \neq Q_{11}/m \tag{3.87}$$

In fact, the difference between E_1 and Q_{11} is shown in Figure 3.17.

Similarly, we can show

$$E_2 = Q_{22} - \nu_{12}Q_{21} + \nu_{62}Q_{26} \tag{3.88}$$

$$E_6 = Q_{66} + \nu_{16}Q_{61} + \nu_{26}Q_{62} \tag{3.89}$$

The difference between E_6 and Q_{66} is also shown in Figure 3.17.

Off-axis engineering constants provide another insight into the nature of anisotropic materials. The highly coupled behavior provides an excellent opportunity of capitalizing on the unique properties of composite materials not possible with conventional materials. Designing with composite materials is no longer an extension of that with conventional materials. If we limit ourselves to orthotropic materials, composite materials can be viewed as a special isotropic material. In the absence of the shear and normal coupling, response of such materials is intuitively similar to isotropic materials. But as we face fully anisotropic materials with shear and normal coupling, the intuition developed from working with isotropic or orthotropic materials is no longer

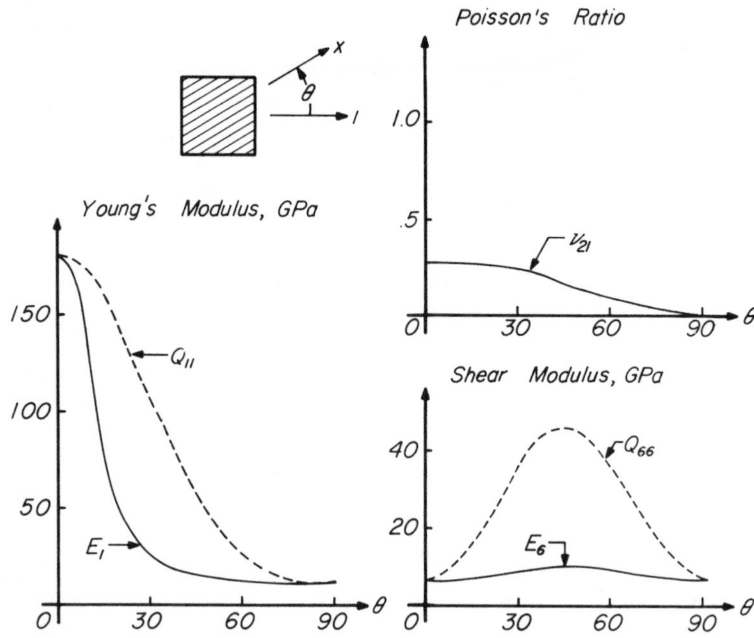

Figure 3.17 Transformed Young's modulus and shear modulus of T-300/5208 unidirectional composite. The data are taken from Table 3.17. Transformed stiffness are also shown for comparison. At 0 and 90 degrees, there is essentially no difference between the two curves. But significant differences exist in off-axis orientations. The differences come from the last two terms in Equations 3.84 and 3.89.

valid. New intuition must be acquired. This is the challenge to designers. We must think composites.

7. conclusions

One of the most important features of composite materials is the variation of properties as the ply orientation changes. The stiffness of an off-axis unidirectional composite is governed by appropriate stress-strain relations as before. The functional relationship remains the same. Although the number of constants have increased from 4 to 6, the number of independent constants remain the same at 4. The two additional constants are related to shear and normal coupling.

The shear and normal coupling does not have a counterpart in the conventional material. The coupling results in more complicated behavior. We should take this as an opportunity unique with composite materials rather than a liability. It will require time, effort and willingness to face new challenges for us to acquire the confidence working with anisotropic materials.

We would like to emphasize again the importance of the sign of ply orientation. Shear and normal coupling changes sign with the ply orientation. A wrong sign will change completely the effect of this coupling.

The concept of the invariants is also very important in composite materials. The magnitude of the components of stiffness and compliance vary as a function of ply orientation, but the area under the transformed components remain constant. This constant value is equal to the particular invariant associated with each component of stiffness or compliance. The invariants therefore represent the potential of a component of stiffness or compliance that is embodied in a unidirectional composite. When we go to the laminated composite consisting of multidirectional plies of a given composite material, we will see that the invariants of a laminated composite remains the same as that of the constituent unidirectional plies. In a sense that lamination only changes the directional properties but does not affect the total potential of stiffness that a composite material can provide. In the limit we recover the isotropic material in which case all directional properties disappear.

Engineering constants are useful but are based on measurements derived from one dimensional tests. Engineering constants are related directly to the components of compliance which in turn can be derived from the inversion of the components of stiffness. For an off-axis unidirectional composite, there is no direct relationship between the engineering constants and the components of stiffness. The components of compliance is the bridge between them. This is different from the on-axis unidirectional composite shown in Chapter 1 where direct link between the engineering constants and the components of stiffness existed.

The variation of the engineering constants as the fiber-orientation changes can be derived from the transformation of the components of compliance. The engineering constants themselves are not covered by any transformation equation. This is a fundamental difference between the engineering constants, which are derived qualities from the compliance, and the compliance components themselves. There are no invariants, for example, associated with the off-axis Young's moduli. The Poisson's ratios, the shear and normal coupling coefficients are dimensionless ratios. They are not governed by any transformation equations; they have no invariants; and they are not symmetric. They are useful for indication of behavior of off-axis unidirectional composites, but their uses are limited because they are fundamentally one dimensional constants. Since composite materials are normally used in two dimensional configurations, engineering constants may not be used directly in many instances.

Effective use of composite materials must not be limited on a replacement or substitution basis. Again, the stress-strain relation is fundamentally the same as the conventional material. It is conceptually simple. We must learn to take advantage of anisotropy. Do not eliminate it for sake of simplicity.

8. homework problems

a. Derive the transformation of stiffness of a unidirectional composite from one off-axis orientation to another.

b. Repeat the process above for the compliance.

c. Draw the generalized Mohr's circles for the stiffness of all the unidirectional materials listed in Table 3.6 on the same scale. Locate the position for the transformed shear modulus Q_{66}. How would aluminum appear?

d. Show that stress-strain relations of an anisotropic material can be expressed in terms of the *p-q-r* and the *U*'s in the following tables:

table 3.18
stress-strain relations in stiffness

	p_ϵ	q_ϵ	r_ϵ
p_σ	$U_{1Q}+U_{4Q}$	U_{2Q}	$2U_{6Q}$
q_σ	U_{2Q}	$2U_{5Q}+2U_{3Q}$	$2U_{7Q}$
r_σ	$2U_{6Q}$	$2U_{7Q}$	$2U_{5Q}-2U_{3Q}$

table 3.19
stress-strain relations in compliance

	p_σ	q_σ	r_σ
p_ϵ	$U_{1S}+U_{4S}$	U_{2S}	U_{6S}
q_ϵ	U_{2S}	$\frac{1}{2}U_{5S}+2U_{3S}$	U_{7S}
r_ϵ	U_{6S}	U_{7S}	$\frac{1}{2}U_{5S}-2U_{3S}$

Show how the tables can be simplified for orthotropic, square-symmetric and isotropic materials.

e. Find the locations of various key points on the transformed stiffness curves in Figure 3.7. These points represent the locations of the inflection points, extremum values and slopes of tangents.

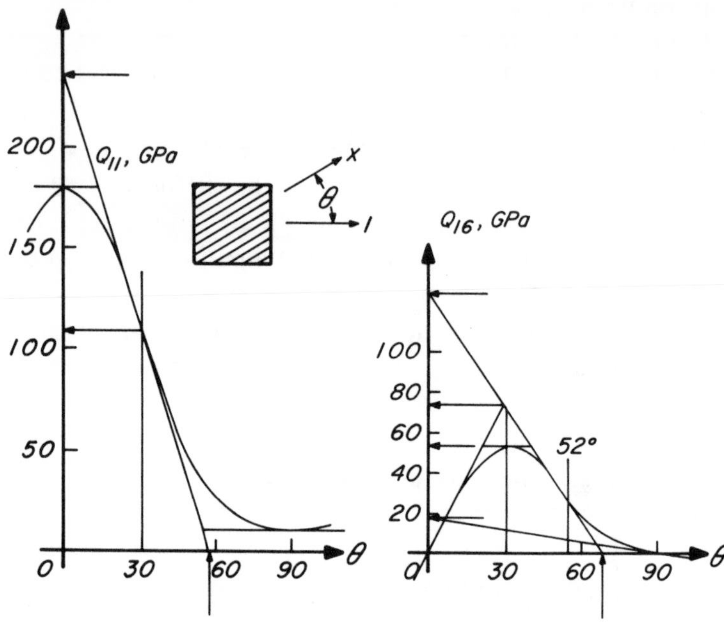

Figure 3.18 Key points in the transformed stiffness of T300/5208.

f. Repeat the process in Problem *e* for the boron-epoxy composite. Are there any unusual features for this composite?

g. Are there bounds for the Poisson's ratio of an off-axis unidirectional composite? What happens if fiber stiffness approaches infinity or matrix stiffness to zero?

h. How can shear or normal coupling coefficients be used to create an apparent infinite stiffness of an off-axis unidirectional composite subjected to a biaxial normal and shear stress components (assuming the other normal component σ_2 is zero)?

1) Derive the condition for zero resulting shear strain $(\epsilon_6 = 0)$. What is the resulting normal strain ϵ_1 for this case?

2) Derive the condition for zero normal strain $(\epsilon_1 = 0)$. What is the resulting shear strain ϵ_6 for this case?

3) If the applied stress components are:

$$\sigma_1 = \sigma_6 = 100 \text{ MPa}, \quad \sigma_2 = 0 \tag{3.90}$$

Is it possible to have infinite shear ($\epsilon_6 = 0$) or infinite normal stiffness ($\epsilon_1 = 0$) for T300/5208? How do you find the ply orientation for each?

4) Will the glass-epoxy composite work equally well for the stresses in Equation 3.90?

5) What principles emerge from this problem? Is complete rigidity under biaxial stress possible?

i. Derive the relationships between the invariants of the stiffness and those of the compliance for an orthotropic material.

j. Are there invariants associated with the transformed engineering constants? An average Young's modulus can be defined from the area under the transformed Young's modulus; e.g., in Figure 3.17. What is the relation between this and that derived from the transformed compliance ($1/U_1$ in Figure 3.12)?

k. What difficulties are involved for testing of an off-axis unidirectional composite? What is the difference in response between the tubular and flat specimens with off-axis ply orientation? Examine the cases of uniaxial extension, pure shear and hydrostatic pressure. What kind of stresses are induced in the load introduction points (the ends)? What load and displacement controls are desired for these tests?

l. A quick estimate (the back of an envelope calculation) of the off-axis stiffness of a unidirectional can be based on only one on-axis stiffness component (the others are zero):

$$U_1 = \frac{3}{8} Q_{xx}$$

$$U_2 = \frac{1}{2} Q_{xx} \qquad (3.91)$$

$$U_3 = U_4 = U_5 = \frac{1}{8} Q_{xx}$$

This was shown in Equation 3.46. The relation is easy to use and can readily be related to the fiber modulus such as:

$$Q_{xx} = v_f E_f \qquad (3.92)$$

where subscript f refers to the fiber. Using T300/5208 data, estimate the off-axis stiffness at $\pi/6$, $\pi/4$ and $\pi/3$ radians. Show the error introduced by this estimate with the exact values listed in Table 3.5. Explain if this estimate works for the compliance components.

nomenclature

a, b, c = Ratios of engineering constants

E_1, E_2 = Young's modulus parallel and transverse to an off-axis unidirectional composite

E_6 = Longitudinal-transverse shear modulus of an off-axis unidirectional composite

I_1, I_2 = Linear or first order invariants of stiffness or compliance

m, n = $\cos\theta$, $\sin\theta$

Q_{ij} = Components of stiffness; $i,j = x,y,s$ or 1,2,6

R_1, R_2 = Quadratic or second order invariants of stiffness or compliance

S_{ij} = Components of compliance; $i,j = x,y,s$ or 1,2,6

U_i = Linear combinations of on-axis stiffness and compliance in the multiple-angle formulation. Same notation but different combinations are used for stiffness and compliance; $i = 1$ to 5

U_i' = Linear combinations of off-axis stiffness and compliance; $i = 1$ to 7

σ_i = On-axis components of stress, $i = x,y,s$; off-axis, $i = 1,2,6$

ϵ_i = On-axis components of strain, $i = x,y,s$; off-axis, $i = 1,2,6$

ν_{21}, ν_{12} = Longitudinal and transverse Poisson's ratios of an off-axis unidirectional composite

ν_{61}, ν_{62} = Longitudinal and transverse shear coupling coefficients of an off-axis unidirectional composite

ν_{16}, ν_{26} = Longitudinal and transverse normal coupling coefficients of an off-axis unidirectional composite

θ = Ply orientation

Δ = Determinant of stiffness or compliance

chapter 4
in-plane stiffness of symmetric laminates

The stiffness of multidirectional laminates consisting of plies and ply groups with arbitrary orientations will be described. Those laminates with midplane symmetry will behave like homogeneous anisotropic plates. The stiffness modulus of the laminate is simply the arithmetic average of the stiffness of the constituent plies. We will also show that the stiffness properties of bidirectional laminates can vary significantly and be fundamentally different from conventional materials. Like unidirectional composites, laminated composites can be described by three sets of elastic constants. The set consisting of stiffness components would be the easiest to use because of the simple property relationship between the laminate and the constituent plies.

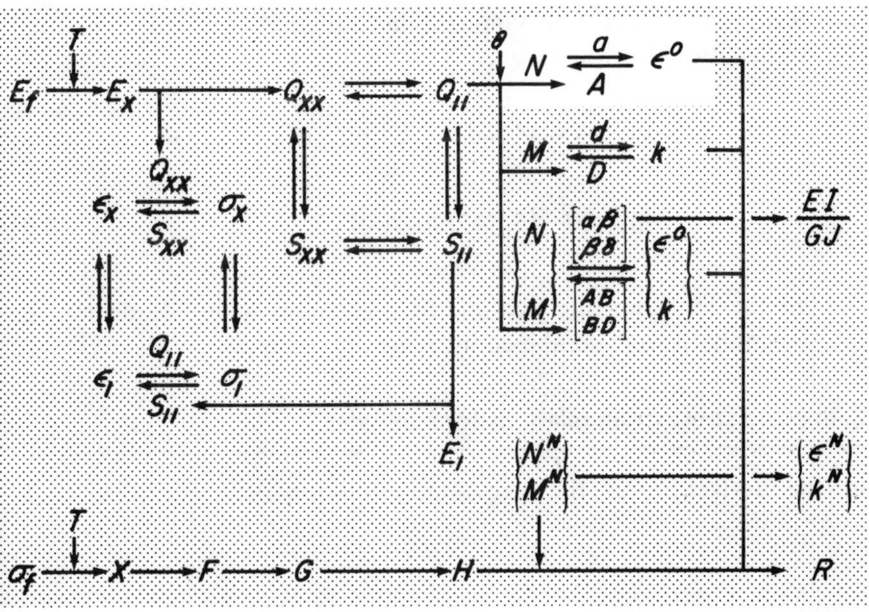

1. laminate code

A multidirectional composite laminate is defined by the following code to designate the stacking sequence of ply groups:

$$[0_3/90_2/45/-45_3]_S \tag{4.1}$$

This code is represented diagrammatically in Figures 4.1 and 4.2, and contains the following features:

- Starting from the bottom of the plate, at $z = -h/2$, the first ply group has three plies of 0-degree orientation; followed by the next group with two 90-degree plies; followed by one 45-degree ply; and finally the last group with three −45-degree plies. For symmetric plate, the ascending order from the bottom face is identical to a descending order from the top face or $z = h/2$. But for unsymmetric laminates, the ascending order will have the opposite laminate code as the descending order. We have arbitrarily decided the use of ascending order for this book.
- The subscript S denotes that the laminate is symmetric with respect to the midplane or the $z = 0$ plane. The upper half of the laminate is the same as the lower half except the stacking sequence is reversed in order to maintain the midplane symmetry.
- A subscript T is used to designate the total laminate, without any omission of the symmetrical upper portion of the laminate. If we want to describe the laminate in Equation 4.1 using the total designation, we will have

$$[0_3/90_2/45/-45_3/-45_3/45/90_2/0_3]_T \quad \text{or} \tag{4.2}$$

$$[0_3/90_2/45/-45_6/45/90_2/0_3]_T \tag{4.3}$$

In the last step, the two middle ply groups with −45 degree ply orientation were combined into one ply group.

2. in-plane stress-strain relations for laminates

In the derivation of the stress-strain relation of a multidirectional

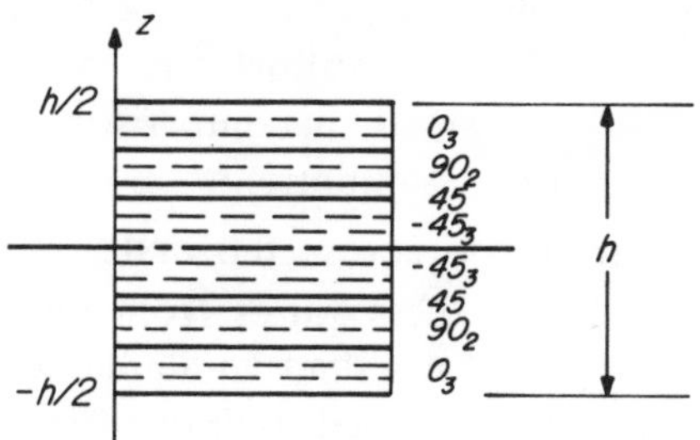

Figure 4.1 Typical stacking sequence of a symmetric laminate. The laminate code as stated in Equation 4.1 follows an ascending order from the bottom ply.

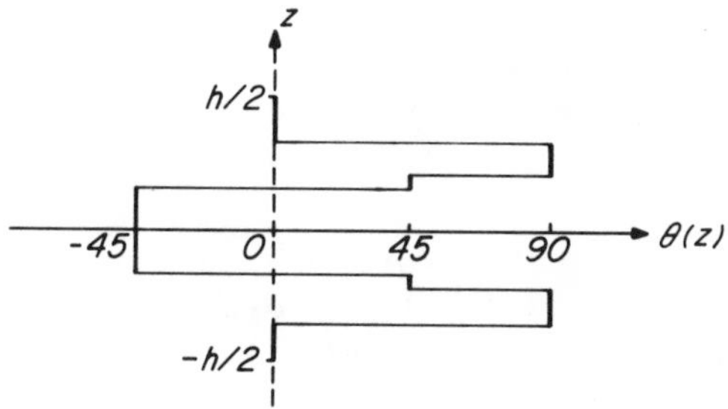

Figure 4.2 Ply orientations as function of z. This is another representation of Figure 4.1.

laminate, we must make the following simplifying assumptions:

- The laminate is symmetric; i.e.,

$$\theta(z) = \theta(-z) \tag{4.4}$$

and

$$Q_{ij}(z) = Q_{ij}(-z) \tag{4.5}$$

Thus, both the ply orientation and the ply material modulus are symmetric with respect to the midplane of the laminate.

- The strain remains constant across the laminate thickness. We will use superscript zero to signify the assumed constant in-plane strain components as follows:

$$\begin{aligned} \epsilon_1(z) &= \epsilon_1^o \\ \epsilon_2(z) &= \epsilon_2^o \\ \epsilon_6(z) &= \epsilon_6^o \end{aligned} \tag{4.6}$$

Note that there is no z-dependency or z-variation across the laminate thickness. This assumption is reasonable when the thickness of the laminate is small in comparison with the length and width and the laminate is symmetric.

Since the stress distribution across the multidirectional laminate is not constant because the stiffness varies from ply to ply, it is much easier to define an average stress than an actual stress across the laminate. This average stress can be used to define the stress-strain relation of the laminate. The stress, in this case, will be the average stress; and the strain, the in-plane strain in Equation 4.6. We can then calculate the stress at any ply within the laminate from the in-plane strain. We will show this later after the stress-strain relation for the laminate is established. The average stress is defined as follows:

$$\bar{\sigma}_1 = \frac{1}{h}\int_{-h/2}^{h/2} \sigma_1 \, dz$$

$$\bar{\sigma}_2 = \frac{1}{h}\int_{-h/2}^{h/2} \sigma_2 \, dz \qquad (4.7)$$

$$\bar{\sigma}_6 = \frac{1}{h}\int_{-h/2}^{h/2} \sigma_6 \, dz$$

In Figure 4.3, we show the relationship between the actual stress from ply to ply and the average stress across the laminate by the averaging process of Equation 4.7.

Substituting the stress-strain relation for any ply orientation into Equation 4.7, we have

$$\bar{\sigma}_1 = \frac{1}{h}\int [Q_{11}\epsilon_1 + Q_{12}\epsilon_2 + Q_{16}\epsilon_6]\, dz \qquad (4.8)$$

Substituting the assumed constant strain in Equation 4.6, we have

$$\bar{\sigma}_1 = \frac{1}{h}\int [Q_{11}\epsilon_1^o + Q_{12}\epsilon_2^o + Q_{16}\epsilon_6^o]\, dz \qquad (4.9)$$

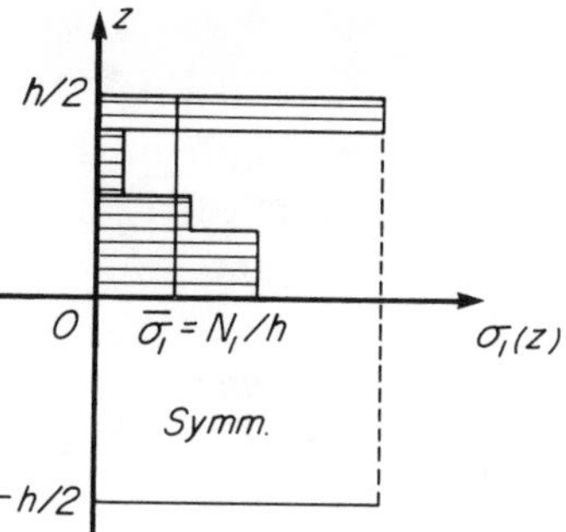

Figure 4.3 Definition of average stress. Comparison between the corresponding components of the actual ply stress and the average laminate stress is shown.

Since the in-plane strain components are independent of z, we can factor them out of the integral signs. Only the stiffness components are left inside the integral because they vary from ply to ply depending on each ply orientation.

$$\bar{\sigma}_1 = \frac{1}{h}\left[\int Q_{11}dz\epsilon_1^o + \int Q_{12}dz\epsilon_2^o + \int Q_{16}dz\epsilon_6^o\right] \tag{4.10}$$

$$= \frac{1}{h}[A_{11}\epsilon_1^o + A_{12}\epsilon_2^o + A_{16}\epsilon_6^o] \tag{4.11}$$

Similarly,

$$\bar{\sigma}_2 = \frac{1}{h}[A_{21}\epsilon_1^o + A_{22}\epsilon_2^o + A_{26}\epsilon_6^o] \tag{4.12}$$

$$\bar{\sigma}_6 = \frac{1}{h}[A_{61}\epsilon_1^o + A_{62}\epsilon_2^o + A_{66}\epsilon_6^o] \tag{4.13}$$

where

$$\begin{aligned} A_{11} &= \int Q_{11}dz, & A_{22} &= \int Q_{22}dz, & A_{12} &= A_{21}, \\ A_{12} &= \int Q_{12}dz, & A_{66} &= \int Q_{66}dz, & A_{16} &= A_{61}, \\ A_{16} &= \int Q_{16}dz, & A_{26} &= \int Q_{26}dz, & A_{26} &= A_{62}. \end{aligned} \tag{4.14}$$

where A_{ij} is the equivalent modulus for a multidirectional laminate. This modulus is simply the average of the stiffness of the constituent plies. There is a difference of a length in the physical dimension of stiffness Q_{ij} in Pa or Nm^{-2}; and that of A_{ij} in Pam or Nm^{-1}.

We can further define stress resultants as

$$N_1 = h\bar{\sigma}_1$$
$$N_2 = h\bar{\sigma}_2 \qquad (4.15)$$
$$N_6 = h\bar{\sigma}_6$$

Note the unit of stress resultant is Pam or Nm^{-1}, which is an integrated stress or force per unit width of a laminate with thickness h. The in-plane stress-strain relation for a laminate is actually the stress resultant versus in-plane strain relation. The latter is derived by combining Equations 4.15 with 4.11 et al. We have:

$$N_1 = A_{11}\epsilon_1^o + A_{12}\epsilon_2^o + A_{16}\epsilon_6^o$$
$$N_2 = A_{21}\epsilon_1^o + A_{22}\epsilon_2^o + A_{26}\epsilon_6^o \qquad (4.16)$$
$$N_6 = A_{61}\epsilon_1^o + A_{62}\epsilon_2^o + A_{66}\epsilon_6^o$$

This set of simultaneous equations can be inverted to yield the in-plane strain in terms of the stress resultant. This process is exactly the same as that described in Chapter 3, Section 5.

Equation 4.16 is based on stiffness of the laminate, and we wish to find the corresponding compliance by inversion such that

$$\epsilon_1^o = a_{11}N_1 + a_{12}N_2 + a_{16}N_6$$
$$\epsilon_2^o = a_{21}N_1 + a_{22}N_2 + a_{26}N_6 \qquad (4.17)$$
$$\epsilon_6^o = a_{61}N_1 + a_{62}N_2 + a_{66}N_6$$

We show both stress-strain relations in Equations 4.16 and 4.17 in matrix multiplication tables as follows:

table 4.1
in-plane stress-strain relation of symmetric laminates in terms of stiffness

	ϵ_1^o	ϵ_2^o	ϵ_6^o
N_1	A_{11}	A_{12}	A_{16}
N_2	A_{21}	A_{22}	A_{26}
N_6	A_{61}	A_{62}	A_{66}

table 4.2
in-plane stress-strain relation of symmetric laminates in terms of compliance

	N_1	N_2	N_6
ϵ_1^o	a_{11}	a_{12}	a_{16}
ϵ_2^o	a_{21}	a_{22}	a_{26}
ϵ_6^o	a_{61}	a_{62}	a_{66}

These stress-strain relations are valid for the in-plane deformation of symmetric laminates. If stress resultants are given, we can find the induced in-plane strain immediately from Table 4.2. Then the on-axis ply strain can be obtained by strain transformation from the initial 1-2 axes to the orientation of a specific ply or ply group. The ply stress is nothing more than the ply strain multiplied by the on-axis stiffness. The complete process going from stress resultants to on-axis ply strain and ply stress is illustrated in Figure 4.4.

From the compliance in Table 4.2, we can calculate the effective engineering constants, following the process used in the off-axis unidirectional composites in Equation 3.68, et al. We will have typically

$$\text{In-plane longitudinal modulus} = E_1^o = \frac{1}{a_{11}h}$$

$$\text{In-plane shear modulus} = E_6^o = \frac{1}{a_{66}h} \quad (4.18)$$

(continues)

$$\text{In-plane Poisson's ratio} = \nu_{21}^{o} = -\frac{a_{21}}{a_{11}}$$

$$\text{In-plane shear coupling coefficient} = \nu_{61}^{o} = \frac{a_{61}}{a_{11}}$$

$$\text{In-plane normal coupling coefficient} = \nu_{16}^{o} = \frac{a_{16}}{a_{66}} \tag{4.18}$$

(concluded)

Again we want to emphasize that engineering constants are the constants associated with simple tests such as uniaxial tensile and compressive tests and simple shear tests. They are the results of 1-dimensional tests and represent 1-dimensional characteristics of laminates. But composites are rarely used in 1-dimensional configuration. The 2-dimensional properties of composites are much different from the 2-dimensional properties of conventional materials. The coupling coefficients are large. Their effects are not always intuitively obvious and can result in opportunities unique with composite materials.

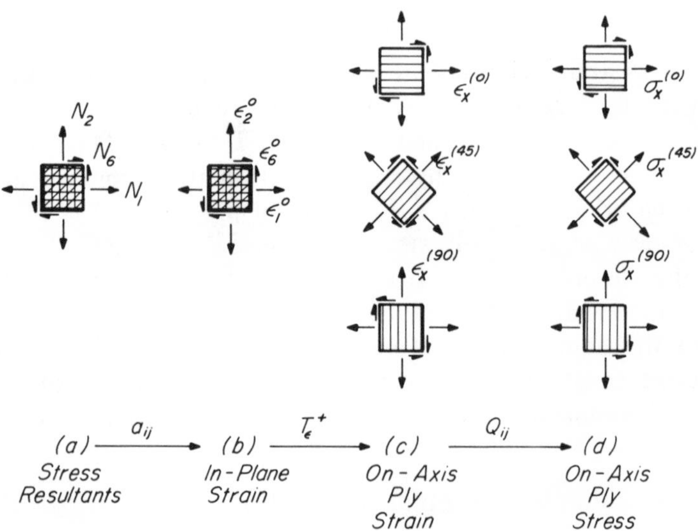

Figure 4.4 On-axis ply strain and stress calculations. From given stress resultants applied to a multidirectional laminate; we can go from (*a*) to (*b*): use in-plane stress-strain relation in laminate compliance from Table 4.2. From (*b*) to (*c*): use positive strain transformation in Tables 2.5 et al. The strain in Table 2.5 shall be replaced by in-plane strain. From (*c*) to (*d*): use the on-axis stress-strain relation of unidirectional composites in terms of stiffness, in Table 1.6.

3. evaluation of in-plane stiffness modulus

We have only mentioned that the in-plane modulus of a multi-directional laminate is the arithmetic average of the off-axis stiffness of the individual plies or ply groups. The averaging process is shown as integrals in Equation 4.14. We will now describe the steps needed to perform the integrations so that the contribution of the ply stiffness to the laminate modulus can be defined.

We want to mention again that in the averaging process of our laminate the modulus in Equation 4.14 is the off-axis stiffness of unidirectional composites. Using the transformed stiffness in Table 3.3:

$$A_{11} = \int Q_{11} dz = \int [U_1 + U_2 \cos 2\theta + U_3 \cos 4\theta]\, dz \tag{4.19}$$

$$= U_1 \int dz + U_2 \int \cos 2\theta\, dz + U_3 \int \cos 4\theta\, dz \tag{4.20}$$

The U's for a given composite remain constant. They can be factored out because they are not dependent on the z-axis.

$$A_{11} = U_1 h + U_2 V_1 + U_3 V_2 \tag{4.21}$$

where the geometric factors are:*

$$V_1 = \int_{-h/2}^{h/2} \cos 2\theta\, dz \tag{4.22}$$

$$V_2 = \int_{-h/2}^{h/2} \cos 4\theta\, dz \tag{4.23}$$

We can repeat the process for other in-plane components.

$$A_{22} = U_1 h - U_2 V_1 + U_3 V_2 \tag{4.24}$$

$$A_{12} = \int Q_{12} dz = \int [U_4 - U_3 \cos 4\theta]\, dz$$

$$= U_4 h - U_3 \int \cos 4\theta\, dz = U_4 h - U_3 V_2 \tag{4.25}$$

*A more general definition can be found in Equations 6.79-6.82.

$$A_{66} = U_5 h - U_3 V_2 \tag{4.26}$$

$$A_{16} = \int Q_{16} dz = \int \left[\frac{1}{2} U_2 \sin 2\theta + U_3 \sin 4\theta \right] dz$$
$$= \frac{1}{2} U_2 V_3 + U_3 V_4 \tag{4.27}$$

$$A_{26} = \frac{1}{2} U_2 V_3 - U_3 V_4 \tag{4.28}$$

where

$$V_3 = \int_{-h/2}^{h/2} \sin 2\theta \, dz \tag{4.29}$$

$$V_4 = \int_{-h/2}^{h/2} \sin 4\theta \, dz \tag{4.30}$$

We have seen that the evaluation of the in-plane modulus is reduced to the evaluation of four geometric factors, defined by the V's. These relations can be summarized in a matrix multiplication table as follows.

table 4.3

formulas for in-plane stiffness modulus of laminates

	h	U_2	U_3
A_{11}	U_1	V_1	V_2
A_{22}	U_1	$-V_1$	V_2
A_{12}	U_4		$-V_2$
A_{66}	U_5		$-V_2$
A_{16}		$\frac{1}{2} V_3$	V_4
A_{26}		$\frac{1}{2} V_3$	$-V_4$

where the V's, the geometric factors, can be defined as follows:

$$V_{[1,2,3,4]} = \int_{-h/2}^{h/2} [\cos 2\theta, \cos 4\theta, \sin 2\theta, \sin 4\theta] \, dz \tag{4.31}$$

There are four V's that will completely determine the six components of the in-plane stiffness of a laminate consisting of constituent plies of the same material; i.e., the same U's in Table 4.3. There are therefore at most four independent variables. This will be illustrated again in Equation 4.33.

A condensed definition of the V's in Equation 4.31 means that the numeral in the bracket on the left-hand side applies to the corresponding term on the right-hand side of the equation. The value of V's is dependent on the variation of ply orientations in the multidirectional laminate. It is implicitly assumed that the laminate consists of plies of the same unidirectional composite. Because sine and cosine functions are bounded between -1 and $+1$, the V's are bounded by the same limits, as we soon shall see. The similarity between Tables 4.3 and 3.3 is the result of using the same transformation equations. In the limit when the laminate has only a ply orientation, we recover Table 3.3 from 4.3 because the integrands in Equation 4.31 are constant. The V's become simply the trigonometric functions times the laminate thickness:

$$\begin{aligned} V_1 &= h \cos 2\theta \\ V_2 &= h \cos 4\theta \\ V_3 &= h \sin 2\theta \\ V_4 &= h \sin 4\theta \end{aligned} \tag{4.32}$$

From the formulas in Table 4.3, we have

$$A_{11} + A_{22} + 2A_{12} = 2[U_1 + U_4]h$$

or

$$A_{12} = h[U_1 + U_4] - \frac{1}{2}[A_{11} + A_{22}] \tag{4.33}$$

Similarly

$$A_{66} = A_{12} + [U_5 - U_4]h$$

Thus of the first four in-plane modulus components, only two are dependent on the stacking sequence. If we know the first two, we can immediately determine the others without integration. The variation of the in-plane modulus is constrained by the invariants of the constituent ply. The number of degrees of freedom are limited to two among the four components of modulus in Equation 4.33.

Let us now evaluate the first V in Equation 4.31. We first normalize it with respect to the laminate thickness:

$$V_1^* = \frac{V_1}{h} = \frac{1}{h} \int_{-h/2}^{h/2} \cos 2\theta \, dz \tag{4.34}$$

Normalization is useful in two aspects; viz., first the V^*'s become dimensionless and are valid for all physical units; SI, English, etc. Secondly, direct comparison can be made between the stiffness of laminate and that of the constituent plies. Table 4.4 can be compared with Table 3.3 component by component. The generalized Mohr's circles for the normalized in-plane modulus can be drawn directly over Figure 3.9 so that the effect of lamination can be illustrated graphically. (This will be done in Figure 4.6.)

If the laminate is symmetric, we only need to evaluate one half of the thickness, say, from $z = 0$ to $z = h/2$. The new limits of integration call for a new interpretation of the laminate code as defined in Equation 4.1. The starting point of the ascending ply sequence has been reversed from the $z = 0$ to $z = h/2$. Only the upper half of a symmetric laminate needs to be evaluated. Thus,

$$V_1^* = \frac{2}{h} \int_0^{h/2} \cos 2\theta \, dz \tag{4.35}$$

Since each ply group is assumed to have the same ply orientation and material, this integration can now be replaced by summation as we move from ply group to ply group:

$$V_1^* = \frac{2}{h} \sum_{i=1}^{m/2} \cos 2\theta_i [z_i - z_{i-1}]$$

$$= \frac{2}{h} \sum_{i=1}^{m/2} \cos 2\theta_i h_i \tag{4.36}$$

where h_i = thickness of i-th ply group; where i begins from the mid-plane. See Figure 4.5 for the definitions of geometric terms.

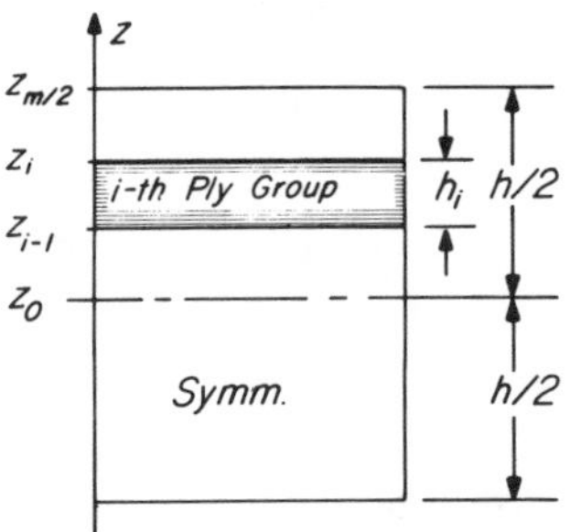

Figure 4.5 Definitions of terms in a symmetric laminate. The index for ply group goes from 0 to $m/2$ when m is the total number of ply groups.

Let $\quad v_i$ = volume fraction of plies with θ_i orientation

$$= 2h_i/h \tag{4.37}$$

If each index i in Equation 4.36 represents a unique ply orientation, we can now substitute Equation 4.37 into 4.36.

$$V_1^* = \sum_{i=1}^{m/2} \cos 2\theta_i v_i$$

$$= v_1 \cos 2\theta_1 + v_2 \cos 2\theta_2 + v_3 \cos 2\theta_3 + \ldots \tag{4.38}$$

where

$$v_1 + v_2 + v_3 + \ldots = 1 \tag{4.39}$$

Thus, V_1^* is simply the rule of mixtures equation, or the weighted average of the $\cos 2\theta$ functions. Since cosine functions can never be greater than unity (or less than minus unity), each term in Equation 4.38 is always equal to or less than the corresponding term in 4.39. We can therefore conclude that V_1^* is bounded as follows:

$$-1 \leqslant V_1^* \leqslant 1 \tag{4.40}$$

By applying the identical process to the remaining V's in Equation 4.31, we can get the following:

$$\begin{aligned} V_2^* &= \nu_1 \cos 4\theta_1 + \nu_2 \cos 4\theta_2 + \nu_3 \cos 4\theta_3 + \ldots \\ V_3^* &= \nu_1 \sin 2\theta_1 + \nu_2 \sin 2\theta_2 + \nu_3 \sin 2\theta_3 + \ldots \\ V_4^* &= \nu_1 \sin 4\theta_1 + \nu_2 \sin 4\theta_2 + \nu_3 \sin 4\theta_3 + \ldots \end{aligned} \tag{4.41}$$

With these simple equations, we can easily compute the in-plane modulus of multidirectional laminates with any ply orientation. The information needed is: the orientation and the volume fraction of each ply group. Then from Equations 4.38 et al. we can calculate the V^*'s. From Table 4.3 we can compute the modulus for any multidirectional laminates.

When normalized V's or V^*'s are used, Table 4.3 can be rewritten in a matrix multiplication table as follows:

table 4.4

formulas for normalized in-plane stiffness modulus

	1	U_2	U_3
A_{11}/h	U_1	V_1^*	V_2^*
A_{22}/h	U_1	$-V_1^*$	V_2^*
A_{12}/h	U_4		$-V_2^*$
A_{66}/h	U_5		$-V_2^*$
A_{16}/h		$\frac{1}{2}V_3^*$	V_4^*
A_{26}/h		$\frac{1}{2}V_3^*$	$-V_4^*$

We can now define the linear combinations of the in-plane modulus, in the same way those of the stiffness of a unidirectional composite are defined in Equations 3.15 and 3.21.

$$
\begin{aligned}
U_{1A}/h &= \frac{1}{8h}[3A_{11} + 3A_{22} + 2A_{12} + 4A_{66}] \\
&= \frac{1}{8}[6U_{1Q} + 2U_{4Q} + 4U_{5Q}] \\
&= U_{1Q}
\end{aligned}
$$

Here subscripts A and Q are added to differentiate the U's for the laminate from those for the unidirectional composite.

Similarly, we can show from Table 4.4

$$
\begin{aligned}
U_{4A}/h &= U_{4Q} \\
U_{5A}/h &= U_{5Q} \\
U_{2A}/h &= \frac{1}{2h}[A_{11} - A_{22}] = V_1^* U_{2Q} \\
U_{3A}/h &= \frac{1}{8h}[A_{11} + A_{22} - 2A_{12} - 4A_{66}] = V_2^* U_{3Q} \qquad (4.42) \\
U_{6A}/h &= \frac{1}{2h}[A_{16} + A_{26}] = \frac{1}{2} V_3^* U_{2Q} \\
U_{7A}/h &= \frac{1}{2h}[A_{16} - A_{26}] = V_4^* U_{3Q}
\end{aligned}
$$

We can now derive the two second-order in-plane invariants from Equations 3.22 and 3.24. These invariants correspond to the radii of the generalized Mohr's circles for a multidirectional laminate analogous to Equations 3.22 and 3.24, respectively:

$$
\begin{aligned}
R_{1A} &= \sqrt{U_{2A}^2 + 4U_{6A}^2} \\
&= \sqrt{V_1^{*2} + V_3^{*2}}\; U_{2Q}\, h \qquad (4.43) \\
R_{1A}/h &= \sqrt{V_1^{*2} + V_3^{*2}}\; R_{1Q}
\end{aligned}
$$

$$
\begin{aligned}
R_{2A} &= \sqrt{U_{3A}^{2} + U_{7A}^{2}} \\
&= \sqrt{V_2^{*2} + V_4^{*2}}\ U_{3Q} h \\
R_{2A}/h &= \sqrt{V_2^{*2} + V_4^{*2}}\ R_{2Q}
\end{aligned}
\tag{4.44}
$$

We can easily show that the square root always has a value equal to or less than unity.* Therefore, the normalized radii of the generalized Mohr's circles for the laminate are always equal to or less than those of the constituent ply. The square root in Equations 4.43 and 4.44 defines the reduction in two radii in Figure 3.9 for a T300/5208 multidirectional laminate. This is another indication of the constraints imposed on laminates by the transformation properties of a unidirectional composite. In-plane modulus cannot be arbitrarily chosen with six degrees of freedom.

Geometric interpretation of Equation 4.43 can be shown using the generalized Mohr's circles for T300/5208 unidirectional composite in Figure 3.9. This is done in Figure 4.6 when normalized components of the in-plane modulus are superposed on those of the unidirectional ply. The radii for the Mohr's circles for the in-plane modulus are less than those of the constituent ply. The degree of anisotropy is reduced. The reduction in the radii is related to the V's which, in turn, are related to the stacking sequence or volume fractions of the constituent plies. The two phase angles in Figure 4.6 specify the starting points in the Mohr's circles. These points, shown in solid dots, are determined by the orientation of the reference coordinates of the laminate. The magnitude of the phase angles can be derived from the geometric relations in Figure 4.6.

$$
\begin{aligned}
\tan 2\delta_1 &= \frac{V_3^*}{V_1^*} \\
\tan 4\delta_2 &= \frac{V_4^*}{V_2^*}
\end{aligned}
\tag{4.45}
$$

* Similar to Equation 4.40, we can show that:

$$V_1^2 + V_2^2 + V_3^2 + \ldots + 2V_1V_2\cos 2[\theta_1 - \theta_2] + \ldots \leqslant [V_1 + V_2 + V_3 + \ldots]^2 \leqslant 1$$

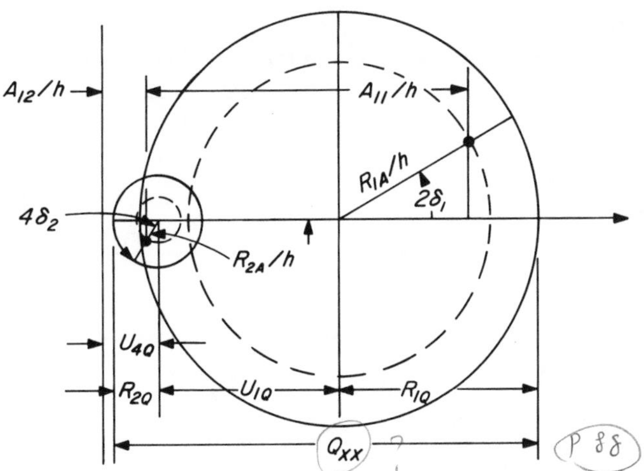

Figure 4.6 Generalized Mohr's circles for T300/5208 unidirectional composite and normalized in-plane modulus of a laminate.

The relations here are also shown in Appendix A. The existence of a symmetry axis is very important to the behavior of composite materials. For unidirectional composites orthotropic symmetry exists when both shear coupling terms vanish simultaneously. This occurs along the horizontal axis in Figure 3.9. For the in-plane modulus of a laminate to have orthotropic symmetry, the phase angles above must be equal. Then by rotating the reference coordinate axes we can always have the starting points in Figure 4.6 along the horizontal axis. The most obvious case for the in-plane modulus to be orthotropic is for the phase angles equal to zero, then

$$V_3^* = V_4^* = 0$$

4. cross-ply laminates

We will now examine some commonly encountered symmetric laminates and determine the values of their in-plane stiffness and compliance.

First, we will study cross-ply composites. The ply orientations are limited to 0 and 90 degrees. In Table 4.5, all the values of the trigonometric functions which will be needed are listed.

table 4.5
values of trigonometric functions for cross-ply laminates

θ_i	$\cos 2\theta_i$	$\cos 4\theta_i$	$\sin 2\theta_i$	$\sin 4\theta_i$
0	1	1	0	0
90	−1	1	0	0

Substituting these trigonometric functions into Equation 4.38, et al., we have

$$V_1^* = v_0 - v_{90}$$

$$V_2^* = v_0 + v_{90} = 1$$

$$V_3^* = V_4^* = 0$$

Based on the condition specified in Equation 4.45, this laminate is orthotropic. By taking these values and putting them into Table 4.4, we will have the normalized in-plane modulus for cross-ply composites as functions of volume ratios. This is done in Table 4.6 where matrix multiplication is implied.

table 4.6
formulas for in-plane stiffness modulus for cross-ply composites

	1	U_2	U_3
A_{11}/h	U_1	$v_0 - v_{90}$	1
A_{22}/h	U_1	$v_{90} - v_0$	1
A_{12}/h	U_4		-1
A_{66}/h	U_5		-1

$A_{16} = A_{26} = 0$

Note that only the first two components are affected by the volume fractions of the constituent plies in the laminate. The remaining four components are constant or zero. Cross-ply laminates are orthotropic because the shear coupling terms are zero. If we substitute the definitions of the U's from Equation 3.15 into the last two equations in Table 4.6, we will have:

$$A_{12}/h = U_4 - U_3 = Q_{xy}$$
$$A_{66}/h = U_5 - U_3 = Q_{ss} \qquad (4.46)$$

Following are some numerical examples of the in-plane stiffness and compliance of cross-ply laminates. T300/5208 will be used as our sample material. The elastic modulus in terms of the U's is listed in Table 3.6. Combining the modulus data with the formulas in Table 4.6, we arrive at the following expressions in GPa:

$$A_{11}/h = 76.37 + [v_0 - v_{90}]85.73 + 19.71$$
$$A_{22}/h = 76.37 - [v_0 - v_{90}]85.73 + 19.71$$
$$A_{12}/h = 22.61 - 19.71 = 2.90 \qquad (4.47)$$
$$A_{66}/h = 26.88 - 19.71 = 7.17$$
$$A_{16} = A_{26} = 0$$

The results from these equations are plotted in Figure 4.7, using the volume fraction of 90-degree plies as the abscissa. Note that the rule of mixtures relations apply in the first two components. Both are linear. The other two nonzero components in Equation 4.47 are constant for all volume fractions.

We calculate the compliance components by inverting the stiffness at a given volume fraction. The inversion process though must be repeated for each fraction. Let us take the volume fraction 50 percent. The modulus components are, in GPa:

$$A_{11}/h = A_{22}/h = 96.08, A_{12}/h = 2.90, A_{66}/h = 7.17 \qquad (4.48)$$

Using the matrix inversion method described in Equations 3.64 and 3.65, we have the following solutions where we substituted the in-plane

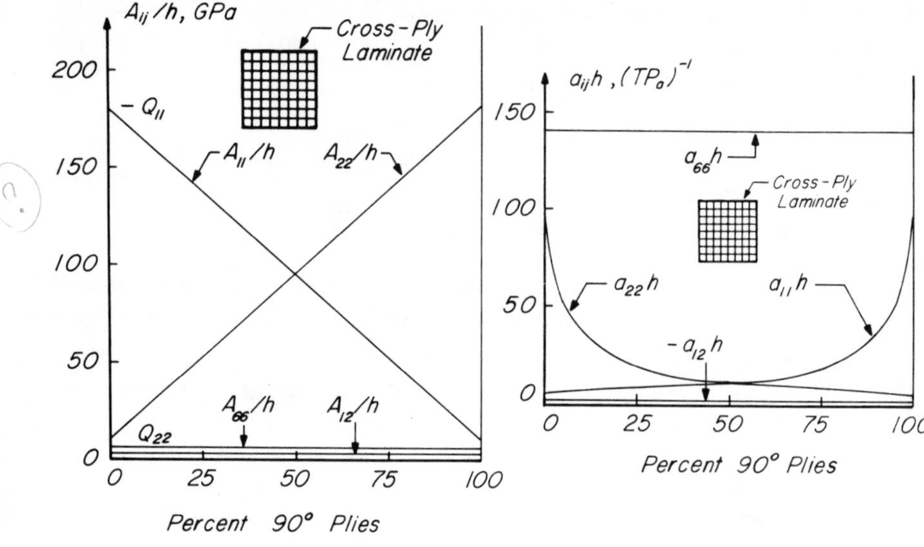

Figure 4.7 In-plane stiffness and compliance of cross-ply composites. The Poisson and shear components are constant and independent of volume fractions; the normal stiffness components are linear but the compliance components are not.

components of the laminate for those of the unidirectional:

$$\begin{aligned}
\Delta A_{ij} &= 66.126 \times 10^{30} h^3 \,(\text{Pa})^3 \\
a_{11}h &= a_{22}h = 10.41 \,(\text{TPa})^{-1} \\
a_{12}h &= -0.3141 \,(\text{TPa})^{-1} \\
a_{66}h &= 139.44 \,(\text{TPa})^{-1} \\
a_{16} &= a_{26} = 0
\end{aligned} \tag{4.49}$$

Typical values of cross-ply laminates are listed in Table 4.7. If our laminate has 16 plies with the following unit ply thickness of:

$$\begin{aligned}
h_o &= 125 \times 10^{-6} \text{ m} \\
16h_o &= 2 \times 10^{-3} \text{ m}
\end{aligned} \tag{4.50}$$

Representative unnormalized components of the in-plane stiffness and compliance of this laminate are:

$$A_{11} = 96.08h = 192.16 \text{ MNm}^{-1}$$
$$a_{11} = 10.41/h = 5.205 \text{ (GN/m)}^{-1} \tag{4.51}$$

table 4.7
normalized in-plane stiffness and compliance of T300/5208 cross-ply laminates, GPa and (TPa)$^{-1}$, respectively.

	A_{11}/h	A_{22}/h	A_{12}/h	A_{66}/h	$A_{16}=A_{26}$
[0/90]	96.08	96.08	2.89	7.17	0
[0_2/90]	124.65	67.50	2.89	7.17	0
[0_4/90]	147.51	44.63	2.89	7.17	0
[0_8/90]	162.75	29.39	2.89	7.17	0

	$a_{11}h$	$a_{22}h$	$a_{12}h$	$a_{66}h$	$a_{16}=a_{26}$
[0/90]	10.41	10.41	−.314	139.47	0
[0_2/90]	8.03	14.82	−.344	139.47	0
[0_4/90]	6.78	22.43	−.440	139.47	0
[0_8/90]	6.15	34.07	−.606	139.47	0

Because of the constraints imposed on the in-plane modulus by the invariants of the constituent ply cited in Equation 4.33, cross-ply laminates can have only one variable. Of the six possible components of in-plane modulus, the shear coupling terms are zero. Of the remaining four, the Poisson and shear components are fixed by the respective components of the constituent ply. Of the remaining two normal components, only one can be free because the sum of these components must be invariant; i.e.,

$$[A_{11} + A_{22}]/h = 96.08 + 96.08 = 124.65 + 67.50 \ldots = 192.16$$

Thus the only degree of freedom is the value of one of the components above. There is an additional constraint implicit for all laminates; i.e.,

$$Q_{yy} \leqslant A_{11}/h \leqslant Q_{xx}$$

From the compliance components, we can get the following engineering constants using Equation 4.18:

$$
\begin{aligned}
E_1^o &= 1/10.41 = 96.0\text{ GPa} \\
E_6^o &= 1/139 = 7.17\text{ GPa} \\
\nu_{21}^o &= 0.3141/10.41 = 0.0301
\end{aligned}
\tag{4.52}
$$

Since cross-ply laminates are orthotropic, we could have calculated and obtained the same engineering constants in Equation 4.52 from those in-plane modulus components in Equation 4.48 directly using the relations in Equation 1.13.

$$E_1^o = \frac{Q_{11}}{m} = \frac{A_{11}/h}{m}$$

$$\text{where } m = \left[1 - \frac{A_{12}\,A_{21}}{A_{11}\,A_{22}}\right]^{-1} = \left[1 - \frac{2.90^2}{96.08^2}\right]^{-1} = 1.001$$

$$\text{then } E_1^o = \frac{96.08}{1.001} = 96.0\text{ GPa} \tag{4.53}$$

This agrees with the longitudinal modulus in Equation 4.52.

The constraining effect of the 90 degree ply is responsible for the low Poisson's ratio. The value of m is almost unity, therefore

$$A_{11}/h \simeq E_1^o$$

This is true only for cross-ply composites. As we have seen Equation 4.47 follows the rule of mixtures relation. We can then say the E_1^o will follow approximately the same relation. This simple relation will hold for laminates only if they are orthotropic and have very small effective Poisson's ratios.

5. angle-ply laminates

Angle-ply laminates form another very common class that deserves special attention. In this class, there are only two ply orientations which have the same magnitude but opposite signs. The laminate is balanced when there are equal numbers of plies with positive and negative orientations.

Thus for angle-ply laminates we have:

$$\theta_1 = +\phi, \theta_2 = -\phi \tag{4.54}$$

and $$v_1 = v_2 = \frac{1}{2}$$

Substituting these values into Equation 4.38 et al.,

$$V_1^* = \frac{1}{2}(\cos 2\phi + \cos 2\phi) = \cos 2\phi$$

$$V_2^* = \cos 4\phi \tag{4.55}$$

$$V_3^* = V_4^* = 0$$

This laminate is orthotropic because of Equation 4.45.

The formulas for the in-plane modulus for angle-ply laminates are listed in Table 4.8, where matrix multiplication is implied. They are obtained by substituting the values from Equation 4.55 into Table 4.4.

Note that the first four rows of this table are identical to those in Table 3.3 for the unidirectional modulus except where the ply orientation θ in Table 3.3 is replaced by the angle ϕ in the angle-ply laminate. The shear coupling terms vanish for the angle-ply laminate because of the last line in Equation 4.55.

table 4.8
formulas for in-plane stiffness modulus for angle-ply laminates

	I	U_2	U_3
A_{11}/h	U_1	$\cos 2\phi$	$\cos 4\phi$
A_{22}/h	U_1	$-\cos 2\phi$	$\cos 4\phi$
A_{12}/h	U_4		$-\cos 4\phi$
A_{66}/h	U_5		$-\cos 4\phi$

$A_{16} = A_{26} = 0$

Because of the similarity between this table and Table 3.3, we can immediately convert the numerical results for T300/5208 in Figure 3.5 by simply replacing θ by $\pm\phi$ and deleting the shear coupling terms. The comparable components of in-plane modulus for T300/5208 angle-ply laminates as functions of $\pm\phi$ are equal to those in the first four columns in Table 3.5. The last two columns are zero.

For [45/−45] angle ply, we have

$$\cos 2\phi = 0, \ \cos 4\phi = -1$$

Using the data for T300/5208 from Table 3.6, we have from the formulas for in-plane modulus in Table 4.8

$$\begin{aligned} A_{11}/h &= A_{22}/h = 76.37 - 19.71 = 56.66 \text{ GPa} \\ A_{12}/h &= 22.61 + 19.71 = 42.32 \text{ GPa} \\ A_{66}/h &= 26.88 + 19.71 = 46.59 \text{ GPa} \\ A_{16} &= A_{26} = 0 \end{aligned} \tag{4.56}$$

With the exception of these shear coupling terms, these values are identical to those for $\theta = 45$ degrees in Table 3.5. The in-plane modulus of angle-ply laminates is listed in Table 4.9 and plotted in Figure 4.8. Using the inversion method applied in Equation 3.65, we have for the [45/−45] from Equation 4.56:

$$\begin{aligned} \Delta A_{ij} &= 66.126 \times 10^{30} h^3 \text{ (Pa)}^3 \\ a_{11}h &= a_{22}h = 39.91 \text{ (TPa)}^{-1} \\ a_{12}h &= -29.81 \text{ (TPa)}^{-1} \\ a_{66}h &= 21.46 \text{ (TPa)}^{-1} \\ a_{16} &= a_{26} = 0 \text{ (TPa)}^{-1} \end{aligned} \tag{4.57}$$

These compliance values together with those for other values of ϕ are listed in Table 4.9 and are shown as solid lines in Figure 4.9. The dashed lines are the transformed components of the compliance of unidirectional composites taken from Figure 3.11. The big difference between the two lines (dashed versus solid) is due to the matrix inversion with or without the shear coupling terms. The off-axis unidirectional composite is anisotropic and the angle-ply laminate is orthotropic.

table 4.9
normalized in-plane stiffness and compliance of T300/5208 angle-ply laminates, GPa and (TPa)$^{-1}$

$\pm\phi$	A_{11}/h	A_{22}/h	A_{12}/h	A_{66}/h	$A_{16}=A_{26}$
0	181.8	10.3	2.90	7.17	0
15	160.4	11.9	12.75	17.02	0
30	109.3	23.6	32.46	36.73	0
45	56.6	56.6	42.32	46.59	0
60	23.6	109.3	32.46	36.73	0
75	11.9	160.4	12.75	17.02	0
90	10.3	181.8	2.90	7.17	0

$\pm\phi$	$a_{11}h$	$a_{22}h$	$a_{12}h$	$a_{66}h$	$a_{16}=a_{26}$
0	5.52	97.08	− 1.54	139.47	0
15	6.80	91.21	− 7.24	58.73	0
30	15.42	71.36	−21.18	27.22	0
45	39.91	39.91	−29.81	21.46	0
60	71.36	15.42	−21.18	27.22	0
75	91.21	6.80	− 7.24	58.73	0
90	97.08	5.52	1.54	139.47	0

The engineering constants for off-axis unidirectional composites will also be completely different from those for angle-ply laminates. The engineering constants for ϕ = 45 degrees, which is simply the ±45-degree angle ply, can be obtained directly from the results in Equation 4.57 using the relations in Equation 4.18:

$$E_1^o = E_2^o = 1/a_{11}h = 39.91^{-1} = 25.05 \text{ GPa}$$

$$E_6^o = 1/a_{66}h = 21.46^{-1} = 46.59 \text{ GPa} \tag{4.58}$$

$$\nu_{21}^o = -a_{12}/a_{11} = 29.81/39.91 = 0.746$$

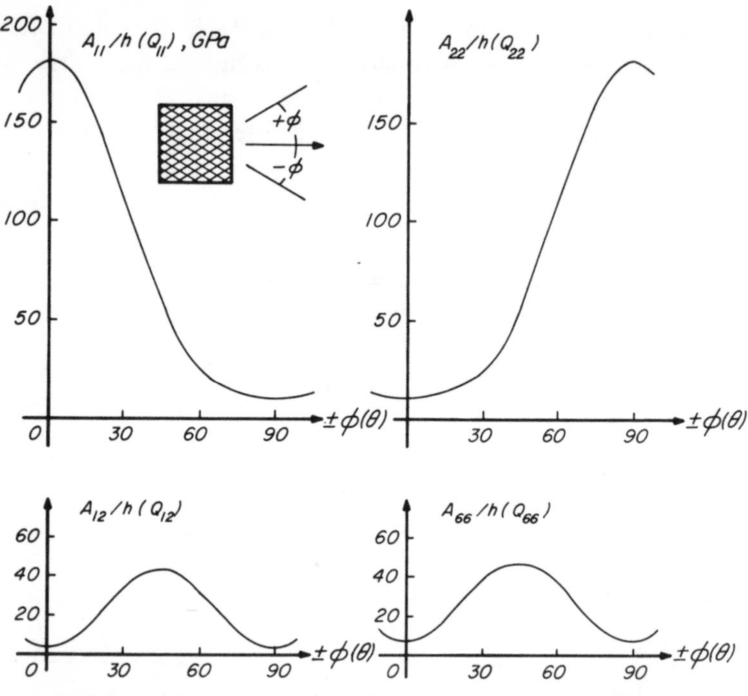

Figure 4.8 In-plane stiffness of angle-ply laminate of T300/5208 composite. With the exception of the non-zero shear coupling components, the curves above are identical to the transformed modulus of unidirectional T300/5208 shown in Figure 3.5 where the coordinates are defined in the parenthesis. The curves here are the laminate stiffness as a function of lamination angle ϕ.

The corresponding engineering constants for an off-axis unidirectional T300/5208 were calculated from the data in Table 3.12:

$$
\begin{aligned}
E_1 &= E_2 = 59.75^{-1} = 16.73 \text{ GPa} \\
E_6 &= 105.7^{-1} = 9.46 \text{ GPa} \\
\nu_{21} &= 9.99/59.75 = 0.167
\end{aligned}
\tag{4.59}
$$

Compare like constants in Equations 4.58 and 4.59; we see that the values for angle-ply laminates are much higher than the off-axis unidirectional. The Poisson's ratios of 0.746 exceed the upper limit for isotropic materials, which is 1/2. This is theoretically admissible for

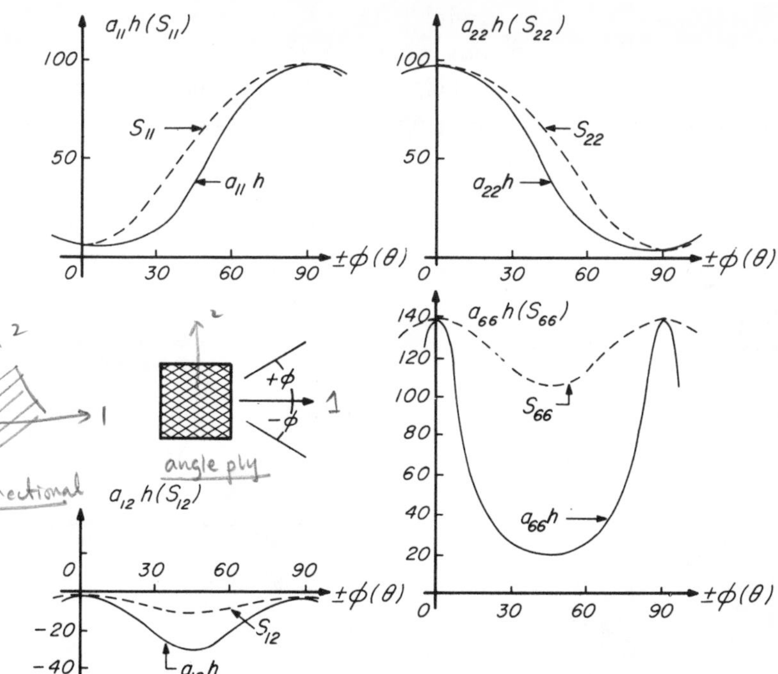

Figure 4.9 In-plane compliance of T300/5208 angle-ply laminates. The solid lines are based on the data in Table 4.8, and are not transformation curves. The dashed lines, taken from Figure 3.11 are the transformed components of compliance of the same unidirectional composite, with the coordinates defined in parenthesis; i.e., θ vs S_{ij}. Close similarity between A_{ij}/h and Q_{ij} does not exist between $a_{ij}h$ and S_{ij}. The shear coupling components are responsible for the differences between the solid and dashed lines.

non-isotropic materials. The comparison of these engineering constants between off-axis unidirectional and angle-ply laminates as functions of ply orientation θ and lamination angle $\pm\phi$ are shown in Figure 4.10. The significant increase in the angle-ply laminates over that of the off-axis unidirectional over the entire range of angles is very apparent. The increase is caused by the constraining influence imposed on each ply within a laminate. The plies are bonded together and are not free to deform independently.

There is an important message here. Laminates are governed by very rigorous conditions such as those in Equation 4.6. Laminates cannot be modeled by a one-dimensional arrangement of parallel springs. Plies within a laminate are constrained and interact with one another. It is not always intuitively obvious when we add or subtract plies from a laminate if we are actually helping or hurting the stiffness and the strength. The key is to understand the ply-to-ply interaction and to adhere to the mathematical models faithfully.

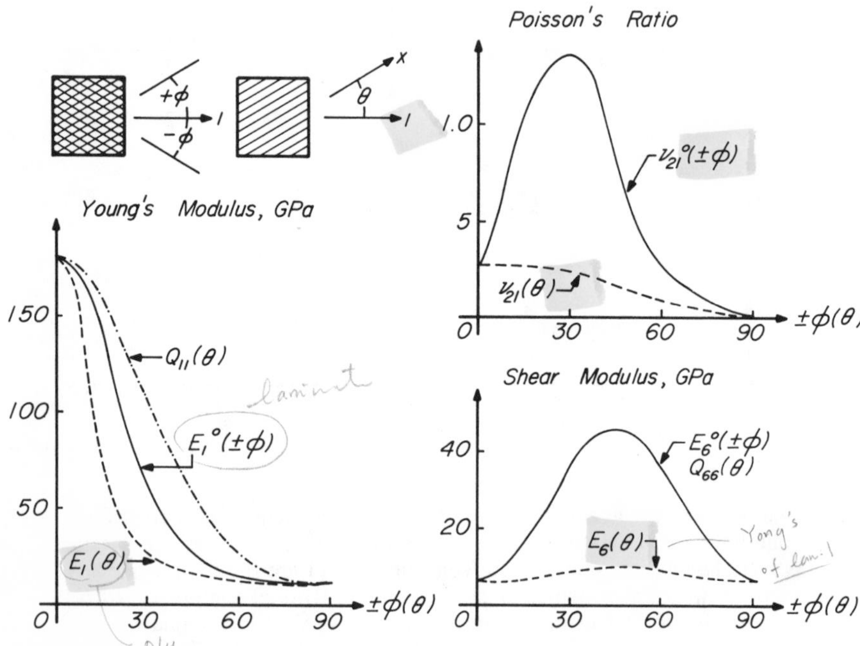

Figure 4.10 Comparison between engineering constants of angle-ply and unidirectional composites. The variations of these constants are shown as dashed and solid lines. The dashed lines are identical to the solid lines in Figure 3.17.

6. quasi-isotropic laminates

If the following conditions for the in-plane modulus of a laminate are satisfied, the laminate is quasi-isotropic:

$$
\begin{aligned}
A_{11} &= A_{22} \\
A_{16} &= A_{26} = 0 \\
A_{66} &= \frac{1}{2}[A_{11} - A_{12}]
\end{aligned}
\tag{4.60}
$$

The last relation is analogous to that for the on-axis stiffness in Equation 1.23. With the last constraint we have one less degree of freedom than the cross-ply laminate cited in Table 4.7. We have no freedom at all! If we construct a quasi-isotropic laminate out of a given material its normalized in-plane properties are predetermined. We cannot change them unless we use a different material. There are numerous stacking sequences for this laminate.

Intuitively, if ply orientations are random, we would expect an isotropic laminate. Directionality would disappear. For example, chopped fiber composites are quasi-isotropic. If we examine Equation 4.31, which defines the V's, it is reasonable to expect these geometric factors to vanish. Physically, when there is equal probability of fibers oriented in any direction, or there is a continuous variation in fiber orientation, the cyclic terms in Table 4.3 as defined by the V's will vanish. The in-plane modulus components will converge toward the invariants in the first column of the table. The in-plane modulus becomes:

$$\begin{cases} A_{11}/h = A_{22}/h = U_1 \\ A_{12}/h = U_4 \\ A_{66}/h = U_5 \end{cases} \tag{4.61}$$

The conditions for isotropy in Equation 4.60 are satisfied because of the relations between the invariants as described in Equation 3.20. We have a quasi-isotropic material.

Since this is an isotropic material, we can find the quasi-isotropic engineering constants, as follows:

From Equation 1.13,

$$\begin{aligned} \nu_y &= \nu_x = \nu = \frac{A_{12}}{A_{11}} = \frac{U_4}{U_1} \\ E_s &= G = A_{66}/h = U_5 \end{aligned} \tag{4.62}$$

From Equation 1.23,

$$E = 2(1 + \nu)G = 2\left[1 + \frac{U_4}{U_1}\right]U_5$$

If we use the values of U's for T300/5208 from Table 3.6, we have

$$\nu = 22.61/76.37 = 0.296$$

$$G = 26.88 \text{ GPa} \tag{4.63}$$

$$E = 2(1 + 0.296)26.88 = 69.67 \text{ GPa}$$

There are other than random orientations that will produce quasi-isotropic laminates. Let us examine the following two laminates:

$$[0/60/-60]_S, \text{ and } [0/90/45/-45]_S$$

For the first laminate, we have from the definitions of V^*'s in Equation 4.38 et al.,

$$\begin{aligned}
v_1 &= v_2 = v_3 = \frac{1}{3} \\
V_1^* &= \frac{1}{3}[\cos 0 + \cos 120 + \cos(-120)] = 0 \\
V_2^* &= \frac{1}{3}[\cos 0 + \cos 240 + \cos(-240)] = 0 \\
V_3^* &= \frac{1}{3}[\sin 0 + \sin 120 + \sin(-120)] = 0 \\
V_4^* &= \frac{1}{3}[\sin 0 + \sin 240 + \sin(-240)] = 0
\end{aligned} \tag{4.64}$$

For the second laminate, we have:

$$v_1 = v_2 = v_3 = v_4 = \frac{1}{4}$$

$$V_1^* = \frac{1}{4}[\cos 0 + \cos 180 + \cos 90 + \cos(-90)] = 0 \tag{4.65}$$

(continued)

$$V_2^* = \frac{1}{4}[\cos 0 + \cos 360 + \cos 180 + \cos(-180)] = 0$$

$$V_3^* = \frac{1}{4}[\sin 0 + \sin 180 + \sin 90 + \sin(-90)] = 0$$

$$V_4^* = \frac{1}{4}[\sin 0 + \sin 360 + \sin 180 + \sin(-180)] = 0 \qquad (4.65)$$

(concluded)

Since all the V's are zero, the laminates are quasi-isotropic. In fact, we can generalize that any laminate with "m" ply groups spaced at ply orientations of Pi/m radian will be quasi-isotropic. In the first case we had $m = 3$; in the second case, $m = 4$. Moreover, with symmetric laminates we must double the number of ply groups within a quasi-isotropic laminate. The minimum number of plies are 6 and 8, respectively. The first laminate is also called Pi/3; and the second Pi/4. For quasi-isotropic laminates we have no freedom because the stacking sequence is fixed and the resulting stiffness is also fixed.

There is a very practical reason for quasi-isotropic laminates beyond being isotropic like conventional materials. This configuration represents the minimum performance that we can expect from a composite laminate. If we are uncomfortable in dealing with directionally varying properties, we can always use the quasi-isotropic laminate. A direct substitution of this laminate for the conventional material can be done without hesitation because this substitution is no different from the substitution of conventional materials.

The quasi-isotropic Young's modulus of T300/5208, as listed in Equation 4.63, is equal to the Young's modulus of aluminum. But there is a minimum of 40 percent savings in weight. When directionality is judiciously added, the advantages of composites are overwhelming.

The quasi-isotropic laminates can be used as the starting point of optimization of ply orientation. If minimum weight is a criterion, the quasi-isotropic laminate should be the upper bound of the weight. An optimized material taking full advantage of the directionality of properties should only have lower weight than the quasi-isotropic configuration.

7. general Pi/4 laminates

This is a family of laminates having four ply orientations spaced at 45-degree intervals. The normal Pi/4 laminates have four ply groups with equal thickness and are therefore quasi-isotropic. General Pi/4 laminates refer to those with arbitrary thicknesses in ply groups, including the limiting cases of zero thickness for one or more ply groups. We will now list all the trigonometric functions and their values for our ply orientations in Table 4.10.

table 4.10
values of trigonometric functions for in-plane modulus of general Pi/4 laminates

θ_i	$\cos 2\theta_i$	$\cos 4\theta_i$	$\sin 2\theta_i$	$\sin 4\theta_i$
0	1	1	0	0
90	−1	1	0	0
45	0	−1	1	0
−45	0	−1	−1	0

Substituting these values into Equation 4.38 et al., we have

$$
\begin{aligned}
V_1^* &= v_0 - v_{90} \\
V_2^* &= v_0 + v_{90} - v_{45} - v_{-45} \\
V_3^* &= v_{45} - v_{-45} \\
V_4^* &= 0
\end{aligned}
\tag{4.66}
$$

With these values, the formulas for in-plane modulus in Table 4.4 can be specialized for our general Pi/4 laminates. This is done in a matrix multiplication table as follows. Note when all the v's are equal, we recover the quasi-isotropic laminates. When the ±45-degree plies are zero, we recover the formulas for cross-ply laminates listed in Table 4.6. When we have a special angle-ply laminate with the lamination angle ϕ equal to 45 degrees, we recover from Table 4.11 the special ±45 laminate from Table 4.8. Finally, Pi/4 laminates are orthotropic when the

45-degree plies are balanced, or when

$$\nu_{45} = \nu_{-45} \tag{4.67}$$

When this is true, the shear coupling components become zero.

table 4.11
formulas for in-plane stiffness modulus of general Pi/4 laminates

	I	U_2	U_3
A_{11}/h	U_1	$v_0 - v_{90}$	$v_0 + v_{90} - v_{45} - v_{-45}$
A_{22}/h	U_1	$-v_0 + v_{90}$	$v_0 + v_{90} - v_{45} - v_{-45}$
A_{12}/h	U_4		$-v_0 - v_{90} + v_{45} + v_{-45}$
A_{66}/h	U_5		$-v_0 - v_{90} + v_{45} + v_{-45}$
A_{16}/h		$\frac{1}{2}[v_{45} - v_{-45}]$	
A_{26}/h		$\frac{1}{2}[v_{45} - v_{-45}]$	

The formulas in Table 4.11 can be represented by a series of diagrams or plots. First of all, the components of stiffness are all linear functions of the ply fractions. The components are proportional to four linear combinations of the ply fractions; viz., ν_0, ν_{90}, $\nu_{45} + \nu_{-45}$ and $\nu_{45} - \nu_{-45}$. In Figure 4.11 we show the in-plane modulus of general Pi/4 laminates for T300/5208 composite. The first chart shows component A_{11}/h. This chart is valid for balanced as well as unbalanced laminates; i.e., independent of the values of ν_{45} and ν_{-45}. The same can be applied to the second chart on components A_{12}/h and A_{66}/h. Component A_{22}/h is not shown because it can be found by interchanging ν_0 with ν_{90} from the A_{11}/h chart. Only in unbalanced laminate (that is the +45 has a different number of plies from the −45) will the last chart in Figure 4.11 become necessary.

The open hexagon in each diagram represents the properties of the quasi-isotropic laminate. Note all relationships in Table 4.11 are described by straight lines. This did not happen by accident. In fact, it is important to choose the correct parameters so linear relationships exist between the ply and the laminate properties. The correct property set

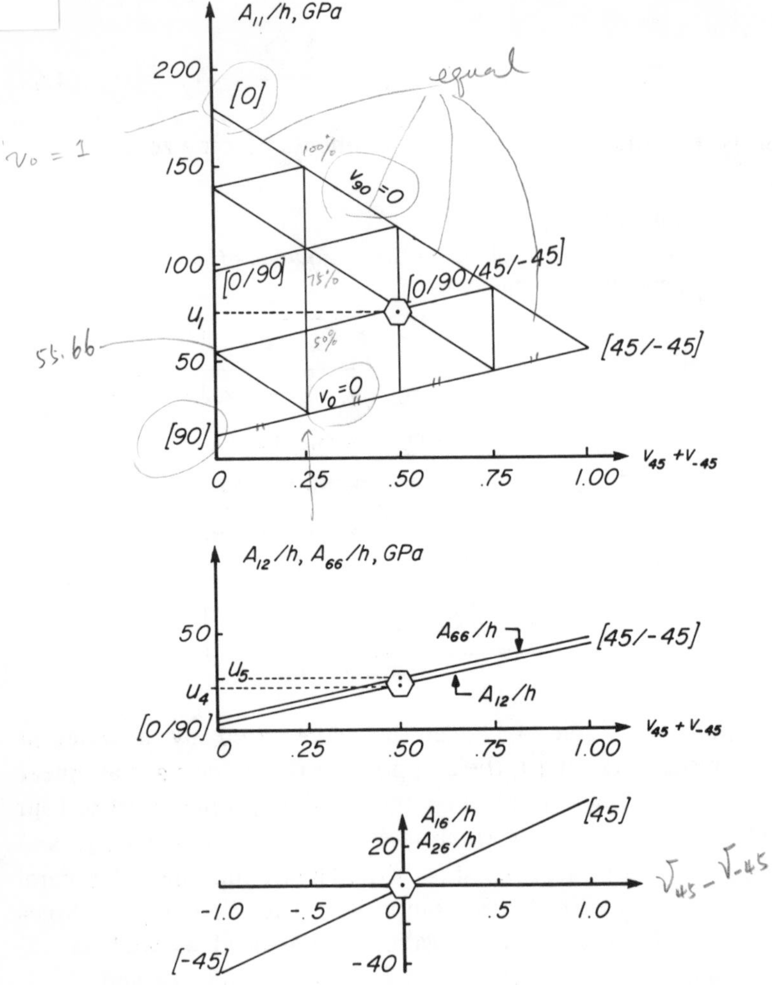

Figure 4.11 In-plane modulus of general Pi/4 laminates of T300/5208 composite. Quasi-isotropic points are shown as open hexagons.

for the stiffness of a laminate is the stiffness modulus of the unidirectional composites. If another property set is chosen, nonlinear relationship would result. For example, when the property set of engineering constants is chosen, the straight lines in Figure 4.11 will be replaced by curved lines. In fact the curved lines have been referred to by a trade name: the carpet plot. The moral of the story is that stiffness modulus is the simplest property for the description of the stiffness of laminated composites. The carpet plot is an unnecessarily complicated way of showing properties of composites.

8. general bidirectional laminates

We have seen earlier two classes of bidirectional laminates, viz., cross-ply and angle-ply laminates. Both laminate classes are orthotropic. We have seen unique properties of laminates that do not have a counterpart in conventional materials. For example, the in-plane Poisson's ratio, shown in Figure 4.10, extends beyond the upper limit of 1/2 imposed on isotropic materials. While orthotropic materials can be viewed as a simple extension of conventional materials, nonorthotropic materials, however, must be viewed from a completely different viewpoint. We must understand the unique properties of anisotropic materials and learn to capitalize on these properties to perform functions not possible with conventional materials. In this section, we will illustrate how simple, unique properties can be derived from general bidirectional laminates.

A general bidirectional laminate consists of two arbitrary ply orientations and ply ratios. In Figure 4.12 we show the two orientations. The two orthogonal bisectors would be the symmetry axes if the two ply orientations are balanced; i.e., the ply ratio is unity.

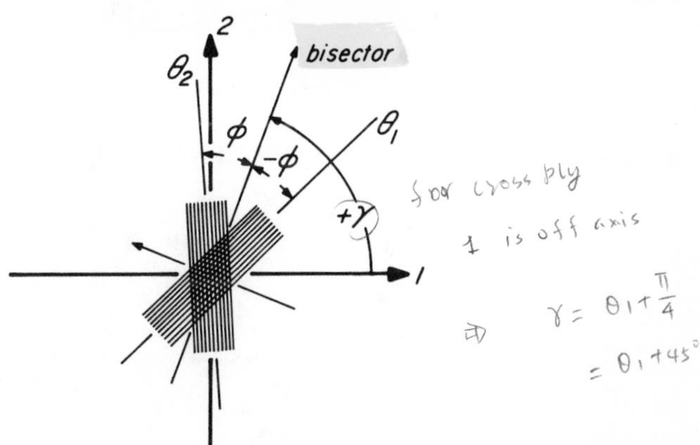

Figure 4.12 Orientations of general bidirectional laminates.

In terms of the angles shown in Figure 4.12, we have

$$\begin{aligned} \theta_1 &= \gamma - \phi \\ \theta_2 &= \gamma + \phi \end{aligned} \tag{4.68}$$

We will introduce another variable:

$$p = \text{ply ratio} = \frac{v_1}{v_2} \tag{4.69}$$

where the v's are the volume fractions of ply orientations. This was defined in Equation 4.37. From Equations 4.38 et al., we can immediately define for all bidirectional laminates.

$$\begin{aligned} V_1^* &= v_1 \cos 2\theta_1 + v_2 \cos 2\theta_2 \\ V_2^* &= v_1 \cos 4\theta_1 + v_2 \cos 4\theta_2 \\ V_3^* &= v_1 \sin 2\theta_1 + v_2 \sin 2\theta_2 \\ V_4^* &= v_1 \sin 4\theta_1 + v_2 \sin 4\theta_2 \end{aligned} \tag{4.70}$$

We have a general cross-ply laminate when:

$$\phi = 45 \text{ degrees} \tag{4.71}$$

By combining this with Equations 4.68–4.70, we have

$$\begin{aligned} V_1^* &= (v_1 - v_2)\sin 2\gamma \\ V_2^* &= -\cos 4\gamma \\ V_3^* &= -(v_1 - v_2)\cos 2\gamma \\ V_4^* &= -\sin 4\gamma \end{aligned} \tag{4.72}$$

This is an off-axis cross-ply laminate. The rigid body rotation of the laminate is specified by γ. This laminate becomes the usual cross-ply laminate when

$$\gamma = 45 \text{ degrees} \tag{4.73}$$

Equation (4.72) becomes:

$$\begin{aligned} V_1^* &= \nu_1 - \nu_2 \\ V_2^* &= 1 \\ V_3^* &= V_4^* = 0 \end{aligned} \tag{4.74}$$

We have recovered the formulas in Table 4.6.

We have a general angle-ply laminate when:

$$\gamma = 0 \tag{4.75}$$

We can easily show by combining this with Equation 4.68 et al.

$$\begin{aligned} V_1^* &= \cos 2\phi \\ V_2^* &= \cos 4\phi \\ V_3^* &= -(\nu_1 - \nu_2)\sin 2\phi \\ V_4^* &= -(\nu_1 - \nu_2)\sin 4\phi \end{aligned} \tag{4.76}$$

This is a general angle-ply laminate when it is not balanced. When we have a balanced laminate,

$$\nu_1 - \nu_2 = 0 \tag{4.77}$$

Then Equation 4.76 becomes the same for the usual angle-ply laminate shown in Table 4.8. It is intuitively obvious that the magnitude of the shear and normal coupling is related to the degree of the imbalance. Equation 4.77 is still a rule of mixtures relation that goes from +1 to −1 with zero at the midpoint.

As an illustration of the range of properties for a general bidirectional laminate, we take:

$$\begin{aligned} \phi &= 30 \text{ degrees} \\ \gamma &= -90 \text{ to } 90 \text{ degrees} \\ p &= 0, \frac{1}{9}, \frac{1}{4}, 1, 4, 9, \infty \end{aligned} \tag{4.78}$$

Note the two limiting values of the ply ratio correspond to unidirectional composites with +30 and −30 degrees orientations, respectively.

For the complete characterization of the stiffness of a symmetric laminate, the following material constants are useful for various purposes.

1. Six normalized components of stiffness: A_{ij}/h
2. Six normalized components of compliance: $a_{ij}h$
3. Two Young's modulus: E_1^o, E_2^o
4. Shear modulus: E_6^o
5. Two Poisson's ratios: ν_{21}^o and ν_{12}^o
6. Two shear coupling coefficients: ν_{61}^o and ν_{62}^o
7. Two normal coupling coefficients: ν_{16}^o and ν_{26}^o
8. One ratio of Young's moduli: $a = E_1^o/E_2^o$
9. Two ratios of Young's to shear moduli:

We must keep in mind that not all the elastic constants above are independent. From Equation 4.33 we can conclude that there are at most four independent constants among the six components of the in-plane modulus. Thus, it is safe to say that we cannot manipulate the elastic constants of a laminated plate at will. There are constraints. When we increase one constant such change will induce other changes in accordance with the law of transformation and its invariants.

We will show the variations of some typical elastic constants of general bidirectional laminates. The material used is T300/5208. The definition of the angles follow those in Figure 4.12. The ply ratio p as defined in the Equation 4.69 is also shown of each of the following figures:

1. A typical component of stiffness is shown in Figure 4.13 (top). This is a shear/normal coupling component. The two limiting cases which correspond to the ply ratios of infinity and zero are the upper/lower bounds of the component of the stiffness. Furthermore, the usual lever rule for phase diagrams applies. This is not surprising because the stiffness component of a laminated composite is obtained by a straight averaging or the rule of mixtures relation.

2. A typical component of the in-plane compliance is shown in Figure 4.13(bottom). The same notation used in the previous figure applies. Note that the compliance components as a result of the changes in the ply ratio are essentially bounded by the two limiting

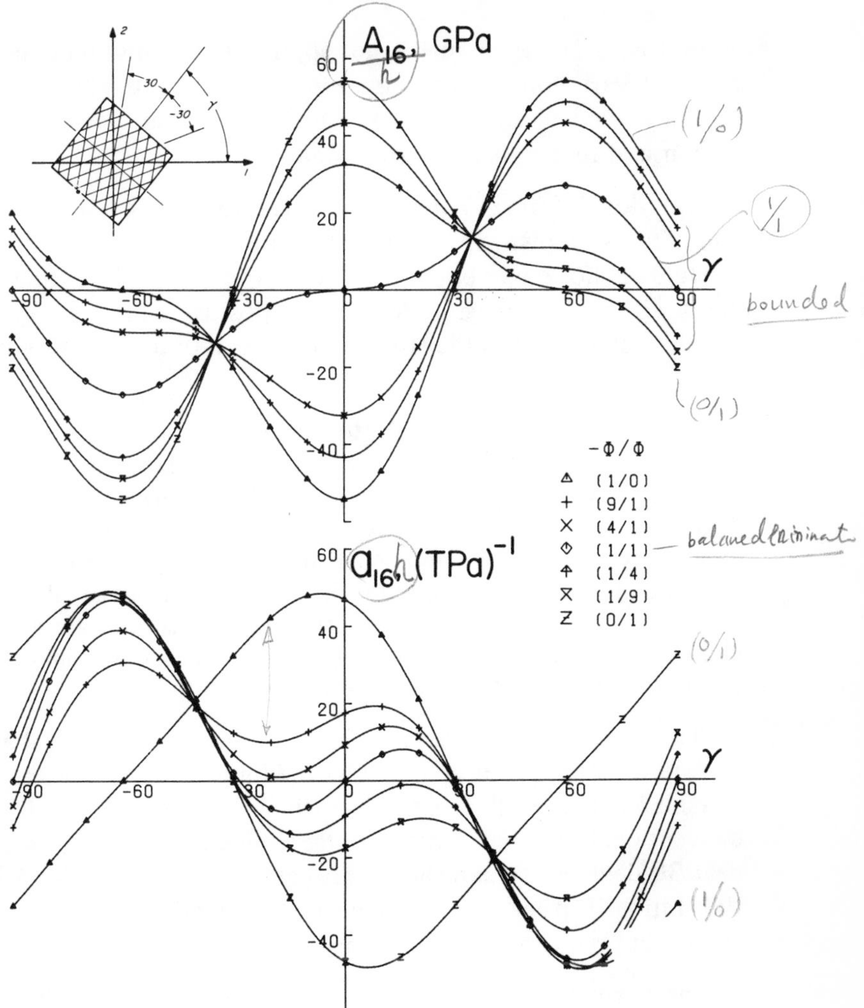

Figure 4.13 Typical components of stiffness and compliance of T300/5208 bidirectional laminates.

cases. There are exceptions but the difference is very small for this material. There is, however, a very drastic change in the magnitude of the component of compliance as one goes from all −30 degree unidirectional ply (p = infinity) to the case where the ply ratio becomes 9/1. In other words, a 10 percent change in the ply orientation can result in over 100 percent changes in the resulting components of compliance. The lever rule is not applicable for this component. Th range of variation for this component covers both the positive an the negative values. This can provide complete reversals in th response of materials if it is so desired.

3. The longitudinal Young's modulus for a general bidirection laminate is shown in Figure 4.14(top). The magnitude of th modulus is no longer bounded. Needless to say the lever rule will nc apply either. The magnitude remains positive which is required fc the diagonal terms in the stiffness matrix in order to insure materia stability.

4. The ratio of the Young's to shear moduli is shown in Figur 4.14(bottom). This ratio is a measure of the bending stiffness t shear rigidity. This ratio is often referred to as the EI/GJ where I an J are moments of inertia of the cross-section of a structure, and and G are the Young's and shear moduli of the material. Note tha the ratio is essentially bounded by the limiting cases although th lever rule does not apply. Wide variations of this ratio are possible b changing the ply ratios or ply orientations. This component remain positive as its required from the stability standpoint.

5. The longitudinal Poisson's ratios is shown in Figure 4.15(top This ratio is no longer bounded by the limiting cases. Poisson's ratio can be zero as well as negative which is not permissible for th conventional material. Note the steep descend of the Poisson's rati near 30 degrees, a small change in angle can result in great changes i this ratio. Further, the ratio near this angle is insensitive to th change in ply ratios as we go from 1/1 to 9/1.

6. The normal coupling coefficient is shown in Figure 4.15(bottom This coefficient is essentially bounded by the limiting cases. Th lever rule however does not apply. There is, again, a wide variation o the magnitude of this coefficient covering both the positive an negative values. Again, near 30 degrees this coefficient is insensitiv to the changes in ply ratios, for example, as we go from zero t infinity.

We have just displayed some typical examples of the elastic constant of general bidirectional laminates. It appears that only the component of the stiffness can be readily anticipated in that the rule of mixture equation and the normal lever rule apply. The "carpet plot' is

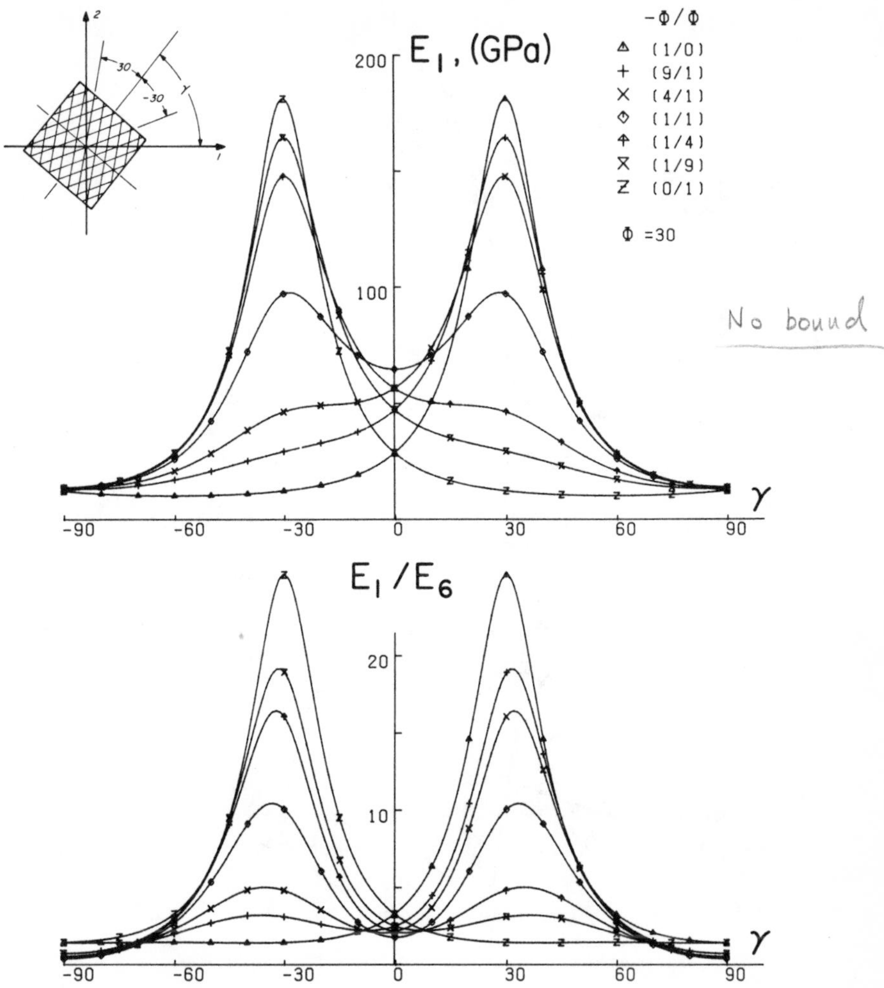

Figure 4.14 Longitudinal Young's modulus and the ratio of extension to shear modulus of T300/5208 bidirectional laminates.

linearized as shown in Figure 4.11. Such simple, linear relationships no longer exist in the case of the components of the compliance and the engineering constants. It is very difficult to bound or visualize the magnitude of the change as we change ply orientation or ply ratios. Whatever methodology or algorithm is used for the design and optimization, we must appreciate the possible variations of the elastic behavior as the laminate becomes fully anisotropic. Both ply orientation and

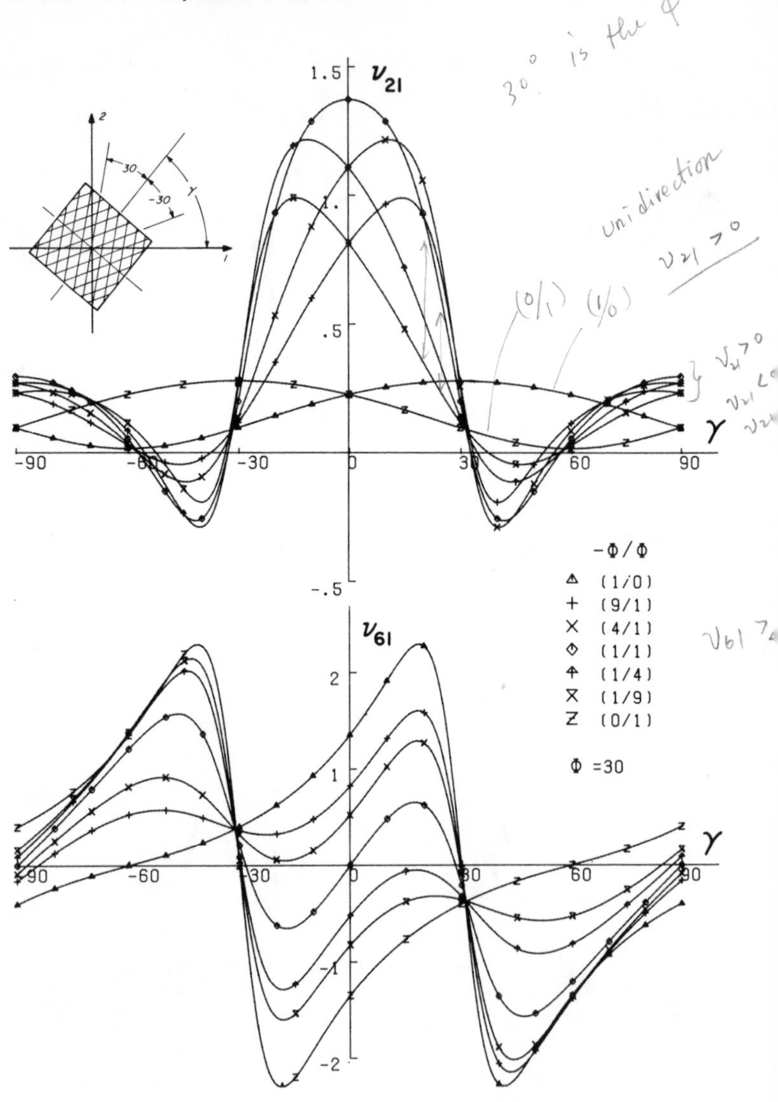

Figure 4.15 Typical coupling coefficients of T300/5208 bidirectional laminates.

ply ratio are variables. These figures are intended to demonstrate the range of variability and the sensitivity as a function of stacking sequence for typical bidirectional laminates.

It is difficult to characterize multidirectional laminates with three or more distinct ply orientations. It appears safe to say that as the number of ply orientations increase, the resulting elastic constants will approach those of the quasi-isotropic laminates. Should this be the case bidirectional laminates are unique because they can provide the widest variation of properties. These laminates can have properties with greater variations than the unidirectional constituent plies.

9. ply stress and ply strain analysis

Following the process outlined in Figure 4.4 we can readily calculate the stress and strain at any ply within a symmetric laminate under arbitrarily applied stress resultants. The motivation for the determination of the stress and strain at a ply level is for the assessment of ply failure. In a multidirectional laminate subject to in-plane strains or stress resultants it is necessary to examine ply by ply or ply group by ply group to determine if any of them has failed or about to fail. The failure criterion for the plies will be covered in Chapter 7. We are only outlining the method for the ply stress and ply strain analysis which is a prerequisite for the eventual failure determination.

We will now list a few simple examples.

1. Determine the ply stress and ply strain of a T300/5208 $[0_4/90_4]_S$ laminate subjected to a uniaxial stress resultant of 1 MN/m.

   From Equation 4.50 and Table 4.7, using unit ply thickness of 125×10^{-6}m.

$$h = 16 \times 125 \times 10^{-6} = 2 \times 10^{-3} \text{ m}$$

$$\begin{aligned} a_{11} &= 10.41/2 \times 10^{-3} = 5.21\ (\text{GN/m})^{-1} \\ a_{21} &= -.314/2 \times 10^{-3} = -.157\ (\text{GN/m})^{-1} \\ a_{61} &= 0 \end{aligned} \tag{4.79}$$

   From the stress-strain relation of Table 4.2

$$\begin{aligned} \epsilon_1^o &= 5.21 \times 10^{-3} \\ \epsilon_2^o &= -.157 \times 10^{-3} \\ \epsilon_6^o &= 0 \end{aligned} \tag{4.80}$$

For the 0-degree ply, the on-axis stress and strain are:

$$\begin{aligned} \epsilon_x &= 5.21 \times 10^{-3} \\ \epsilon_y &= -.157 \times 10^{-3} \\ \epsilon_s &= 0 \end{aligned} \tag{4.81}$$

From the on-axis stress-strain relation of Table 1.6 and the modulus in Table 1.9,

$$\begin{aligned} \sigma_x &= 181.8 \times 5.21 - 2.89 \times .157 = 946.7 \text{ MPa} \\ \sigma_y &= 2.89 \times 5.21 - 10.34 \times .157 = 13.4 \text{ MPa} \\ \sigma_s &= 0 \end{aligned} \tag{4.82}$$

For the 90-degree ply, the normal strains are interchanged:

$$\begin{aligned} \epsilon_x &= -.157 \times 10^{-3} \\ \epsilon_y &= 5.21 \times 10^{-3} \\ \epsilon_s &= 0 \end{aligned} \tag{4.83}$$

$$\begin{aligned} \sigma_x &= -181.8 \times .157 + 2.89 \times 5.21 = -13.4 \text{ MPa} \\ \sigma_y &= -2.89 \times .157 + 10.34 \times 5.21 = 53.41 \text{ MPa} \\ \sigma_s &= 0 \end{aligned} \tag{4.84}$$

We can compute the average stress to verify equilibrium:

$$\begin{aligned} \overline{\sigma}_1 &= \frac{1}{2}(946.7 + 53.41) \simeq 500 \text{ MPa} \\ \overline{\sigma}_2 &= 0 \\ \overline{\sigma}_6 &= 0 \end{aligned} \tag{4.85}$$

We can recover the stress resultants given originally:

$$\begin{aligned} N_1 &= \bar{\sigma}_1 h = 500 \times 2 = 1 \text{ MN/m} \\ N_2 &= N_6 = 0 \end{aligned} \tag{4.86}$$

If we use the maximum strain failure criterion (Problem h in Chapter 1) which states that failure occurs when:

$$\begin{aligned} &1. \quad \epsilon_x > 8 \times 10^{-3} \quad \text{or} \\ &2. \quad \epsilon_y > 4 \times 10^{-3} \end{aligned} \tag{4.87}$$

Examining the on-axis strains, we can conclude that the 90-degree ply has failure because

$$5.21 > 4 \tag{4.88}$$

In fact, the 90-degree ply failed when the applied stress resultant is

$$N_{1(\text{FPF})} = 1 \times \frac{4}{5.21} = .767 \text{ MN/m} \tag{4.89}$$

where subscript FPF refers to the "first ply failure" stress level.

The second or ultimate ply failure will occur when the 0-degree ply fails. Using the same failure criterion, the ultimate stress resultant is

$$N_{1(\max)} = 1 \times \frac{8}{5.21} = 1.53 \text{ MN/m} \tag{4.90}$$

2. Examine the same cross-ply laminate under compressive loads and determine the first and ultimate failure stress resultants or average stresses.

Assuming that the maximum strain failure criterion

$$\begin{aligned} &\left| -\epsilon_x \right| > 7 \times 10^{-3} \quad \text{(slightly less than tensile strain)} \\ &\left| -\epsilon_y \right| > 20 \times 10^{-3} \quad \text{(much more than tensile strain)} \end{aligned} \tag{4.91}$$

For the 90-degree ply, the induced strains have opposite signs,

$$
\begin{aligned}
N_1 &= 1 \times \frac{8}{.157} = 50.9 \text{ MN/m, (tension), or} \\
N_1 &= 1 \times \frac{20}{5.21} = 3.83 \text{ MN/m (compression)}
\end{aligned}
\tag{4.92}
$$

For the 0-degree ply,

$$
\begin{aligned}
N_1 &= 1 \times \frac{7}{5.21} = 1.34 \text{ MN/m, (compression), or} \\
N_1 &= 1 \times \frac{4}{.157} = 25.4 \text{ MN/m (tension)}
\end{aligned}
\tag{4.93}
$$

Note the first-ply failure now occurs in the 0-degree ply at an average stress of

$$
\begin{aligned}
\overline{\sigma}_{1\,(\text{FPF})} &= 1.34/2 \times 10^{-3} = 672 \text{ MPa} \\
\overline{\sigma}_{1\,(\text{max})} &= 3.83/2 \times 10^{-3} = 1910 \text{ MPa}
\end{aligned}
\tag{4.94}
$$

3. Determination of shear stiffness from a 45-degree angle-ply laminate.

For our laminate, we know from Equation 4.57,

$$
\begin{aligned}
a_{11}h &= 39.91 \text{ (TPa)}^{-1} \\
a_{12}h &= -29.81 \text{ (TPa)}^{-1} \\
a_{61} &= 0
\end{aligned}
\tag{4.95}
$$

Under a uniaxial stress of

$$
\overline{\sigma}_1 = 100 \text{ MPa} \tag{4.96}
$$

$$
\begin{aligned}
\epsilon_1^o &= 39.91 \times 100 = 3.99 \times 10^{-3} \\
\epsilon_2^o &= -29.81 \times 100 = -2.98 \times 10^{-3} \\
\epsilon_6^o &= 0
\end{aligned}
\tag{4.97}
$$

If we transform both the stress and strain −45 degrees, and call the new axes 1′ and 2′, from Table 2.1,

$$\bar{\sigma}_1{}' = \bar{\sigma}_2{}' = 50 \text{ MPa}$$
$$\bar{\sigma}_6{}' = 50 \text{ MPa} \tag{4.98}$$

From Table 2.5,

$$\epsilon_1^{o\prime} = \frac{1}{2}(3.99 - 2.98) = .50 \times 10^{-3}$$
$$\epsilon_2^{o\prime} = \epsilon_1^{o\prime} \tag{4.99}$$
$$\epsilon_6^{o\prime} = 3.99 + 2.98 = 6.97 \times 10^{-3}$$

In the 1′-2′ coordinate system, our laminate is a cross-ply laminate subjected to the stress and strain above. Since the stress-strain relation is orthotropic, the shear components in stiffness and compliance are uncoupled. A direct relation exists. The resulting shear modulus is,

$$A_{66}/h = \frac{\sigma_6{}'}{\epsilon_6^{o\prime}} = \frac{50}{6.97 \times 10^{-3}} = 7.17 \text{ GPa} \tag{4.100}$$

This agrees with the result of Equation 4.47. We can take advantage of the transformation properties of stress and strain to convert the uniaxial tensile or compressive stress applied to a 45-degree angle-ply to a shear test. In terms of the applied stress and measured strain in the symmetry axes of the angle-ply, the 1-2 coordinate system,

$$Q_{ss} = A_{66}/h = \frac{\bar{\sigma}_1/2}{\epsilon_1^o - \epsilon_2^o} \tag{4.101}$$

When the applied stress is positive,

$$\epsilon_1^o > 0$$
$$\epsilon_2^o < 0 \tag{4.102}$$

The denominator is the sum of the longitudinal and Poisson's strain if the negative sign of Poisson's strain is ignored.

10. conclusions

The in-plane stiffness of a laminated composite can be obtained directly by applying the rule of mixtures equation to the stiffness of the unidirectional composite. The in-plane compliance is simply the inverse of the in-plane stiffness. Finally, if engineering constants are desired, they can be obtained from the components of the in-plane compliance. The process described above is straight-forward and is applicable to the flexural stiffness of laminated composites as well. The relationship between engineering constants of the constituent plies to those of a laminate is very complicated. In place of simple linear rule of mixtures equations, we have highly non-linear relationships. These non-linear relations are responsible for the curves in "carpet plots."

Matrix inversion is required to obtain the components of compliance from those of stiffness. In the process, all the components of the stiffness participate in the determination of each component of compliance. It is therefore difficult to visualize the impact of a change in the stiffness to the change in compliance. Simple ratios or linear relationship no longer exists. The effect of such change in the stiffness on the resulting engineering constants become even less obvious. This is shown in the general bidirectional laminates where Poisson's ratios are no longer bounded by the limits imposed on isotropic materials. They can be greater than one-half or less than zero. This presents a challenge to design formulas intended for the conventional material. The tendency to make composite materials orthotropic or quasi-isotropic is understandable, but the designer may be depriving himself of the opportunity of an optimum design.

It should be emphasized again that laminated composite materials are governed by analogous stress-strain relations to those of unidirectional composites. The in-plane stiffness of a laminate is bounded by the stiffness of the constituent plies. But the compliance and the associated engineering constants are not always bounded by those of the constituent plies. When material constants are not bounded, it is very difficult to rely on intuition. This is why discipline is so important when we work with composite materials. We must keep track of the signs; we must know how the properties of the constituent plies are translated into the stiffness of a laminate. In this respect, the components of stiffness are preferred because their variations with the ply properties and stacking sequence are governed by simple, explicit, linear relations. It is also important to know how many degrees of freedom are available as we change ply orientations, or add or subtract plies in order to achieve an optimum laminate.

10. homework problems

a. Find stiffness compliance, and engineering constants of T300/5208 angle-ply laminates with $\pm\pi/8$?

b. Find stiffness, compliance, and engineering constants of hybrid cross-ply laminates with 0-degree T300/5208 and 90-degree Scotch-ply 1002 for ply ratios of 1, 2, 4 and 8 as in Table 4.7. Write down the elastic constants for ply ratios of 1/2, 1/4 and 1/8. Which constants of the hybrid are bounded? Which are unbounded?

c. How do we determine the off-axis properties of cross-ply laminates? Rotate the entire laminate by angle γ as that in Figure 4.12. Show the coupling coefficients for the laminates in Table 4.7.

d. What is the Young's modulus of an 8-ply 45-degree T300/5208 slender body? What is the Young's modulus of two parallel but unbonded slender bodies, one with +45 degree, the other −45 degree? What is the Young's modulus of a symmetric laminate containing the same ply angles; i.e., $[\pm 45_2]_S$? Explain the difference between the unbonded and the laminate (bonded).

e. The in-plane modulus of a symmetric laminate follow the rule-of-mixtures relation using the ply modulus; e.g.,

$$A_{11}/h = \sum Q_{11}^{(i)} v_i \tag{4.103}$$

Is the following rule-of-mixtures relation using the Young's modulus valid?

$$E^o = \sum E_1^{(i)} v_i \tag{4.104}$$

Explain the different results of T300/5208 [0/90] and [±45] laminates derived from the two mixtures equations.

f. Are there bounds on the in-plane Poisson's ratio as fiber stiffness approaches infinity or matrix stiffness to zero for cross-ply and angle-ply laminates? Are there bounds on shear and normal coupling coefficients?

g. What devices can use a material with negative Poisson's ratio?

h. Can a quick estimate be made of the in-plane modulus of laminates following the homework problem *l* in Chapter 3? Compare the quick estimate with the exact results in Tables 4.7 and 4.9 for various cross-ply and angle-ply laminates.

i. We observe in Table 4.7 that the sum of the first two columns (or all four columns) of modulus is independent of the ply ratio. Why does the sum for the compliance components change with ply ratios?

j. What quantity of the average stress components remains constant when a hydrostatic extension ($\epsilon_1^o = \epsilon_2^o = p_\epsilon$) is imposed on a cross-ply laminate? Does this quantity depend on the ply ratio?

k. What quantity of the strain components remains constant when a hydrostatic tension or compression ($N_1 = N_2 = \pm p_\sigma h$) is imposed on the same cross-ply laminate? Does this quantity depend on the ply ratio? Does this quantity change if the entire laminate is rotated? How many invariants in Table 4.7 are equal to those in Table 4.9?

l. In sizing a conventional material, thickness change is the only option. A thickness increase will reduce linearly the stress and strain. All components of stress and strain will change proportionally. A change in thickness is equivalent to proportional loading or unloading. In sizing composite laminates, proportional increase in stress and strain is possible if the ply ratio or ply orientations remain constant. But, in general, the ply orientations change as we seek an optimum laminate. What principles and constraints are involved in achieving an optimum laminate from the standpoint of stiffness only (not strength)? We will be concerned with a point within a large structure, and it is assumed that the stress resultants are given and do not change with ply orientation.

m. Calculate quasi-isotropic constants using Equation 4.62 for all unidirectional composites listed in Table 3.6. These constants represent the lower bound performance that we can expect from laminates made from these composite materials. How do they compare with aluminum on the specific stiffness basis (stiffness E divided by density or specific weight)?

nomenclature

- A_{ij}, a_{ij} = In-plane stiffness and compliance of multidirectional symmetric laminates; in Nm^{-1}, and $N^{-1}m$, respectively
- E_i^o = In-plane engineering constants; $i = 1,2,6$
- h = Total thickness of a laminate
- h_i = $n_i h_o$ = Thickness of i-th ply group; $i = 1$ to m
- h_o = h/n = Unit ply thickness, in m
- m = Total number of ply groups in a laminate
- N_i = Stress resultant, in Nm^{-1}; $i = 1,2,6$
- n = Total number of plies in a laminate
- n_i = h_i/h_o = Number of plies in the i-th ply group
- p = Ply ratio of a bidirectional laminate
- $Q_{ij}^{(\theta)}$ = Stiffness of ply group with θ ply orientation
- $R_{iQ,A}$ = Radii of generalized Mohr's circle for unidirectional and laminated composites; $i = 1,2$
- U_i = Linear combinations of stiffness; $i = 1$ to 5
- V_i = Geometric factors; $i = 1$ to 4
- v_θ = Volume fraction of ply group with θ orientation
- γ = Rigid body rotation of a laminate
- $\bar{\sigma}_i$ = N_i/h = Average stress across thickness of a laminate
- ϵ_i^o = In-plane strain; $i = 1,2,6$
- ϕ = The angle of a $\pm\phi$ angle-ply laminate
- ν_{ij}^o = ν_{21}^o, ν_{12}^o Poisson's ratios; ν_{61}^o, ν_{62}^o shear coupling coefficients, ν_{16}^o, ν_{26}^o normal coupling coefficients
- θ = Ply orientation

chapter 5
flexural stiffness of symmetric sandwich laminates

'he flexural stiffness of symmetric sandwich laminates with honey-omb core and multidirectional composite facing will be covered. The train in the laminate is assumed to be proportional to the curvature. 'lexural stiffness can then be defined in terms of the modulus and ompliance and the moment-curvature relation. The contribution of the ore and the effect of stacking sequence on the flexural modulus can be lescribed by explicit formulas. For design optimization the com-onents of modulus are therefore easier to use than those of com-liance and equivalent engineering constants.

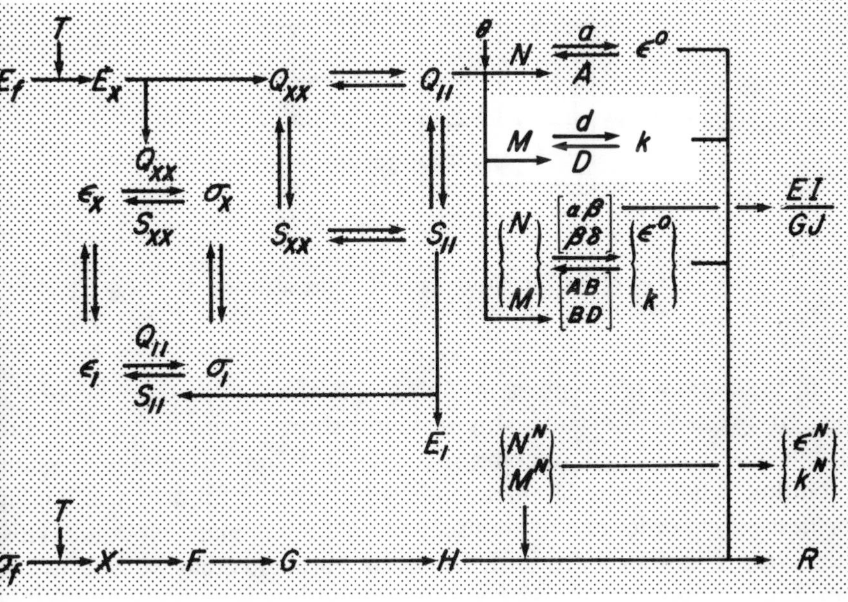

1. laminate code

The same laminate code convention as that used for the in-plane modulus in Equation 4.1 will be followed for the flexural modulus. For symmetric laminates we can add the half depth of the core in the code for example:

$$[0_3/90_2/45/-45_3/z_c]_S \tag{5.1}$$

The orientations, ply groups and the core for this laminate are shown in Figure 5.1. The plies are arranged in an ascending order from the bottom or the $z = -h/2$ face. This again can be a source of confusion. The code in Equation 5.1 applies to the lower half of a symmetric laminate starting from the bottom face. The stacking sequence in the upper half of the laminate is in reverse order of the code. The actual integration for the calculation of the flexural modulus for symmetric laminates is applied over the upper half of the laminate which extends from $z = 0$ to $z = h/2$.

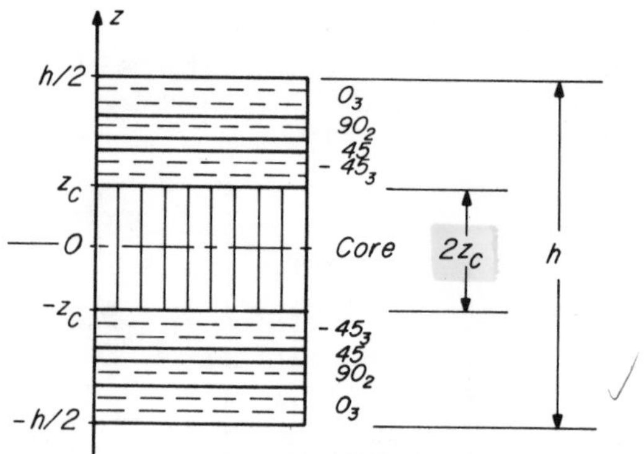

Figure 5.1 Dimensions and stacking sequence of symmetric sandwich laminates.

In the case of the in-plane modulus, only the volume fractions of the ply groups are important. This is clearly shown in Equation 4.38 and Table 4.4. The actual stacking sequence does not affect the in-plane modulus. Whether the laminate code is intended to follow an ascending or descending order is of no consequence to the in-plane modulus. This, however, is no longer true for the flexural modulus that we will discuss in this chapter. The positions of ply groups in a laminate have direct effect on the flexural modulus. That is why we are discussing the laminate code again.

2. moment-curvature relations

In the flexural behavior of laminates, moment and curvature are the key variables, similar to the role of stress resultant and in-plane strain in the in-plane behavior of the last chapter. The counterpart of the stress-strain relation for the in-plane behavior is the moment-curvature relation for the flexural behavior. The elastic constants for the latter relation will be called the flexural stiffness and flexural compliance. It is the purpose of this chapter to develop definitions of moment and curvature, and their relationship to each other.

The distribution of ply stresses can be symmetric and anti-symmetric with respect to the midplane. In Chapter 4, the stress distribution was symmetric and this was shown in Figure 4.3. In Figure 5.2 we will repeat the symmetric distribution of Figure 4.3, and we will also show the case of anti-symmetric distribution.

As the result of symmetric stress distribution in Figure 5.2(*a*), we can represent the variable stress by an average stress and a stress resultant, shown in Equations 4.7 and 4.15, respectively. The in-plane behavior of symmetric laminates can be characterized using the average stress or stress resultant. When the stress distribution is anti-symmetric, as shown in Figure 5.2(*b*), the average stress across the entire laminate thickness is zero. One approach of dealing with the anti-symmetric stress distribution is to define a new quantity: the moment, to take the place of the stress resultant. The simplest or first moment has three components:

$$
\begin{aligned}
M_1 &= \int_{-h/2}^{h/2} \sigma_1 z dz \\
M_2 &= \int_{-h/2}^{h/2} \sigma_2 z dz \qquad (5.2)\\
M_6 &= \int_{-h/2}^{h/2} \sigma_6 z dz
\end{aligned}
$$

The unit of moment is N, or Nm/m; i.e., a moment per unit width of a laminate with thickness h and width b.

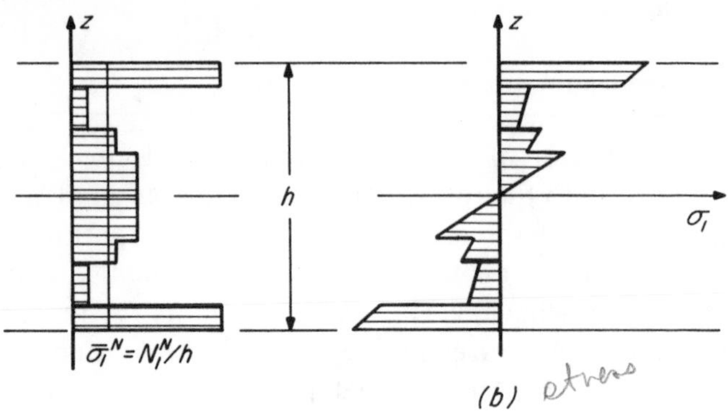

Figure 5.2 Stress variations across laminates. Illustration of symmetric ply stresses in (a), and anti-symmetric ply stresses in (b).

The sign of the components of moment is also critical. The bending components of moment, like the normal components of stress and strain, are easy to rationalize and readily defined. A bending moment is positive if the average induced stress in the upper half of the laminate is positive. In Figure 5.3(a) we define the positive component for M_1; in Figure 5.3(b), the positive M_2. When M_1 or M_2 is negative, the average induced stress in the upper half of the laminate will be negative. We use average stress here because in a laminated material it is possible to have both positive and negative stresses in each half of the laminate. Figure 5.2(b) shows this possibility.

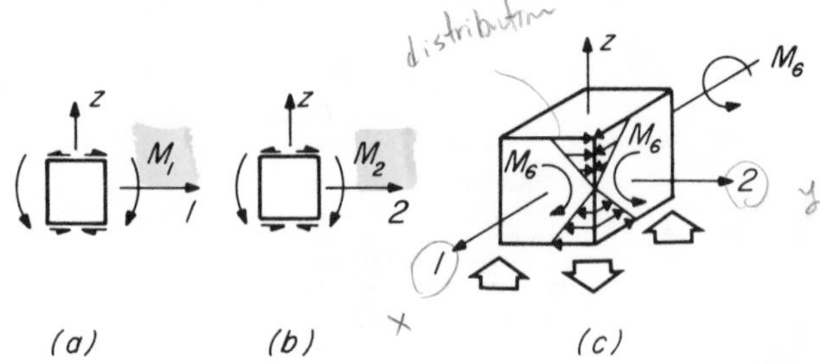

Figure 5.3 The positive directions of components of moment. Bending moments are shown in (a) and (b). In (c), positive twisting moment appears as clockwise torque on the positive 1-axis face; counterclockwise on the positive 2-axis face. The effect of the positive twisting moment can be duplicated by four self-equilibrating forces acting at the corners as shown.

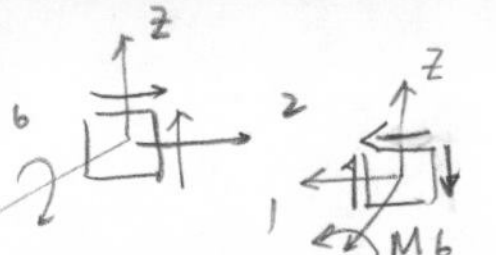

The sign convention for twisting moment follows the same rule; viz., a positive shear stress on the upper half of the laminate is associated with the positive twisting moment. The positive shear stress component is defined in Figure 1.6. Figure 5.3(*c*) shows the result of positive twisting moment and the induced shear stress distribution. All the arrows will reverse their directions if the twisting moment is negative. We are not imposing the right-hand rule for the sign convention except the coordinates and angle of rotation. If the right-hand rule is followed, as shown by Timoshenko,* we must distinguish the twisting moment on the 1-axis face as M_{12}, and that on the 2-axis as M_{21}; or M_{xy} and M_{yx}, respectively.

Then we have the following relations:

$$\begin{aligned} M_{12} &= M_{xy} = -M_6 \\ M_{21} &= M_{yx} = M_6 \end{aligned} \tag{5.3}$$

Therefore $$M_{12} = -M_{21} \tag{5.4}$$

The important issue here is not what sign convention we use. We must understand the rationale and be consistent. Again we would like to mention how critical signs are when we work with composite materials. A wrong guess is often inconsequential for conventional materials, but can be disastrous for composites. The signs for shear stress, shear strain, twisting moment shown here and twisting curvature, which we will introduce presently, are all sources of uncertainty and error.

We will now derive the strain-displacement relation for the bending of a plate similar to that for the in-plane stretching of a plate in Equations 1.1 and 1.4. We will assume that the plate is initially flat as shown in Figure 5.4(*a*). After bending, the plate can be described by a function w where:

$$w = w(x, y) \tag{5.5}$$

It is implied that the vertical displacement of each point does not vary in the z-direction. The normal to the plate does not stretch or deform.

*S. Timoshenko and S. Woinowsky-Krieger, *Theory of Plates and Shells*, McGraw-Hill, 1959, p. 80

It only rotates as the plate is bent or twisted. Figure 5.4(b) is an illustration of a bent plate.

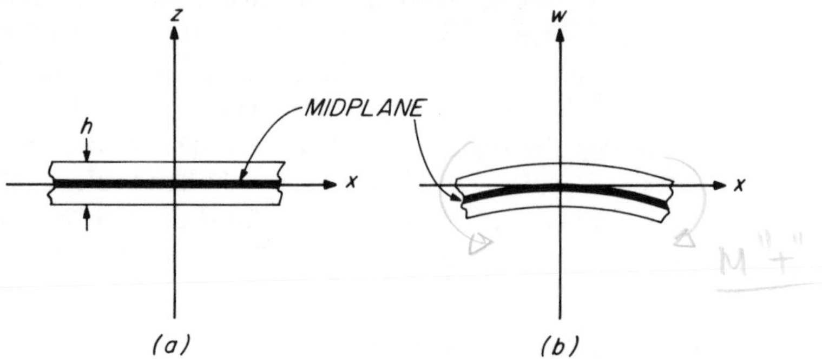

Figure 5.4 Definition of a plate or laminate before and after bending. The deformed midplane is described by a function $w(x, y)$.

The rotation of the normal to the midplane can be directly related to the first derivative at the same point in the plate. This is shown in Figure 5.5 where two cases of the bent plane are shown for the purpose of establishing the sign convention. Consistency between Figures 5.3 and 5.4 is maintained if we use a negative sign in the displacement derivative relation as follows:

$$u = -z\theta = -z\frac{\partial w}{\partial x} \tag{5.6}$$

Similarly, we can derive the displacement along the y-axis as:

$$v = -z\frac{\partial w}{\partial y} \tag{5.7}$$

From the last two equations, and the strain-displacement relations of Equations 1.1 and 1.4, we can show that:

$$\begin{aligned} \epsilon_1 &= \frac{\partial u}{\partial x} = -z\frac{\partial^2 w}{\partial x^2} \\ \epsilon_2 &= \frac{\partial v}{\partial y} = -z\frac{\partial^2 w}{\partial y^2} \\ \epsilon_6 &= \frac{\partial u}{\partial y} + \frac{\partial v}{\partial x} = -2z\frac{\partial^2 w}{\partial x \partial y} \end{aligned} \tag{5.8}$$

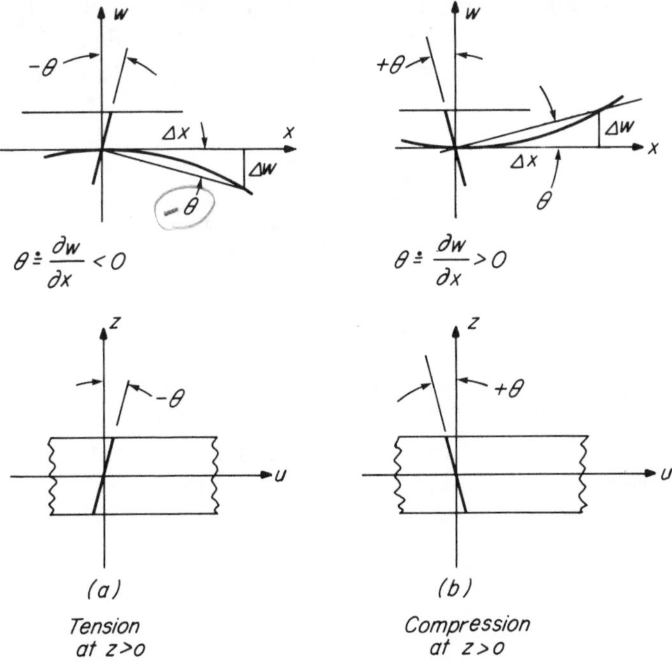

Figure 5.5 Sign convention of midplane displacements. For a concave downward deformation in (*a*), the derivative of *w* is negative, and a negative sign must be added to the displacement-derivative relation in Equation 5.6. When the curvature is reversed in (*b*), the derivative of *w* is now positive.

From elementary calculus, we can relate the second derivatives to curvatures k's as follows:

$$
\begin{aligned}
k_1 &\cong -\frac{\partial^2 w}{\partial x^2} \\
k_2 &\cong -\frac{\partial^2 w}{\partial y^2} \\
k_6 &\cong -2\frac{\partial^2 w}{\partial x \partial y}
\end{aligned}
\tag{5.9}
$$

Negative signs are used here in order to maintain consistency with the definition of moments established in Figure 5.3. The twisting curvature is difficult to illustrate and is not normally covered in elementary text. We derived our relation through the use of the strain-displacement relation in Equation 5.8. Substituting the definitions in Equation 5.9 into 5.8, we have:

$$\begin{aligned} \epsilon_1(z) &= zk_1 \\ \epsilon_2(z) &= zk_2 \\ \epsilon_6(z) &= zk_6 \end{aligned} \tag{5.10}$$

This assumed linear strain distribution is shown in Figure 5.6. A more general assumed state of strain than both Equations 4.6, and 5.10 would be the sum of the two. This combined strain will be used as the basis of general, unsymmetrical laminates which we will cover in Chapter 6.

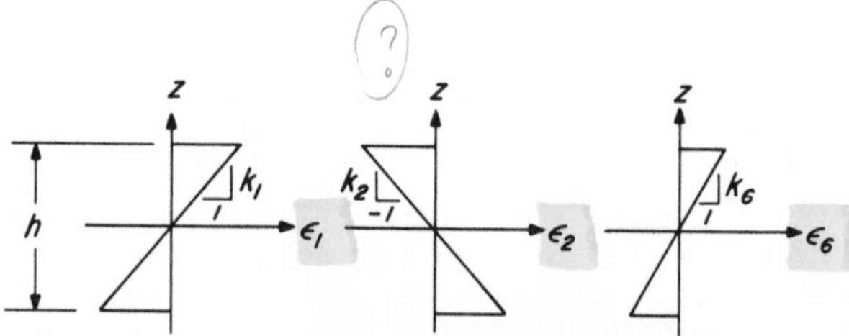

Figure 5.6 Assumed linear strain distribution across laminate thickness. Maximum strain values are reached at the upper and lower faces. They are equal but opposite in signs when the laminate is symmetric.

We can now derive the moment-curvature relations by substituting the assumed strain into the definition of moment in Equation 5.2. We must first, however, use the off-axis stress-strain relations listed in Table 3.1 for this substitution. This will express the stress components in terms of the strain components.

From Equation 5.2

$$M_1 = \int \sigma_1 z dz \tag{5.11}$$

From Table 3.1

$$M_1 = \int [Q_{11}\epsilon_1 + Q_{12}\epsilon_2 + Q_{16}\epsilon_6] z dz \tag{5.12}$$

From Equation 5.10

$$= \int [Q_{11}k_1 + Q_{12}k_2 + Q_{16}k_6] z^2 dz \tag{5.13}$$

Since curvatures are constant, not dependent on z, they can be factored out,

$$M_1 = \left[\int Q_{11} z^2 dz\right] k_1 + \left[\int Q_{12} z^2 dz\right] k_2 + \left[\int Q_{16} z^2 dz\right] k_6$$

$$M_1 = D_{11}k_1 + D_{12}k_2 + D_{16}k_6$$

Similarly

$$\begin{aligned} M_2 &= D_{21}k_1 + D_{22}k_2 + D_{26}k_6 \\ M_6 &= D_{61}k_1 + D_{62}k_2 + D_{66}k_6 \end{aligned} \tag{5.14}$$

where

$$\begin{aligned} D_{11} &= \int Q_{11} z^2 dz, \quad D_{22} = \int Q_{22} z^2 dz, \quad D_{12} = D_{21}, \\ D_{12} &= \int Q_{12} z^2 dz, \quad D_{66} = \int Q_{66} z^2 dz, \quad D_{16} = D_{61}, \\ D_{16} &= \int Q_{16} z^2 dz, \quad D_{26} = \int Q_{26} z^2 dz, \quad D_{26} = D_{62}. \end{aligned} \tag{5.15}$$

We have thus derived the moment-curvature relation in Equation 5.14 and defined the flexural modulus in Equation 5.15.

Inverting the moment-curvature relation we can obtain the following relation in terms of flexural compliance, duplicating the same steps used in the inversion in Chapter 3.

$$k_1 = d_{11}M_1 + d_{12}M_2 + d_{16}M_6$$

$$k_2 = d_{21}M_1 + d_{22}M_2 + d_{26}M_6 \tag{5.16}$$

$$k_6 = d_{61}M_1 + d_{62}M_2 + d_{66}M_6$$

The relationship above can be presented in matrix multiplication tables as follows:

table 5.1

moment-curvature relation of symmetric laminates in terms of stiffness

	k_1	k_2	k_6
M_1	D_{11}	D_{12}	D_{16}
M_2	D_{21}	D_{22}	D_{26}
M_6	D_{61}	D_{62}	D_{66}

table 5.2

moment-curvature relation of symmetric laminates in terms of compliance

	M_1	M_2	M_6
k_1	d_{11}	d_{12}	d_{16}
k_2	d_{21}	d_{22}	d_{26}
k_6	d_{61}	d_{62}	d_{66}

We can now define the effective flexural engineering constants. From the compliance relation in Table 5.2, we know that under simple bending of M relative to the 1-axis only, the resulting curvature along the 1-axis is:

$$k_1 = d_{11}M_1 = d_{11}\frac{M}{b} \tag{5.17}$$

where b is the finite width of a beam or plate; M is the total moment and is equal to $M_1 b$. From elementary theory, we know that rigidity of a homogeneous beam is:

$$\text{Rigidity} = EI \tag{5.18}$$

$$= \frac{M}{k_1} \tag{5.19}$$

where E is the homogeneous Young's modulus; and I is the moment of inertia. By combining the two relationships, we have:

$$EI = b/d_{11} \tag{5.20}$$

or

$$E = E_1^f = b/Id_{11} = 12/h^3 d_{11} = 1/I^* d_{11} \tag{5.21}$$

where $I = bh^3/12$, $I^* = I/b = h^3/12$

E_1^f = Effective Young's modulus along the 1-axis

Similarly, we can show:

$$\begin{aligned} E_2^f &= 12/h^3 d_{22} = 1/I^* d_{22} \\ E_6^f &= 12/h^3 d_{66} = 1/I^* d_{66} \end{aligned} \tag{5.22}$$

The superscript f denotes effective flexural engineering constants. These are the constants if the beam or plate of our multidirectional laminates is treated like a homogeneous material. Other dimensionless engineering constants analogous to those for off-axis unidirectional composites in Chapter 3 and to those for in-plane anisotropic behavior in Chapter 4 are:

$$\nu_{21}^f = -\frac{d_{21}}{d_{11}}, \quad \nu_{12}^f = -\frac{d_{12}}{d_{22}} \tag{5.23}$$

$$\nu_{61}^f = \frac{d_{61}}{d_{11}}, \quad \nu_{16}^f = \frac{d_{16}}{d_{66}} \tag{5.24}$$

$$\nu_{62}^f = \frac{d_{62}}{d_{22}}, \quad \nu_{26}^f = \frac{d_{26}}{d_{66}} \tag{5.25}$$

3. evaluation of flexural stiffness modulus

We will now evaluate the components of flexural modulus by performing the integration of the components in Equation 5.15. Similar to the case of in-plane modulus in Chapter 4, we will first substitute the off-axis stiffness of the unidirectional composites using the multiple-angle transformation relations listed in Table 3.3.

From Equation 5.15

$$D_{11} = \int Q_{11} z^2 \, dz \tag{5.26}$$

From Table 3.3

$$= \int [U_1 + U_2 \cos 2\theta + U_3 \cos 4\theta] z^2 \, dz \tag{5.27}$$

Since the U's are independent of z for a laminate with the same unidirectional composite,

$$D_{11} = U_1 \int z^2 \, dz + U_2 \int \cos 2\theta z^2 \, dz + U_3 \int \cos 4\theta z^2 \, dz \tag{5.28}$$

$$= U_1 h^* + U_2 V_1 + U_3 V_2 \tag{5.29}$$

where

$$h^* = \int_{-h/2}^{h/2} z^2 \, dz = 2 \int_{z_c}^{h/2} z^2 \, dz$$

$$= \frac{h^3}{12} \left[1 - \left(\frac{z_c}{h/2} \right)^3 \right] = \frac{h^3}{12} \left[1 - z_c^{*3} \right] \tag{5.30}$$

$$= I^* [1 - z_c^{*3}]$$

$$z_c^* = \text{Volume fraction of core} = 2z_c/h \tag{5.31}$$

$$V_1 = \int_{-h/2}^{h/2} \cos 2\theta z^2 \, dz = 2\int_{z_c}^{h/2} \cos 2\theta z^2 \, dz$$

$$V_2 = 2\int_{z_c}^{h/2} \cos 4\theta z^2 \, dz \tag{5.32}*$$

t is assumed that the honeycomb core has no stiffness in the 1-2 coordinate system. That is the reason the lower limit of integration is et at the half depth of the core.

Similarly,

$$D_{22} = U_1 h^* - U_2 V_1 + U_3 V_2 \tag{5.33}$$

$$D_{12} = U_4 h^* - U_3 V_2 \tag{5.34}$$

$$D_{66} = U_5 h^* - U_3 V_2 \tag{5.35}$$

$$D_{16} = \frac{1}{2} U_2 V_3 + U_3 V_4 \tag{5.36}$$

$$D_{26} = \frac{1}{2} U_2 V_3 - U_3 V_4 \tag{5.37}$$

where

$$V_3 = 2\int_{z_c}^{h/2} \sin 2\theta z^2 \, dz \tag{5.38}$$

$$V_4 = 2\int_{z_c}^{h/2} \sin 4\theta z^2 \, dz \tag{5.39}$$

*A more general definition can be found in Equations 6.79-6.82.

Here again, the evaluation of the flexural modulus reduces to the evaluation of the four geometric factors, the V's. Similar to Equation 4.31 we can combine the definitions of the V's into one expression:

$$V_{[1,2,3,4]} = 2 \int_{z_c}^{h/2} [\cos 2\theta, \cos 4\theta, \sin 2\theta, \sin 4\theta] z^2 \, dz \tag{5.40}$$

We can also put all the formulas for the components of the flexural modulus into a matrix multiplication table as in Table 5.3. Note the similarity between this table and the formulas for in-plane modulus in Table 4.3. The definitions of the V's, however, are different. Again the geometric factors are separated from the material property. For the same material, the U's stay constant and the V's change from laminate to laminate. For the same laminate but with different material, only new U's are needed.

table 5.3
formulas for flexural modulus of symmetric sandwich laminates

	h^*	U_2	U_3
D_{11}	U_1	V_1	V_2
D_{22}	U_1	$-V_1$	V_2
D_{12}	U_4		$-V_2$
D_{66}	U_5		$-V_2$
D_{16}		$\frac{1}{2}V_3$	V_4
D_{26}		$\frac{1}{2}V_3$	$-V_4$

where $h^* = (1-z_c^{*3})h^3/12 = (1-z_c^{*3})I^*$

Analogous to the in-plane modulus, the number of flexural modulus dependent on the stacking sequence are four, not six. Two linear invariants can be derived from Table 5.3:

$$\begin{aligned} D_{11} + D_{22} + 2D_{12} &= 2[U_1 + U_4]h^* \\ D_{66} - D_{12} &= [U_5 - U_4]h^* \end{aligned} \tag{5.41}$$

The core and thickness correction factor which appear here and in the first column of Table 5.3 will reduce the invariant terms.

Normalized flexural modulus can also be represented by the generalized Mohr's circles like those for the unidirectional composite in Figure 3.9 and the normalized in-plane modulus in Figure 4.6. In the process of lamination, the V's either maintain or reduce the length of the radii of the Mohr's circles, similar to the factors in Equations 4.43 and 4.44 for the in-plane modulus. Honeycomb core will reduce the distance between the Mohr's circles by a magnitude of h^*/I^*. This core, however, will not affect the radii of the Mohr's circles. The use of core provides a degree of freedom in addition to and independent of the stacking sequence of the facing material.

Let us try to evaluate the first term in Equation 5.40.

$$V_1 = 2\int_{z_c}^{h/2} \cos 2\theta z^2 \, dz \tag{5.42}$$

If each ply group would have the same unidirectional material, the integration can be replaced by a summation. See Figure 5.7 for the definitions of indices of summation.

$$V_1 = \frac{2}{3}\sum_{i=c+1}^{m/2} \cos 2\theta_i [z_i^3 - z_{i-1}^3] \tag{5.43}$$

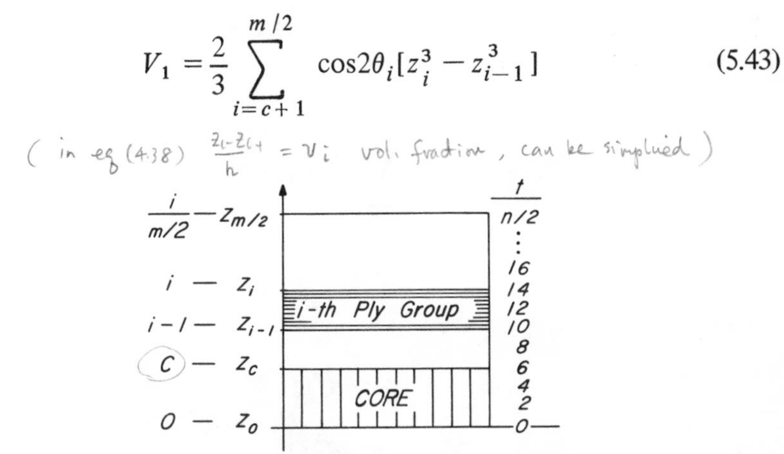

Figure 5.7 Schematic diagram of a symmetric sandwich laminate. There are m ply groups and n plies in the laminate using indices i and t, respectively. Assuming the half depth of the core is equal to a multiple of unit plies, the half depth can be designated by $i = c = 6$ in this figure.

Simplification of this summation in terms of volume fractions such as that for the in-plane modulus in Equation 4.38 is not possible because of the cubic relation here instead of the linear relation. Some simplification is possible if the half thickness of the core is a multiple of the unit ply thickness; i.e.,

$$c = \frac{z_c}{h_o} = \text{an integer}$$

This is assumed in Figure 5.7. Then the z coordinates in Equation 5.43 can be replaced by ply numbers as follows:

$$\begin{aligned} z_c &= c\,h_o \\ z_1 &= (c + n_1)h_o \\ z_2 &= (c + n_1 + n_2)h_o \\ &\ldots\ldots \end{aligned} \tag{5.44}$$

where n_i equals the number of plies in the i-th ply group. In terms of Equation 5.43, this can be rewritten as

$$V_1 = \frac{2h_o^3}{3} \sum_{i=c+1}^{m/2} \cos 2\theta_i [t_i^3 - t_{i-1}^3] \tag{5.45}$$

where

$$t_i = \frac{z_i}{h_o} \tag{5.46}$$

Let

$$V_1^* = \frac{12}{h^3} V_1 = V_1/I^*$$

Substituting Equation 5.45 into Equation 5.46, we obtain

$$V_1^* = \frac{8}{n^3} \sum_{i=c+1}^{m/2} \cos 2\theta_i [t_i^3 - t_{i-1}^3] \tag{5.47}$$

where n equals the total number of plies including the core thickness expressed in equivalent number of plies. The variables in the bracket can be expressed in terms of plies using Equation 5.44. The formulas for the other three V's will take the same form. Only the trigonometric function changes; i.e., cosine in Equation 5.47 is replaced by sine, etc. This bracketed quantity in Equation 5.47 is therefore a weighting factor. In the case of the in-plane modulus, the weighting factor was the volume fraction of each ply orientation; we had the rule-of-mixtures relation. In the case of flexural modulus, this weighting factor put heavier emphasis on the outer plies as the result of a cubic relation. Again, if we assume that all plies have the same thickness, and the core depth is a double multiple of the unit plies, we can establish the numerical values of this weighting factor starting with the midplane as zero and move upward toward the top surface where the $n/2$-th ply is located. The value of this weighting factor is listed in Table 5.4. Equation 5.47 can be rewritten as follows:

$$V_1^* = \frac{8}{n^3} \sum_{t=c+1}^{n/2} \cos 2\theta_t [t^3 - (t-1)^3] \tag{5.48}$$

The index t is used here to distinguish from the index i in Equation 5.47. The latter index is intended for the number of ply groups; and the former index, the number of individual plies. The two indices will be equal if each ply group has only one ply.

The weighting factor above can be applied directly to Equation 5.26, in which case analogous to Equation 5.48 we have:

$$D_{ij}^* = \frac{1}{I^*} D_{ij} = \frac{8}{n^3} \sum_{t=c+1}^{n/2} Q_{ij}^{(t)} [t^3 - (t-1)^3] \tag{5.49}$$

Using the numerical values listed in Table 5.4, Equation 5.48 can now be written as:

$$V_1^* = \frac{8}{n^3} [\cos 2\theta_1 + 7\cos 2\theta_2 + 19\cos 2\theta_3 + 37\cos 2\theta_4 + \ldots] \tag{5.50}$$

table 5.4
weighting factors for the flexural modulus of symmetric sandwich laminates

Ply order number, t	$t-1$	$t^3-(t-1)^3$
1	0	1
2	1	7
3	2	19
4	3	37
5	4	61
6	5	91
7	6	127
8	7	169
9	8	217
10	9	271
11	10	331
12	11	397
13	12	469
14	13	547
15	14	631
16	15	721

If adjacent plies have the same ply orientation, we have for example ply groups with two plies each,

$$\theta_1 = \theta_2, \theta_3 = \theta_4 \ldots \quad (5.51)$$

Then

$$V_1^* = \frac{8}{n^3}[8\cos 2\theta_1 + 56\cos 2\theta_3 + 152\cos 2\theta_5 + \ldots] \quad (5.52)$$

If we have a honeycomb core with a half-depth of 4-ply thickness, or

$$z_c = 4h_o$$

the first ply or ply group for the facing will start with $t = 5$ in Table 5.4.

$$V_1^* = \frac{8}{n^3}[61\cos 2\theta_1 + 91\cos 2\theta_2 + 127\cos 2\theta_3 + 169\cos 2\theta_4 + \ldots] \quad (5.53)$$

here the total number of plies n must include the half-depth of the ›re which is equal to 4 plies. If we have a 3-ply facing, the value of $n/2$ 7.

. flexural behavior of unidirectional laminates

f our laminate is unidirectional, the ply orientation is fixed, independ-nt of the z coordinate. The trigonometric functions in Equation 5.40 an be taken outside of the integrals. The resulting V's are:

$$V_{[1,2,3,4]} = [\cos 2\theta, \cos 4\theta, \sin 2\theta, \sin 4\theta]\, h^* \qquad (5.54)$$

vhere h^* is defined in Equation 5.30. For this specialized case, the ›ormulas for the flexural modulus are as follows:

table 5.5
formulas for the flexural modulus of unidirectional composites

	h^*/I^*	$(h^*/I^*)U_2$	$(h^*/I^*)U_3$
D^*_{11}	U_1	$\cos 2\theta$	$\cos 4\theta$
D^*_{22}	U_1	$-\cos 2\theta$	$\cos 4\theta$
D^*_{12}	U_4		$-\cos 4\theta$
D^*_{66}	U_5		$-\cos 4\theta$
D^*_{16}		$\frac{1}{2}\sin 2\theta$	$\sin 4\theta$
D^*_{26}		$\frac{1}{2}\sin 2\theta$	$-\sin 4\theta$

where $D^*_{ij} = \frac{12}{h^3} D_{ij} = D_{ij}/I^*$, $h^*/I^* = 1 - z_c^{*3}$

Note that the constants in this table are identical to those of the trans-›ormed in-plane modulus of unidirectional composites in Table 3.3. The ›nly difference is the normalizing factor h^* needed for the flexural nodulus. Thus, we can obtain the normalized off-axis flexural modulus lirectly from the off-axis modulus of a unidirectional composite. This s shown in Figure 5.8. All the remarks about the transformed modulus ›f unidirectional composites following Figure 3.5 are equally applicable

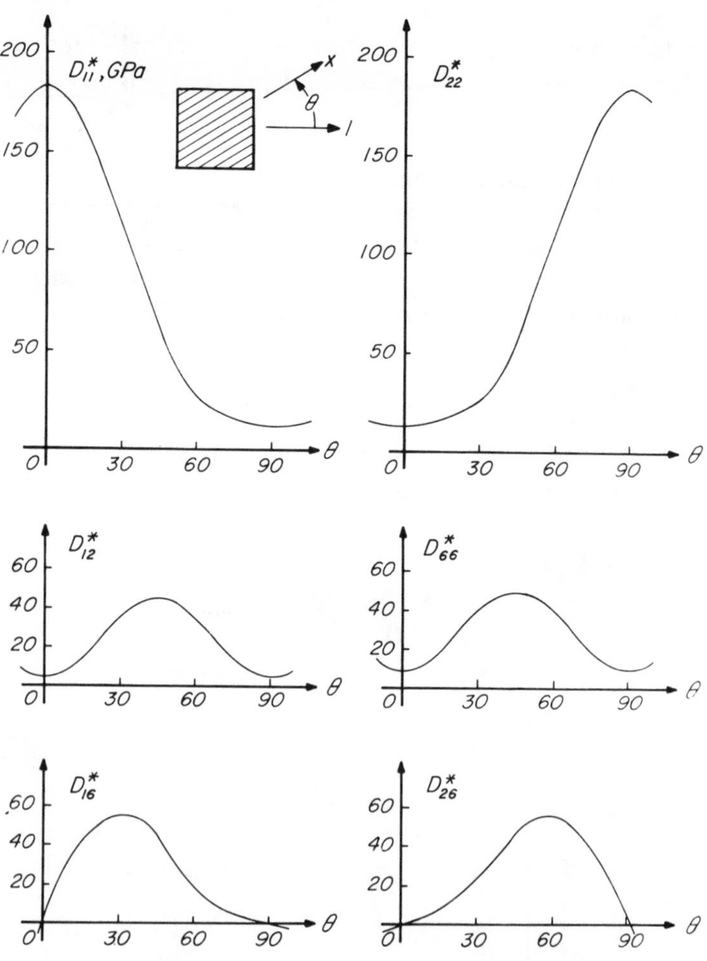

Figure 5.8 Transformed flexural modulus of unidirectional T300/5208. These are the same curves as those in Figure 3.5 with the exception of the normalizing factor in Table 5.5.

to the flexural modulus. The generalized Mohr's circles in Figure 3.9 ar valid for the flexural modulus of T300/5208 if normalized component are used.

By incorporating a sandwich core into Tables 5.5 we have:

$$D_{11} = \frac{h^3\,[1-z_c^{*3}]}{12}\,Q_{11} = h^* Q_{11} \tag{5.55}$$

or

$$D^*_{11} = D_{11}/I^* = [1-z_c^{*3}]\,Q_{11}$$

dentical factor shall be applied to the other components of the flexural nodulus. From this simple relation we can immediately write down the lexural compliance by using the same normalizing factor. We have ıow:

$$d_{11} = \frac{12}{h^3[1-z_c^{*3}]} S_{11} = S_{11}/h^* \qquad (5.56)$$

or

$$d_{11}^* = d_{11} I^* = S_{11}/[1-z_c^{*3}]$$

vhere the transformed compliance can be found from Table 3.12 and 'igure 3.11. The latter is repeated in Figure 5.9 where the normalizing actor has been added. So long as a sandwich beam or plate consists of ymmetric, homogeneous facings, its flexural stiffness and compliance an be obtained directly from the stiffness and compliance of uni- lirectional composites. We only need to know the normalizing factor, s shown in Equations 5.55 and 5.56. We can make the following emarks about the flexural rigidity of beams and plates using the xpressions in Equations 5.55 and 5.56

- Rigidity is highly dependent on the thickness h. If we double the thickness, we will get a cubic increase in return, or 8 times the rigidity.
- Removal of materials near the midplane is a very effective way of reducing the weight without much sacrifice in the rigidity. If one-third of the material at the center is removed; i.e., $z_c^* = 1/3$, the loss in rigidity as measured by z_c^{*3} is only 1/27 of the solid beam or plate.

3oth remarks are valid for composite and conventional materials so ong as the facing material is homogeneous. If multidirectional com- osites are used for the facing, the remarks above are true only qualita- ively. We will discuss this later in this chapter.

For an off-axis unidirectional composite facing, the beam will twist ınder pure bending. This is the equivalent of the shear coupling in the n-plane behavior of off-axis materials. From Table 5.2, we can relate he induced curvatures to bending moments. For example, if we apply bending moment to our off-axis beam as shown in Figure 5.10, from 'able 5.2:

$$k_1 = d_{11} M_1$$
$$k_2 = d_{21} M_1 \qquad (5.57)$$
$$k_6 = d_{61} M_1$$

The first curvature is due to normal bending; the second, the Poisso coupling; and the third, the twisting coupling. The question now is ho will the twist occur: how much, and in what direction. This is th recurring question associated with shear stress and shear strain. Agai we must pay attention to the sign convention. This was illustrated i Figure 3.15 for the in-plane behavior.

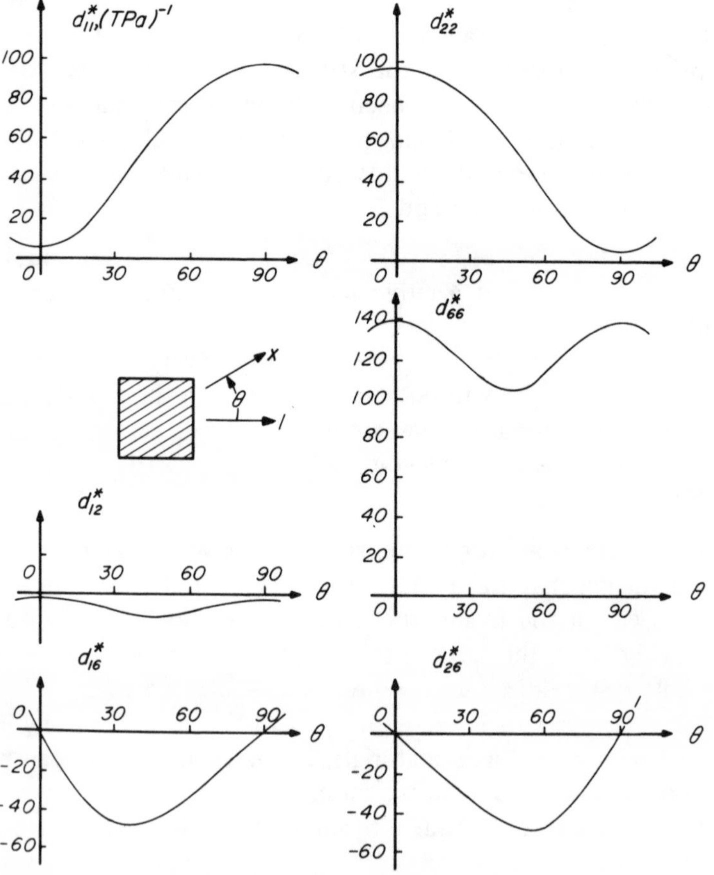

Figure 5.9 Transformed flexural compliance of unidirectional T300/5208. This is the same as Figure 3.11 for the off-axis compliance except the normalizing factor h^* in Equation 5.56 has been added.

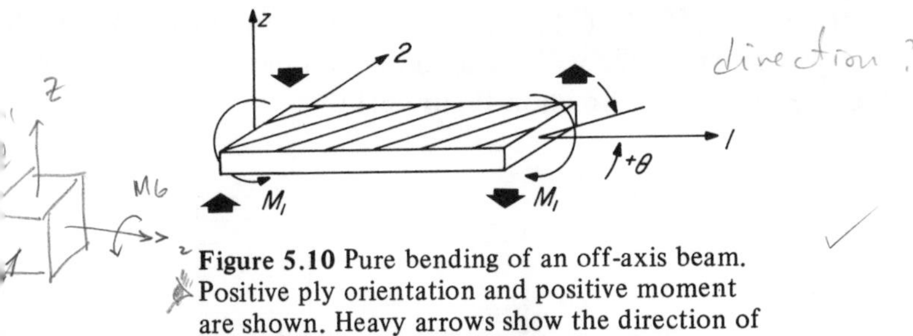

Figure 5.10 Pure bending of an off-axis beam. Positive ply orientation and positive moment are shown. Heavy arrows show the direction of movements of the four corners, similar to Figure 5.3(*c*).

'e know from Figure 5.9 that the shear coupling terms for T300/5208 nd for most practical composites are negative for positive ply angles. ince the moment in Figure 5.10 is also positive, we know from Equa- on 5.57 that the twisting curvature must be negative. Now refer to 'igure 5.3(*c*) where we showed the effect of a positive twisting moment n the stress distribution and the possible directions of displacements ıdicated by heavy arrows. Hence a positive curvature will be a clockwise otation about the 1-axis. For our beam in Figure 5.10, we have nega- ive curvature. Therefore, the twisting curvature caused by the bending ıoment will be a counterclockwise rotation along the 1-axis. This otation is represented by the heavy arrows shown at the corners.

. flexural modulus of cross-ply laminates

'ross-ply laminates are the simplest multidirectional laminates. Repeat- ng the values of the trigonometric functions in Table 4.5, we have the ollowing:

table 5.6
values of trigonometric functions for cross-ply laminates

θ_i	$\cos 2\theta_i$	$\cos 4\theta_i$	$\sin 2\theta_i$	$\sin 4\theta_i$
0	1	1	0	0
90	–1	1	0	0

Let us study the effect of stacking sequence on the flexural modulu of symmetric laminates. We will use a 16-ply laminate with three diffe ent stacking sequences as shown in Figure 5.11.

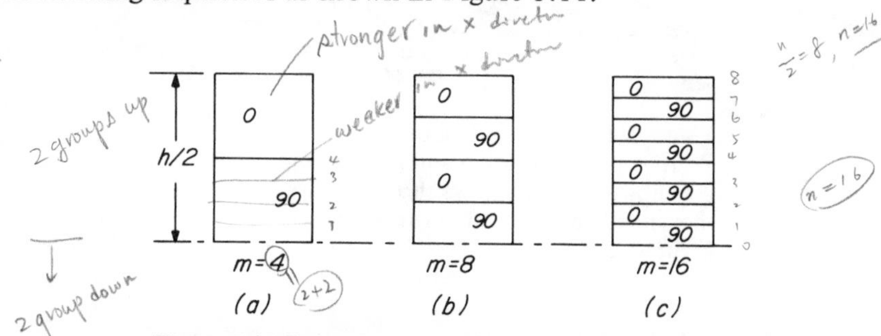

Figure 5.11 Cross-ply laminates with 16 plies but different number of ply groups; viz., $m = 4$, 8 and 16. Because of symmetry only upper half of the laminate is shown.

From the second column of Table 5.6, we know that

$$\cos 4\theta_1 = \cos 4\theta_2 = 1 \tag{5.58}$$

Following the pattern of Equation 5.50 for V_1^*, we can immediatel write down the analogous relation for V_2^*.

$$V_2^* = \frac{8}{n^3}(1 + 7 + 19 + 37 + 61 + 91 + 127 + 169) = \frac{512}{512} = 1 \tag{5.59}$$

where $n = 16$ was used. Because of the special angles in Equation 5.58 this V_2^* will remain constant, independent of the stacking sequence shown in Figure 5.11.

Knowing the values from the first column of Table 5.6, we ca substitute the values into Equation 5.50 for the case of $m = 4$ or Figur 5.11(a).

$$V_1^* = \frac{8}{16^3}(-1 - 7 - 19 - 37 + 61 + 91 + 127 + 169) = \frac{394}{512} = \frac{3}{4} \tag{5.60}$$

Note that the first ply from the midplane upward is a 90-degree ply. We have mentioned before that there is a difference between the laminate code as defined by Equation 5.1 which follows an ascending order from the bottom surface of the laminate. The stacking sequence starting from the midplane is the opposite of the laminate code for all symmetric laminates, as shown in Figure 5.11.

The case of $m = 8$ shown in Figure 5.11(b), and that of $m = 16$ in Figure 5.11(c) are listed below, respectively

For $m = 8$,

$$V_1^* = \frac{8}{16^3}(-1 - 7 + 19 + 37 - 61 - 91 + 127 + 169) = \frac{192}{512} = \frac{3}{8} \tag{5.61}$$

For $m = 16$,

$$V_1^* = \frac{8}{16^3}(-1 + 7 - 19 + 37 - 61 + 91 - 127 + 169) = \frac{96}{512} = \frac{3}{16} \tag{5.62}$$

It appears that a pattern has been established for cross-ply symmetric laminates with increased ply groups

$$V_1^* = \frac{3}{m}, \quad m = 4, 8, 16, \ldots \tag{5.63}$$

Summarizing the results for this family of cross-ply laminates in which the total number of ply groups is a variable, we can enter the values of V's into Table 5.3 and arrive at Table 5.7. Care must be exercised in the proper use of normalizing factors. Note that only V_1^* is affected by the stacking sequence. We only showed the case of changing the number of ply groups. Other stacking sequences than those shown in Figure 5.11 are, of course, possible; an example of which may be $[0_2/90_4/0_2]_S$. The value for V_1 will be different from that shown in Equation 5.60 and Table 5.7. The effect of V_1 on the flexural modulus is the degree of anisotropy, or the difference between D_{11} and D_{22}. In the limit when we have an infinite number of alternating plies, our laminate will become quasi-homogeneous. The property of the laminate will be square symmetric, but not isotropic. This difference between

table 5.7

formulas for flexural modulus of a solid symmetric [0/90] cross-ply laminate

	1	U_2	U_3
D_{11}^*	U_1	$\frac{3}{m}$	1
D_{22}^*	U_1	$-\frac{3}{m}$	1
D_{12}^*	U_4		-1
D_{66}^*	U_5		-1

$D_{16} = D_{26} = 0,\ Z_c^* = 0,\ D_{ij}^* = D_{ij}/I^*$

square symmetric and isotropy was illustrated in Equations 1.22 and 1.23.

Let us calculate the flexural modulus of cross-ply laminates shown in Figure 5.11. Using the data for T300/5208, we have for 16-ply laminates

$$h = 16h_o = 16 \times 125 \times 10^{-6} = 2 \times 10^{-3}\,\text{m}$$

$$D_{11} = \frac{h^3}{12} D_{11}^* \tag{5.64}$$

$$= \frac{h^3}{12}\left[U_1 + \frac{3}{m}U_2 + U_3\right]$$

$$= 666 \times 10^{-12}\left[76.37 + \frac{3}{m}85.73 + 19.71\right] \tag{5.65}$$

For ply group m equal to 4

$$D_{11}^* = 160.3\ \text{GPa}$$

$$D_{11} = 106.9\ \text{Nm} \tag{5.66}$$

or for $m = 8$

$$\begin{aligned} D^*_{11} &= 128.2 \text{ GPa} \\ D_{11} &= 85.4 \text{ Nm} \end{aligned} \tag{5.67}$$

or for $m = 16$

$$\begin{aligned} D^*_{11} &= 112.1 \text{ GPa} \\ D_{11} &= 74.7 \text{ Nm} \end{aligned} \tag{5.68}$$

or for $m = \infty$ in Table 5.7, the laminate becomes quasi-homogeneous.*

$$\begin{aligned} D^*_{11} &= D^*_{22} = 96.0 \text{ GPa} \\ D_{11} &= D_{22} = 64.0 \text{ Nm} \end{aligned} \tag{5.69}$$

We can repeat the process above and obtain all the components of modulus with or without normalization, and the corresponding components of compliance. We purposely list both the normalized and unnormalized components because they serve different purposes.

$m = 4$, $[0_4/90_4]_S$, Figure 5.11(a) (5.70)

$$D^*_{ij} = \begin{bmatrix} 160.37 & 2.89 & \\ 2.89 & 31.77 & \\ & & 7.17 \end{bmatrix} \text{GPa} \qquad d^*_{ij} = \begin{bmatrix} 6.24 & -.569 & \\ -.569 & 31.51 & \\ & & 139.4 \end{bmatrix} (\text{TPa})^{-1}$$

$$D_{ij} = \begin{bmatrix} 106.9 & 1.93 & \\ 1.93 & 21.18 & \\ & & 4.78 \end{bmatrix} \text{Nm} \qquad d_{ij} = \begin{bmatrix} 9.36 & -.85 & \\ -.85 & 47.27 & \\ & & 209.2 \end{bmatrix} (\text{kNm})^{-1} \tag{5.71}$$

*This occurs when normalized flexural modulus is equal to the in-plane modulus in Table 4.11.

- $m = 8$, $[0_2/90_2/0_2/90_2]_S$, Figure 5.11(*b*) (5.72)

$$D^*_{ij} = \begin{bmatrix} 128.22 & 2.89 & \\ 2.89 & 63.92 & \\ & & 7.17 \end{bmatrix} \text{GPa} \qquad d^*_{ij} = \begin{bmatrix} 7.80 & -.353 & \\ -.353 & 15.65 & \\ & & 139.4 \end{bmatrix} (\text{TPa})^{-1}$$

$$D_{ij} = \begin{bmatrix} 85.48 & 1.93 & \\ 1.93 & 42.61 & \\ & & 4.78 \end{bmatrix} \text{Nm} \qquad d_{ij} = \begin{bmatrix} 11.70 & -.530 & \\ -.530 & 23.48 & \\ & & 209.2 \end{bmatrix} (\text{kNm})^{-1}$$

(5.73)

- $m = \infty$ or $[0/90\ldots]_s$ (This is a quasi-homogeneous laminate.)

$$D^*_{ij} = \begin{bmatrix} 96.08 & 2.89 & \\ 2.89 & 96.08 & \\ & & 7.17 \end{bmatrix} \text{GPa} \qquad d^*_{ij} = \begin{bmatrix} 10.41 & -.313 & \\ -.313 & 10.41 & \\ & & 139.4 \end{bmatrix} (\text{TPa})^{-1}$$

$$D_{ij} = \begin{bmatrix} 64.05 & 1.93 & \\ 1.93 & 64.05 & \\ & & 4.78 \end{bmatrix} \text{Nm} \qquad d_{ij} = \begin{bmatrix} 15.62 & -.471 & \\ -.471 & 15.62 & \\ & & 209.2 \end{bmatrix} (\text{kNm})^{-1}$$

(5.74)

Based on the components above, we can say:

1. The shear components are uncoupled from the other four non-zero components. The shear compliance is simply the reciprocal of the shear modulus; i.e.,

$$7.17 \times 0.1394 = 4.78 \times .2092 = 1.000$$

The shear component is independent of the ply groups.

2. The Poisson component of modulus remain constant as the ply groups change. But the Poisson component of compliance changes with m.
3. Because of the constant Poisson component and the invariant constraint of Equation 5.41, the sum of the two normal components of the modulus with or without normalization must be constant; i.e.,

$$160.37 + 31.77 = 128.22 + 63.92 = 2 \times 96.08 = 192.16$$

$$106.9 + 21.18 = 85.48 + 42.61 = 2 \times 64.05 = 128.1$$

4. When the normalized components in Equation 5.74 are equal to those in Table 4.7, the laminate is quasi-homogeneous.
5. Figure 5.12 shows the degree of convergence of a cross-ply laminate to a quasi-homogeneous square-symmetric laminate.

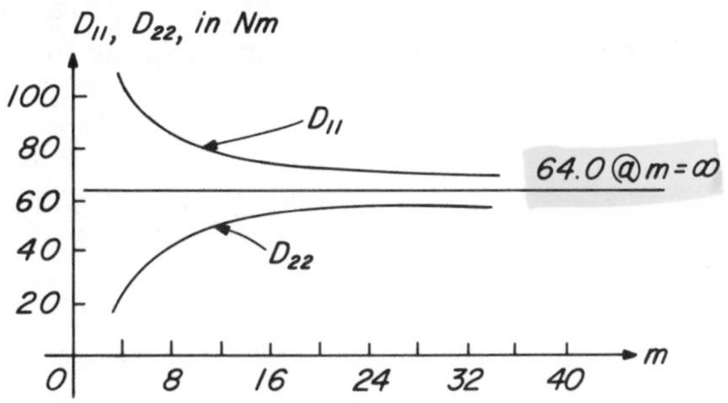

Figure 5.12 Flexural modulus components as functions of ply groups for a T300/5208 laminate. Note that as ply groups m increases, the modulus components approach the modulus of the quasi-homogeneous laminate, although many groups are needed for good convergence.

If we introduce a honeycomb core into our cross-ply laminate, we want to show how the flexural modulus can be calculated. Let us examine three cross-ply laminates in Figure 5.13. These laminates are sandwich constructions with facing materials identical to those solid laminates shown in Figure 5.11. The number of ply groups are different among these laminates. The core half-depth is equal to four plies.

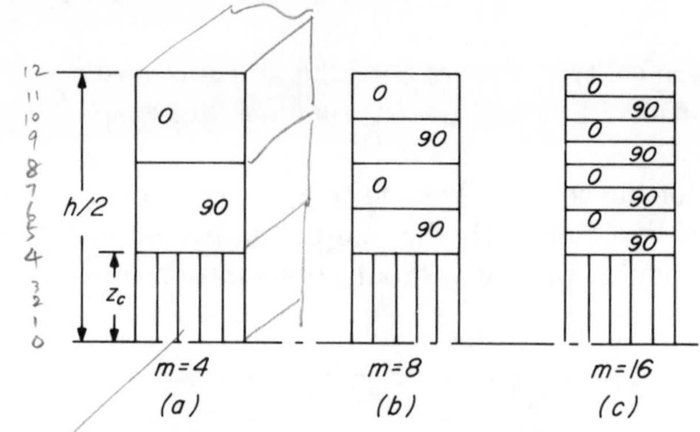

Figure 5.13 Cross-ply sandwich laminates. This symmetric laminate has 2–8 ply facings and 4-ply thick half-depth of core. Total thickness of laminate is 24 equivalent plies. Three different numbers of ply groups are shown; m = 4, 8 and 16. This figure shows the same facing laminates as those solid laminates in Figure 5.11.

The flexural modulus of these sandwich laminates can be readily calculated by substituting the nonzero trigonometric functions into Equation 5.53. For the case of 4-ply group laminate in Figure 5.13(a), or $m = 4$.

$$V_1^* = \frac{8}{24^3}(-61 - 91 - 127 - 169 + 217 + 271 + 331 + 397) = \frac{4}{9} \tag{5.75}$$

or for $m = 8$

$$V_1^* = \frac{8}{24^3}(-61 - 91 + 127 + 169 - 217 - 271 + 331 + 397) = \frac{2}{9} \tag{5.76}$$

or for $m = 16$

$$V_1^* = \frac{8}{24^3}(-61 + 91 - 127 + 169 - 217 + 271 - 331 + 397) = \frac{1}{9} \tag{5.77}$$

We again notice the trend that, as the number of ply groups m increases, the value of V^* decreases by the following relation:

$$V_1^* = \frac{16}{9m} \tag{5.78}$$

The sandwich laminates approach square-symmetric as m increases.

We need the following values before we can use the formulas for the flexural modulus in Table 5.3:

$$V_2^* = \frac{8}{24^3}(61 + 91 + 127 + 169 + 217 + 271 + 331 + 393) = \frac{1664}{1728}$$

$$= \frac{26}{27} \tag{5.79}$$

$$z_c^* = \frac{1}{3} \tag{5.80}$$

$$1 - z_c^{*3} = \frac{26}{27} \tag{5.81}$$

$V_3 = V_4 = 0$ (The laminate is orthotropic.)

Using the same correction factor for the sandwich core is applied to the first column of Table 5.3, and the normalized V's defined in Equation 5.46, we can summarize the results in Equations 5.79 to 5.81 in a matrix multiplication table as follows:

table 5.8

formulas for flexural modulus of a symmetric sandwich laminate with [0/90] facings

	$\frac{26}{27}$	U_2	$\frac{26}{27}U_3$
D_{11}^*	U_1	$\frac{16}{9m}$	1
D_{22}^*	U_1	$-\frac{16}{9m}$	1
D_{12}^*	U_4		-1
D_{66}^*	U_5		-1

$D_{16} = D_{26} = 0$, $D_{ij}^* = D_{ij}^*/I^*$, $h^*/I^* = 26/27$

We are now ready to calculate the flexural modulus of our sandwich laminates assuming the facing material is T300/5208.

$$h = 24h_o = 24 \times 125 \times 10^{-6} = 3 \times 10^{-3}\,\text{m} \tag{5.82}$$

$$z_c = 4h_o = 0.5 \times 10^{-3}\,\text{m} \tag{5.83}$$

From Tables 5.3 and 5.8 for $m = 4$ shown in Figure 5.13(a):

$$D_{11} = \frac{h^3}{12}[26U_1/27 + 16U_2/9m + 26U_3/27] \tag{5.84}$$

$$= 2.25 \times 10^{-9}\,[26 \times 76.37/27 + 16 \times 85.73/9m + 26 \times 19.71/27 \tag{5.85}$$

For $m = 4$,

$$D_{11}^* = 130.62\text{ GPa} \qquad D_{11} = 293.9\text{ Nm} \tag{5.86}$$

for $m = 8$,

$$D_{11}^* = 111.95 \qquad D_{11} = 251.9 \tag{5.87}$$

for $m = 16$,

$$D_{11}^* = 102.04 \qquad D_{11} = 229.6 \tag{5.88}$$

for $m = \infty$,

$$D_{11}^* = 92.52 \qquad D_{11} = 208.1 \tag{5.89}$$

Similarly,

$$D_{22} = \frac{h^3}{12}[26U_1/27 - 16U_2/9m + 26U_3/27] \tag{5.90}$$

For $m = 4$,

$$D_{22}^* = 54.4\text{ GPa} \qquad D_{22} = 122.4\text{ Nm} \tag{5.91}$$

for $m = 8$,

$$D_{22}^* = 73.46 \qquad D_{22} = 165.3 \tag{5.92}$$

For $m = 16$,

$$D_{22}^* = 83.02 \qquad D_{22} = 186.8 \tag{5.93}$$

For $m = \infty$,

$$D_{22}^* = 92.52 \qquad D_{22} = 208.1 \tag{5.94}$$

This last value is the modulus for a quasi-homogeneous laminate. This is the same value as in Equation 5.89. The normalized flexural modulus is equal to the in-plane modulus in Table 4.7 times the core correction factor of 26/27. The absolute flexural modulus is plotted in Figure 5.14.

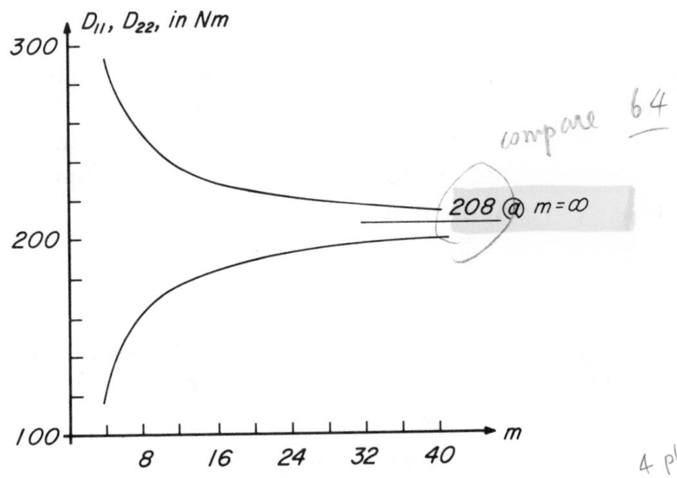

Figure 5.14 Flexural modulus for a sandwich laminate of T300/5208 as functions of the number of ply groups. The convergence toward a square-symmetric laminate is analogous to that in Figure 5.12.

Note the substantial increase in the modulus components of the sandwich construction here over the solid laminates shown in Figure 5.12. First of all, there is a thickness increase from 16 to 24 plies. If our laminate were homogeneous and solid or without a core, the increase in the flexural modulus components will be the cube of the thickness ratio. In our particular case, it will be:

$$(24/16)^3 = 3.375 \tag{5.95}$$

On the other hand, if we have a sandwich construction, there should be a reduction proportional to, as in Equation 5.55,

$$1 - z_c^{*3} = \frac{26}{27} = .962 \tag{5.96}$$

which represents the effect of core if the facings were homogeneous. Assuming the core thickness for our laminate is the same as those in Figure 5.13, the net effect of thickness increases and the presence of core is simply the product of Equations 5.95 and 5.96:

$$3.375 \times \frac{26}{27} = 3.25 \tag{5.97}$$

We can now make direct comparison between the sandwich construction and the solid laminate. This comparison can only be made for the case of quasi-homogeneous material. For example, the ratio of the absolute components of modulus between that in Equation 5.94 and the same component in Equation 5.74 is

$$208.1/64.05 = 3.25$$

This agrees with the result of Equation 5.97. Similarly, we can find the ratio of the normalized components between Equations 5.94 and 5.74:

$$92.52/96.08 = .962$$

This agrees with the result of Equation 5.96. The conclusion is that homogeneous materials with or without honeycomb core can be scaled. Plate thickness and core thickness can be obtained by proper ratios from one construction to another. No such simple scaling will work for laminated composites. They must be assessed on an individual basis. Only in special cases can the calculation of flexural modulus by smearing the laminated facing be approximately accurate. The parallel axis theorem in the next chapter can determine this accuracy.

6. flexural modulus of angle-ply laminates

In the last section we saw that cross-ply laminates are orthotropic, or square symmetric when the number of ply groups approach infinity. We will see in this section that angle-ply laminates with or without core are generally anisotropic. A balanced, symmetric laminate is orthotropic in its in-plane modulus but is anisotropic in its flexural modulus. The reason is that the position of each ply is unique along the z-axis. The shear coupling terms of a $+\theta$ ply cannot be cancelled by those of a $-\theta$ ply unless the positions of these plies are judiciously selected. We will show later that the shear coupling terms can be cancelled if we use antisymmetric laminates. So there are two simple methods of obtaining orthotropic flexural modulus:

- Use on-axis plies only. This is the case of on-axis unidirectional or cross-ply laminates.
- Use antisymmetric laminates. This will be discussed in Chapter 6.

The motivation to make laminates orthotropic (and symmetric) is often driven by the availability of stress analysis tools. Most current tools are limited to orthotropic and homogeneous plates. It is unfortunate that the use of composite materials is limited or penalized by the nonavailability of analytical tools. It is important to understand how anisotropy and nonhomogeneity arise in composite laminates and to what degree they can be manipulated to perform functions not possible with conventional materials.

For angle-ply laminates the ply orientation can be

$$\theta_i = +\phi \text{ or } -\phi \tag{5.98}$$

Figure 5.15 shows three possible ply groups, $m = 4$, 8, and 16 of angle-ply laminates.

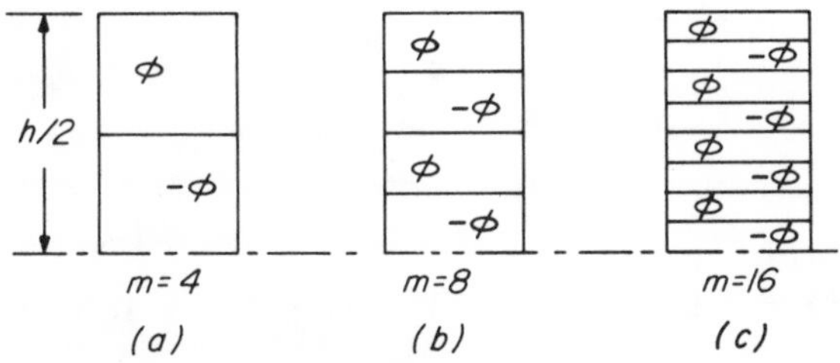

Figure 5.15 Angle-ply laminates with different number of ply groups. Because of symmetry only the upper half of the laminate is shown.

When ply orientations change signs, the cosine functions remain the same, while the sine functions will change signs. The cosine function in Equation 5.48 can be factored out as follows:

$$V_1^* = \frac{8}{n^3} \cos 2\phi \sum [t^3 - (t-1)^3] \tag{5.99}$$

Since $n = 16$, or $t = 8$, from Table 5.4

$$V_1^* = \frac{8}{16^3} \cos 2\phi\,(1 + 7 + 19 + 37 + 61 + 91 + 127 + 169)$$

$$= \cos 2\phi \tag{5.100}$$

Similarly, we can show

$$V_2^* = \cos 4\phi \tag{5.101}$$

For $m = 4$ and the proper sign for the sine functions we have:

$$V_3^* = \frac{8}{16^3} (-1 - 7 - 19 - 37 + 61 + 91 + 127 + 169) \sin 2\phi$$

$$= \frac{3}{4} \sin 2\phi \tag{5.102}$$

Similarly, we can show

$$V_4^* = \frac{3}{4} \sin 4\phi \tag{5.103}$$

For $m = 8$,

$$V_3^* = \frac{8}{16^3} (-1 - 7 + 19 + 37 - 61 - 91 + 127 + 169) \sin 2\phi$$

$$= \frac{3}{8} \sin 2\phi \tag{5.104}$$

For $m = 16$,

$$V_3^* = \frac{8}{16^3}(-1 + 7 - 19 + 37 - 61 + 91 - 127 + 169)\sin 2\phi$$

$$= \frac{3}{16}\sin 2\phi \tag{5.105}$$

It appears that

$$V_3^* = \frac{3\sin 2\phi}{m}, \; m = 4, 8, 16, \ldots \tag{5.106}$$

Similarly,

$$V_4^* = \frac{3\sin 4\phi}{m}, \; m = 4, 8, 16, \ldots \tag{5.107}$$

The formulas for the flexural modulus for angle-ply laminates can now be written in matrix multiplication form in Table 5.9.

table 5.9

formulas for flexural modulus of a solid symmetric angle-ply [$\phi/-\phi$] laminate

	1	U_2	U_3
D_{11}^*	U_1	$\cos 2\phi$	$\cos 4\phi$
D_{22}^*	U_1	$-\cos 2\phi$	$\cos 4\phi$
D_{12}^*	U_4		$-\cos 4\phi$
D_{66}^*	U_5		$-\cos 4\phi$
D_{16}^*		$\frac{3}{2m}\sin 2\phi$	$\frac{3}{m}\sin 4\phi$
D_{26}^*		$\frac{3}{2m}\sin 2\phi$	$-\frac{3}{m}\sin 4\phi$

where $D_{ij}^* = D_{ij}/I^*$

Note the formulas for the flexural modulus of angle-ply laminates are identical to those for in-plane modulus of angle-ply laminates as shown in Table 4.7 with the exception of the shear coupling terms. These terms vanish as the number of ply groups increase. Thus for quasi-homogeneous laminates (as m becomes infinity), the in-plane and flexural moduli are related by:

$$D_{ij}^* = \frac{12}{h^3} D_{ij} = \frac{1}{h} A_{ij} \qquad (5.108)$$

or

$$D_{ij} = \frac{h^2}{12} A_{ij} \qquad (5.109)$$

conversely,

$$d_{ij} = \frac{12}{h^2} a_{ij} \qquad (5.110)$$

We can thus compute the flexural modulus and compliance of a specific angle-ply laminate.

Let

$$\phi = 45 \text{ degrees} \qquad (5.111)$$

Our laminates for $m = 4, 8, 16$ are:

$$[45_4/-45_4]_S,\ [45_2/-45_2]_{2S},\ [45/-45]_{4S} \qquad (5.112)$$

The upper half of these laminates are shown in Figure 5.16. Using these data for T300/5208 and 16 plies, we have

$$h_o = 125 \times 10^{-6}\,\text{m}$$

$$h = 16h_o = 2 \times 10^{-3}\,\text{m} \qquad (5.113)$$

$$D_{16} = D_{26} = \frac{h^3}{12}\frac{3}{2m} U_2 = \frac{1}{m} U_2 \times 10^{-9}$$

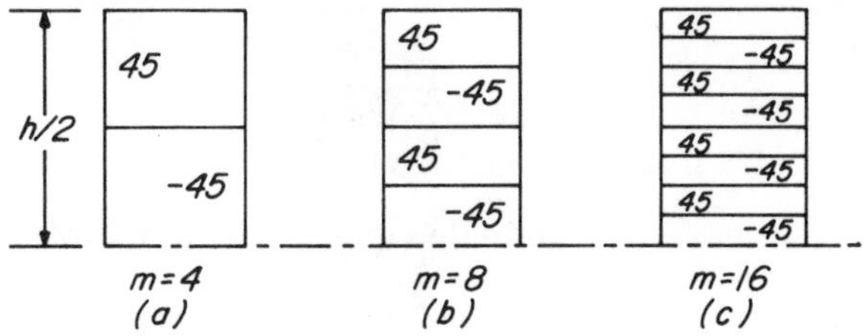

Figure 5.16 Stacking sequence of symmetric angle-ply laminates with 45-degree angles. Ply groups increase from 4 to 16. Only the upper half of the laminate is shown.

The flexural modulus and compliance for various ply groups are shown as follows:

$$D^*_{ij} = \begin{bmatrix} 56.65 & 42.31 & \frac{1}{m}128.5 \\ 42.31 & 56.65 & \frac{1}{m}128.5 \\ \frac{1}{m}128.5 & \frac{1}{m}128.5 & 46.59 \end{bmatrix} \text{GPa} \qquad D_{ij} = \begin{bmatrix} 37.77 & 28.21 & \frac{1}{m}85.73 \\ 28.21 & 37.77 & \frac{1}{m}85.73 \\ \frac{1}{m}85.73 & \frac{1}{m}85.73 & 31.06 \end{bmatrix} \text{Nm} \tag{5.114}$$

Note as m increases, the flexural modulus becomes square symmetric. We can invert the modulus for $m = 4$ and obtain the following compliance, where both the normalized and the unnormalized are included:

$$d^*_{ij} = \begin{bmatrix} 44.02 & -25.70 & -12.63 \\ -25.70 & 44.02 & -12.63 \\ -12.63 & -12.63 & 38.9 \end{bmatrix} (\text{TPa})^{-1} \qquad d_{ij} = \begin{bmatrix} 66.03 & -38.56 & -18.95 \\ -38.56 & 66.03 & -18.95 \\ -18.95 & -18.95 & 58.35 \end{bmatrix} (\text{kNm})^{-1} \tag{5.115}$$

For $m = 8$,

$$d^*_{ij} = \begin{bmatrix} 40.55 & -29.17 & -3.92 \\ -29.17 & 40.55 & -3.92 \\ -3.92 & -3.92 & 24.16 \end{bmatrix} (\text{TPa})^{-1} \qquad d_{ij} = \begin{bmatrix} 60.83 & -43.76 & -5.88 \\ -43.76 & 60.83 & -5.88 \\ -5.88 & -5.88 & 36.25 \end{bmatrix} (\text{kNm})^{-1} \tag{5.116}$$

For $m = \infty$,

$$d_{ij} = \frac{12}{h^2} a_{ij} \tag{5.117}$$

where $a_{ij}h$ for our angle-ply laminate can be found in Table 4.9:

$$d_{11} = \frac{12}{h^3} a_{11} h = \frac{12}{8 \times 10^{-9}} 39.9 \times 10^{-12} = 59.85\ (\text{kNm})^{-1} \tag{5.118}$$

The other components of compliance can be calculated in the same manner. We have

$$d_{ij}^* = \begin{bmatrix} 39.9 & -29.8 & \\ -29.8 & 39.9 & \\ & & 21.46 \end{bmatrix} (\text{TPa})^{-1} \qquad d_{ij} = \begin{bmatrix} \mathbf{59.85} & -44.70 & \\ -44.70 & 59.85 & \\ & & 32.19 \end{bmatrix} (\text{kNm})^{-1} \tag{5.119}$$

As the number of ply groups increase, the shear or normal coupling terms drop rapidly in both the modulus and compliance. The rate of reduction is greater in the compliance than in the modulus. Only when $m = 4$ is the shear coupling significant. The compliance components 11, 22, 12 and 66 do not vary much as the shear coupling components change. This means that the flexural stiffness of the laminate increases as it approaches orthotropy.

7. ply stress and ply strain analysis

The ply stress and ply strain in a symmetric laminate due to flexure can be determined following the procedure for the in-plane stretching. Figure 5.17 is analogous to Figure 4.4 for the in-plane behavior. The process of determining the ply stress and ply strain is straight forward. The motivation is to assess the strength of each ply within the laminate. The strength calculation will be covered in Chapter 7.

The highest z for the i-th ply group in Figure 5.7 will have the highest strain in each ply group by virtue of Equation 5.10. It is the highest strain components that will govern the strength of that ply group. This is obvious if all components of the curvature have the same sign, positive or negative. But it is not so obvious if the signs are mixed. We will show later that regardless of the signs of the curvature components, the highest z will govern their strength. This holds for symmetric laminates under flexural loads.

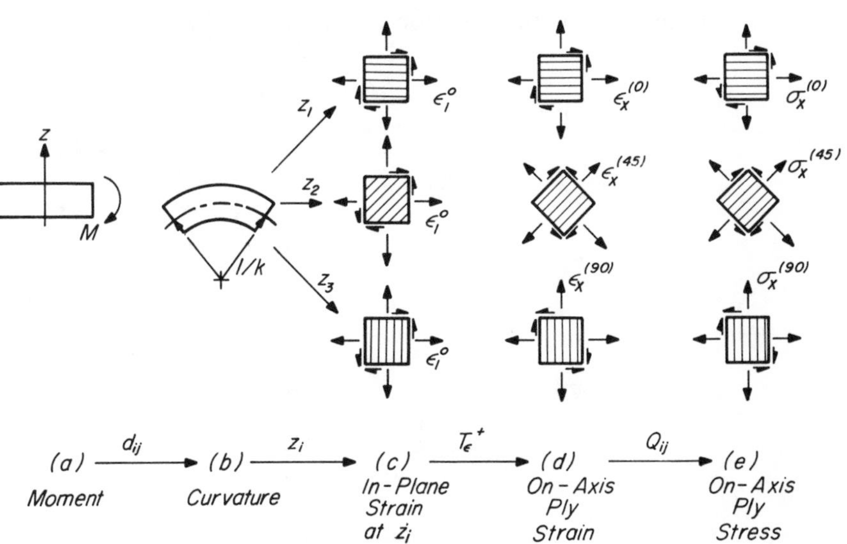

Figure 5.17 Ply stress and strain in a symmetric laminate under flexure:
From (*a*) to (*b*): Use moment-curvature relations in Table 5.2.
From (*b*) to (*c*): Use curvature-strain equation in Equation 5.10. Use top surface of each ply group for the z-value.
From (*c*) to (*d*): Use strain transformation to transform the laminate strain to the on-axis strain.
From (*d*) to (*e*): Use the on-axis stress-strain relation to determine the corresponding on-axis ply stress.

Figure 5.17 outlines the process of going from applied moments to the resulting ply strain and ply stress in a laminate. The initial moments may be obtained a number of ways. Let us assume that the moments are known. The simplest example is the case of a statically determinate structure. One such example is the three-point bend test shown in Figure 5.18, when the load is applied at the midspan. The maximum moment is also at the midspan

$$M = \frac{PL}{4} \tag{5.120}$$

For a beam with a width b, the distributed moments or moments per unit width are

$$M_1 = \frac{M}{b} = \frac{PL}{4b}, M_2 = M_6 = 0 \tag{5.121}$$

From the moment-curvature relation in Equation 5.17,

$$\begin{aligned} k_1 &= d_{11} M_1 = \frac{PL}{4b} d_{11} \\ k_2 &= d_{21} M_1 \\ k_6 &= d_{61} M_1 \end{aligned} \tag{5.122}$$

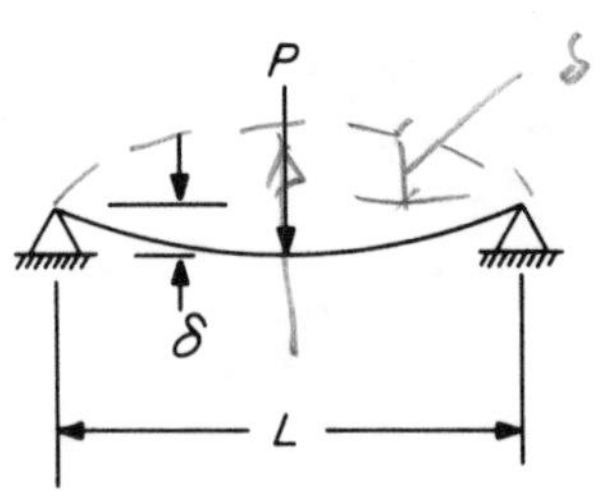

Figure 5.18 Three-point bend test.

Let us assume that the laminate is T300/5208, $[0_4 90_4]_s$:

$$P = 100 \text{ N}$$

$$L = .1 \text{ m}$$

$$b = .01 \text{ m}$$

From the compliance of this laminate listed in Equation 5.71,

$$d_{11} = 9.36 \ (\text{kNm})^{-1}$$

$$d_{21} = -.85 \ (\text{kNm})^{-1}$$

$$d_{61} = 0$$

With the geometric and material properties above, we can immediately calculate the following using Equation 5.122:

$$k_1 = \frac{100 \times .1}{4 \times .01} 9.36 = 2.34 \text{ m}^{-1}$$

$$k_2 = \frac{100 \times .1}{4 \times .01} (-.85) = -.212 \text{ m}^{-1} \qquad (5.123)$$

$$k_6 = 0$$

The strain at the upper face of the beam (top of 0° ply group)

$$z = 8h_o = 1 \times 10^{-3} \text{ m}$$

$$\epsilon_1 = zk_1 = 2.34 \times 10^{-3}$$

$$\epsilon_2 = zk_2 = -.212 \times 10^{-3} \qquad (5.124)$$

$$\epsilon_6 = 0$$

The induced stress components at this upper face which have the 0-degree ply group are:

$$\sigma_1 = \sigma_x^{(0)} = 181.8 \times 2.34 - 2.89 \times .212$$

$$= 424 \text{ MPa}$$

$$\sigma_2 = \sigma_y^{(0)} = 2.89 \times 2.34 - 10.3 \times .212 \qquad (5.125)$$

$$= 4.57 \text{ MPa}$$

$$\sigma_6 = \sigma_s^{(0)} = 0$$

The transverse stress and strain components are negligible compared with the longitudinal components.

The strain and stress at the upper face of the 90-degree ply group is

$$z = 4h_o = .5 \times 10^{-3}\,\text{m}$$

$$\epsilon_1 = 1.17 \times 10^{-3}$$

$$\epsilon_2 = -.106 \times 10^{-3} \qquad (5.126)$$

$$\epsilon_6 = 0$$

$$\sigma_1 = \sigma_y^{(90)} = 10.3 \times 1.17 - 2.89 \times .106$$

$$= 11.7\ \text{MPa}$$

$$\sigma_2 = \sigma_x^{(90)} = 2.89 \times 1.17 - 181.8 \times .106 \qquad (5.127)$$

$$= -15.88\ \text{MPa}$$

$$\sigma_6 = \sigma_s^{(90)} = 0$$

The stress distribution is shown in Figure 5.19.

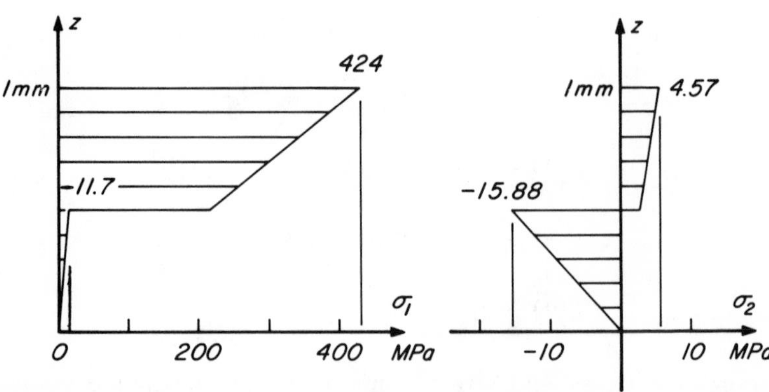

Figure 5.19 Ply stress in a cross-ply beam at the midspan of a three-point bend test.

he deflection δ in Figure 5.18 can be calculated from the beam ormulas in handbooks

$$\begin{aligned}\delta &= \frac{PL^3}{48EI} \\ &= \frac{PL^3}{48b} d_{11} \\ &= \frac{100 \times .1^3}{48 \times .01} 9.36 \\ &= 1.95 \times 10^{-3} \text{ m}\end{aligned} \tag{5.128}$$

The ply stress and ply strain analysis of any determinate structure an be duplicated exactly as the above. New variables are introduced as ollows:

- For different laminates, different compliance must be used.
- For different end conditions, such as a four-point bend test, the moment in Equation 5.120 and the deflection in Equation 5.128 must be changed.

For laminate under complex boundary conditions, the process above emains the same for the ply stress and ply strain determination. But e deflection function w in Equation 5.5 requires a solution based on e theory of plates. No simple relation like those for beams is vailable.

Finally the ply stress and ply strain calculation is only the means for rength determination. Again the appropriate failure criterion which e will cover in Chapter 7 is required for the ply-by-ply examination to scertain the sequence of successive ply failures, from the first (FPF) to e last or the ultimate.

. conclusions

he flexural stiffness of laminated composites can be derived following e pattern for the in-plane stiffness. First the flexural modulus can be elated to the modulus of the constituent plies by some weighting ctors. The composite modulus of the laminate is not simply a linear function of the constituent plies. That was the case for the in-plan modulus, but for the flexural modulus the outer plies will contribut more than the inner ones. Flexural compliance and the equivalen engineering constants can be derived by matrix inversion, and the ratic or reciprocals of the compliance components, respectively.

Light weight core can be used to replace the laminate material nea the mid-plane. This is a very effective method of reducing the tota weight of the laminated plate while sacrificing very little in the flexura stiffness. As the number of ply groups increases, the behavior of th laminated plate approaches that of a quasi-homogeneous laminate. It very difficult performing scaling operation from one laminate confi urations to another. In most instances, such scaling operation can onl be done if the laminates are quasi-homogeneous, with or without core To be safe, flexural modulus should be calculated based on the precis stacking sequence.

A balanced laminate will have zero shear and normal coupling in it in-plane behavior. Since each ply occupies a fixed position along th z-axis, a laminated composite is usually anisotropic in its flexura behavior.

The flexural stiffness of a laminated beam should be derived as special case of a laminated plate. The appropriate modulus of the ind vidual constituent ply must be included in the computation of the tota flexural modulus of the laminated plate. From this modulus, we ca then compute the flexural compliance; from the compliance we ca compute the stiffness of a beam. This process was followed in Equatio 5.21. It is not possible, on the other hand, to compute directly th effective Young's modulus of a beam from the Young's modulus c each constituent layer. A rule of mixtures equation, including th proper weighting factors, will not result in the proper stiffness of th beam. Composite laminates are two-dimensional bodies and only tw dimensional theories are valid for the description of the stiffne behavior.

homework problems

What happens to the flexural modulus and compliance if the definition of displacement in Equations 5.6 and 5.7 does not have a minus sign?

What happens to the flexural modulus and compliance if tensorial shear strain is used instead of the engineering shear strain? The factor 2 in Equation 5.8 must be removed.

How accurate is the following rule-of-mixtures equation for estimating the flexural stiffness of a beam?

$$E_1^f = \frac{8}{n^3} \sum_{t=c+1}^{n/2} E^{(t)} \left[t^3 - (t-1)^3 \right] \qquad (5.129)$$

where the terms are identical to those in Equation 5.48, except $E^{(t)}$ is the Young's modulus of the t-th ply. Compare the flexural stiffness of solid cross-ply and angle-ply beams for various ply groups up to infinity calculated from Equation 5.129 and those in Figure 5.12 and Equation 5.114 et al.

We know from Chapter 4, Section 6, that quasi-isotropic laminates can be obtained from discrete multidirectional laminates ($\pi/3$, $\pi/4$ et al.). We also know that quasi-homogeneous laminates can be obtained from large ply groups or m approaching infinity. But it is possible to approach quasi-homogeneity with finite number of plies. One stacking sequence discovered by Ernest R. Scheyhing (D.Eng thesis, Yale University, 1965) calls for the following 24-ply symmetric laminate:

$$[-60/0/60_2/0/-60/60/0/-60_2/0/60]_S \qquad (5.130)$$

Show, for a T300/5208 laminate, how close homogeneity is satisfied, i.e.,

$$A_{ij} = \frac{12}{h^2} D_{ij} \qquad (5.131)$$

How close is flexural isotropy satisfied, i.e.,

$$D_{66} = \frac{1}{2}(D_{11} - D_{12})$$
$$D_{16} = D_{26} = 0 \tag{5.132}$$

Is in-plane isotropy exact or approximate? Plot the transformed modulus from 0 to 90 degrees.

e. Label the coordinates in Figure 5.20 which shows the flexural modulus, and bending and torsional stiffnesses of 16-ply 45-degree angle-ply laminates of T300/5208. The bending and torsional stiffnesses are:

$$\frac{EI}{b} = \frac{1}{d_{11}}$$
$$\frac{GJ}{b} = \frac{4}{d_{66}} \tag{5.133}$$

The torsional stiffness is based on a wide rectangular cross-section. Also show the asymptotic value when m approaches infinity.

f. Find the load-deflection curve of a three-point bend test (centrally located load) of the following beam:

$$[0_{16}/90_{16}]_S$$
$$\text{T300/5208}$$
$$h = 8 \text{ mm} \tag{5.134}$$
$$L = 10 \text{ cm}$$
$$b = 2 \text{ cm}$$

g. Determine the first ply failure (FPF) and the ultimate using the following maximum strain criteria:

$$\epsilon_x = 8.3 \times 10^{-3}$$
$$\epsilon_y = 3.8 \times 10^{-3} \tag{5.135}$$

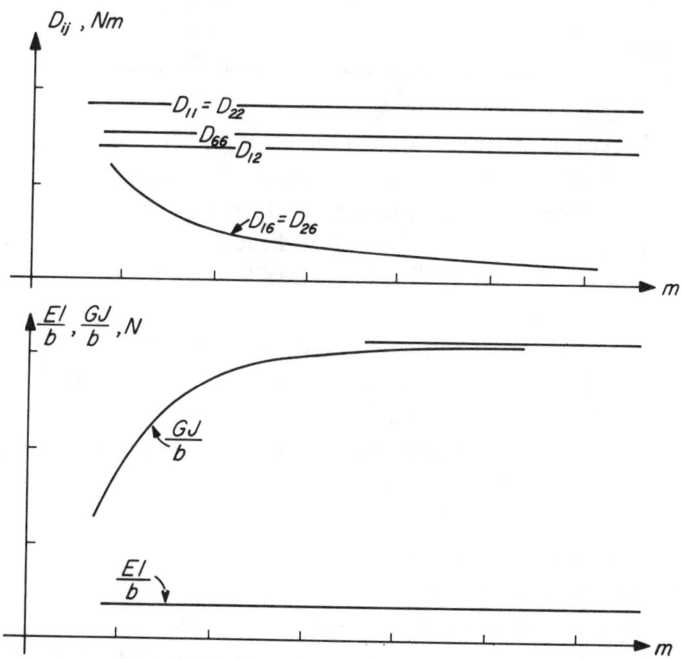

Figure 5.20 Flexural modulus and stiffness of 16-ply 45-degree angle-ply laminates of T300/5208. The abscissa is the number of ply groups. (See Figure 5.15)

h. Repeat the problem above for $[90_{16}/0_{16}]_S$ laminate.

i. What is the natural frequency (the first mode) of the beams in Problems g and h? Use the density from Table 1.7 1600 kgm^{-3}; and the mass per unit length μ = 0.256 kg/m. How close do these frequencies compare with a hinged-hinged beam where

$$\omega_n = \frac{\lambda_1}{L^2} \frac{1}{\sqrt{d_{11}\mu}} \qquad (5.136)$$

where λ_1 = 3.142 (from Handbooks)

The use of hybrid composites is an effective means of optimizing laminates. Hybrid leaf springs with all 0-degree T300/5208 and Scotch-ply 1002 can be made with different ply ratios. Assuming cost ratios of the two materials to be 10, 5 and 2 (T300/5208 is higher), is there a cost-effective ply ratio from the bending stiffness viewpoint? For simplicity, assume a total of 100 plies is needed. All Scotch-ply is located in the core, and all T300/5208 in the facing material.

nomenclature

b	=	Width of beam, in m
D_{ij}	=	Flexural stiffness modulus of multidirectional symmetric laminate including core, in Nm; i,j = 1,2,6
D^*_{ij}	=	D_{ij}/I^*_i or $12D_{ij}/h^3$ when $z_c = 0$, in Pa
d_{ij}	=	Flexural compliance of multidirectional symmetric laminate, and the inverse of D_{ij}, in $(\text{kNm})^{-1}$; i,j = 1,2,6
d^*_{ij}	=	I^*d_{ij}, in Pa^{-1}
E^f_i	=	Flexural engineering constant, in Pa; i = 1,2,6
h	=	Total thickness of laminate, including core, in m
h^*	=	$h^3[1 - z_c^{*3}]/12$, in m^3
h_i	=	$n_i h_o$ = Total thickness of the i-th ply group; i = 1 to m
h_o	=	Unit ply thickness, in m
I	=	Moment of inertia, in m^4
k_i	=	Curvature, in m^{-1}; i = 1,2,6
M_i	=	Moment, in N; i = 1,2,6
m	=	Total number of ply groups in a laminate
n	=	Total number of plies in a laminate including core depth measured in number of plies
n_i	=	Total number of plies in the i-th ply assembly; i = 1 to m
$Q^{(\theta)}_{ij}$	=	Modulus of the ply assembly with θ orientation; i,j = 1,2,6
t_i	=	z_i/h_o or $= i$
U_i	=	Linear combinations of modulus; i = 1 to 7
V_i	=	Geometric factors in formulas for flexural modulus; i = 1 to 4
V^*_i	=	$12V_i/h^3$
z_c	=	Half depth of honeycomb core, in m
z^*_c	=	$2z_c/h$ = Total core to total laminate thickness ratio
z_i	=	Location of ply or ply group, in m
ν^f_{ij}	=	Flexural coupling coefficients

chapter 6

properties of general laminates

eneral laminates are free from midplane symmetry. They can be ymmetric or antisymmetric; and can be of built-up and hybrid conruction. A new coupling between stretching and flexure is introduced. ne modulus and compliance matrices increase from 3 X 3 to 6 X 6. ut the same methodology that governs the symmetric laminates is tended to the general laminates. Unique opportunities not available ith conventional materials can now be exploited to produce novel erformances. The parallel axis theorem is a powerful tool for deterining the modules of general laminates.

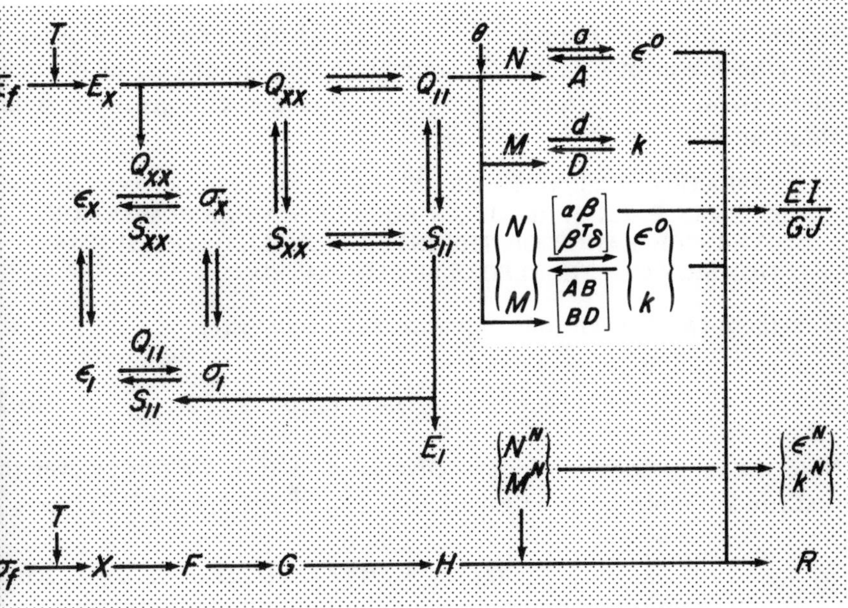

1. index and matrix notations

We have used subscripts for the components of stress, strain, modulu and compliance since the first chapter. We have also used the matri multiplication tables to represent stress-strain relations, transformatio relations, and others. Having had experience with the longhand nota-tion, we will now introduce a shorthand notation that can efficientl represent the equations that we have seen earlier.

For example, in place of the stress-strain relation in Table 3.1 we ca write the same relation in a summation as

$$\sigma_i = \sum_{j=1,2,6} Q_{ij}\epsilon_j, \quad i = 1,2,6 \qquad (6.1$$

There are two types of subscripts in this equation. First, the subscript is called the free index. It assumes values of 1, 2, 6 in this equation. Th rule that governs this subscript or index is called the range conventio defined as follows:

> A FREE INDEX CAN APPEAR ONLY ONCE IN EACH TERM OF AN EQUATION AND ASSUMES A RANGE OF VALUES SPECIFIED.

Secondly, subscript j appears twice on the right-hand side of this equa-tion (the subscript j under the summation sign is not part of the mai relation), we now introduce the summation convention of the inde notation:

> REPEATED SUBSCRIPTS OR INDICES CAN APPEAR ONLY IN PAIRS IN EACH TERM OF AN EQUATION AND A SUMMATION OVER THE RANGE OF THE INDEX IS IMPLIED. THE SUMMA-TION SIGN CAN THEREFORE BE ELIMINATED.

With these two conventions, Equation 6.1 becomes:

$$\sigma_i = Q_{ij}\epsilon_j, \; i,j = 1,2,6 \qquad (6.2$$

Note the range for both indices covers 1, 2 and 6. We can recover th first row of Table 3.1 from Equation 6.2 when $i = 1, j = 1,2,6$; i.e.,

$$\sigma_1 = Q_{11}\epsilon_1 + Q_{12}\epsilon_2 + Q_{16}\epsilon_6 \qquad (6.3$$

Similarly, when $i = 2, j = 1,2,6$, we recover the second row; when $i = 6$, $j = 1,2,6$, we recover the third row. If we let $i = x, j = x,y$, we have,

$$\sigma_x = Q_{xx}\epsilon_x + Q_{xy}\epsilon_y \tag{6.4}$$

This is the first row of Table 1.6.

The index notation is efficient because one equation such as Equation 6.2 can replace three algebraic equations in Table 3.1.

Similarly, the stress-strain relation in terms of compliance is simply

$$\epsilon_i = S_{ij}\sigma_j, \quad i,j = 1,2,6 \text{ or } x,y,s \tag{6.5}$$

For the in-plane behavior of symmetric laminates, we can define

$$N_i = \int \sigma_i dz \tag{6.6}$$

Substituting the stress-strain relation in terms of stiffness modulus,

$$N_i = \int Q_{ij}\epsilon_j \, dz \tag{6.7}$$

If we assume that the in-plane strain is constant, it can be taken out of the integral sign; then we can define the in-plane modulus as:

$$A_{ij} = \int Q_{ij} dz \tag{6.8}$$

Then the in-plane stress-strain relations are:

$$N_i = A_{ij}\epsilon_j^o \tag{6.9}$$

$$\epsilon_i^o = a_{ij}N_j \tag{6.10}$$

These relations are shown in longhand in Tables 4.1 and 4.2, respectively.

For the flexural behavior of symmetric laminates, we can define

$$M_i = \int \sigma_i z dz \tag{6.11}$$

From which we can define the moment-curvature relation:

$$M_i = D_{ij} k_j \tag{6.12}$$

and the flexural modulus:

$$D_{ij} = \int Q_{ij} z^2 dz \tag{6.13}$$

The moment-curvature relation in terms of compliance is:

$$k_i = d_{ij} M_j \tag{6.14}$$

These relations are shown in Tables 5.1 and 5.2, respectively.

The symmetry condition such as

$$Q_{12} = Q_{21}, Q_{16} = Q_{61}, Q_{26} = Q_{62} \tag{6.15}$$

can be expressed as

$$Q_{ij} = Q_{ji} \tag{6.16}$$

Similarly, we have

$$\begin{aligned} S_{ij} = S_{ji}, \; A_{ij} &= A_{ji}, \; D_{ij} = D_{ji}, \\ a_{ij} &= a_{ji}, \; d_{ij} = d_{ji} \end{aligned} \tag{6.17}$$

Instead of the index notation, we can use a matrix notation to express the same relations above. Bold face letters (which can also be represented by the underlined letters) represent matrices. The indices can be eliminated. Equations 6.6 et al. can be rewritten as:

$$\mathbf{N} = \int \boldsymbol{\sigma} dz \tag{6.18}$$

$$\mathbf{A} = \int \mathbf{Q} dz \tag{6.19}$$

$$N = A\,\epsilon^o \tag{6.20}$$

$$\epsilon^o = a\,N \tag{6.21}$$

$$M = \int \sigma z dz \tag{6.22}$$

$$M = D\,k \tag{6.23}$$

$$D = \int Q z^2\,dz \tag{6.24}$$

$$k = d\,M \tag{6.25}$$

Matrix multiplication is implied when two matrices are placed side by side.

Notations are artificial and arbitrary. Symbolically they convey mathematical operations and meanings. Each notation has its advantages and drawbacks. The selection of a notation is often dictated by the particular problem on hand as well as the subjective judgment of the user. Basically, notations are intended to help rather than to hinder communication and understanding. When in doubt, we should resort to the conventional, longhand operations. This will prevent the misapplication or misinterpretation of a notation. In the study of general laminates, we will use both index and matrix notations.

2. stiffness and compliance of general laminates

General laminates are normally unsymmetric. In our context, they can be antisymmetric and hybrid. General laminates have not been used extensively to date for a number of reasons. First, the unsymmetrical laminate will warp after curing and cool down. It may be difficult to meet the dimensional control of a structure. Secondly, the analysis of unsymmetrical plates and shells is more difficult than that for symmetrical structures. Designers feel less experienced working with the unsymmetrical construction and are therefore reluctant in using such unfamiliar construction. But there are many familiar general laminates which include built-up constructions where material cross-sections vary across the depth of a beam or plate. Hybrids are another form of general laminates where plies of different materials or different construction of the same material (woven vs nonwoven) are combined. Then there is a class of antisymmetric laminates which have unique properties.

We intend to show in this chapter that general laminates are no more uncontrollable than the symmetrical, homogeneous laminates. The same theory and material property data control the behavior of all laminates. General laminates have properties which can be effectively utilized to produce unique performance. In many applications, only minimum gage laminates are required. Unsymmetrical laminates can save 50 percent in weight. Other applications may call for predetermined warpage. Use of antisymmetric construction can provide unique coupling. We are therefore not in a position to write off general laminates just because they are more difficult to analyze than symmetric laminates.

The key feature of general laminates lies in the additional degree of coupling, as we will see presently. The basic behavior of this class of laminates is governed by the strain distribution across the thickness of the laminate. By combining the previously assumed strain for both the in-plane and the flexural deformation, that is by taking the strain distribution up to the linear term, we will have

$$\begin{aligned} \epsilon_1(z) &= \epsilon_1^o + zk_1 \\ \epsilon_2(z) &= \epsilon_2^o + zk_2 \\ \epsilon_6(z) &= \epsilon_6^o + zk_6 \end{aligned} \qquad (6.26)$$

In index notation, we have

$$\epsilon_i(z) = \epsilon_i^o + zk_i \qquad (627)$$

Unless otherwise stated, the range of index is always 1, 2, and 6, or $i =$ 1,2,6. The assumed strain components are shown in Figure 6.1. No reference is made concerning the material property. The strain, as always, is defined by geometry with no direct connection to equilibrium or material property. Stress, on the other hand, must satisfy equilibrium; stress-strain relations must reflect material behavior and property. The assumed strain is applicable to all materials, homogeneous and hybrid composites.

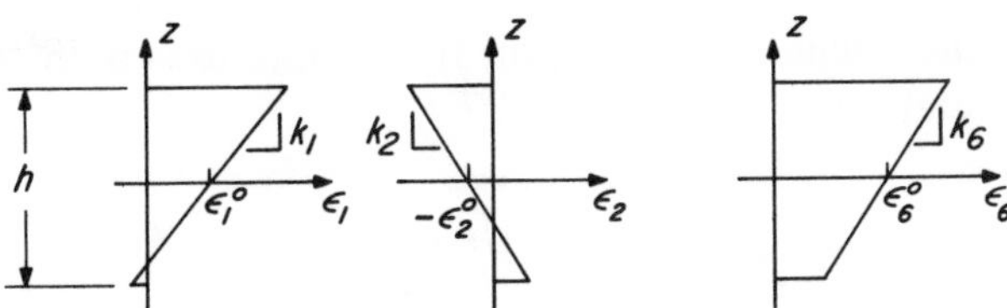

Figure 6.1 Assumed linear strain distributions for general laminates.

We will now substitute the assumed strain in Equation 6.26 into the definition of stress resultant, we have

$$N_1 = \int (Q_{11}[\epsilon_1^o + zk_1] + Q_{12}[\epsilon_2^o + zk_2] + Q_{16}[\epsilon_6^o + zk_6])dz \tag{6.28}$$

$$= \int Q_{11}dz\epsilon_1^o + \int Q_{12}dz\epsilon_2^o + \int Q_{16}dz\epsilon_6^o$$
$$+ \int Q_{11}zdzk_1 + \int Q_{12}zkzk_2 + \int Q_{16}zdzk_6 \tag{6.29}$$

$$= A_{11}\epsilon_1^o + A_{12}\epsilon_2^o + A_{16}\epsilon_6^o + B_{11}k_1 + B_{12}k_2 + B_{16}k_6 \tag{6.30}$$

Similarly

$$N_2 = A_{21}\epsilon_1^o + A_{22}\epsilon_2^o + A_{26}\epsilon_6^o + B_{21}k_1 + B_{22}k_2 + B_{26}k_6 \tag{6.31}$$

$$N_6 = A_{61}\epsilon_1^o + A_{62}\epsilon_2^o + A_{66}\epsilon_6^o + B_{61}k_1 + B_{62}k_2 + B_{66}k_6 \tag{6.32}$$

where the components of the new coupling modulus are:

$$B_{11} = \int Q_{11}zdz,\ B_{22} = \int Q_{22}zdz,\ B_{12} = \int Q_{12}zdz$$
$$B_{66} = \int Q_{66}zdz,\ B_{16} = \int Q_{16}zdz,\ B_{26} = \int Q_{26}zdz \tag{6.33}$$

In place of the longhand derivation from Equations 6.28 to 6.33, we have from Equations 6.7, 6.2 and 6.27:

$$N_i = \int Q_{ij}(\epsilon_j^o + zk_j)dz \tag{6.34}$$

$$= \int Q_{ij}\epsilon_j^o\,dz + \int Q_{ij}k_j z\,dz \tag{6.35}$$

Since ϵ^o and k are independent of z, they can be taken out of the integral signs,

$$N_i = \left[\int Q_{ij}dz\right]\epsilon_j^o + \left[\int Q_{ij}z\,dz\right]k_j \tag{6.36}$$

$$= A_{ij}\epsilon_j^o + B_{ij}k_j \tag{6.37}$$

where

$$B_{ij} = \int Q_{ij}z\,dz \tag{6.38}$$

This is the coupling modulus, which links curvature to stress resultant. In symmetric laminates, we have by definition:

$$Q_{ij}(z) = Q_{ij}(-z) \tag{6.39}$$

This can be seen in Figures 5.1 and 5.2. The ply orientation is symmetric with respect to the midplane. The modulus is an even function in z. We will first split the integration in Equation 6.39 in two parts:

$$B_{ij} = \int_{-h/2}^{0} Q_{ij}z\,dz + \int_{0}^{h/2} Q_{ij}z\,dz \tag{6.40}$$

$$= -\int_{0}^{-h/2} Q_{ij}z\,dz + \int_{0}^{h/2} Q_{ij}z\,dz \tag{6.41}$$

By virtue of the symmetry conditions of the modulus in Equation 6.39, we can change z to $-z$ in the first integral in Equation 6.41, then

$$B_{ij} = -\int_0^{h/2} Q_{ij} z dz + \int_0^{h/2} Q_{ij} z dz = 0 \qquad (6.42)$$

We can also show a coupling between in-plane strain to moment as follows:

From Equation 6.11

$$M_i = \int \sigma_i z dz \qquad (6.43)$$

Substituting Equations 6.2 and 6.27

$$M_i = \int Q_{ij} [\epsilon_j^o + z k_j] z dz \qquad (6.44)$$

$$= \left[\int Q_{ij} z dz \right] \epsilon_j^o + \left[\int Q_{ij} z^2 dz \right] k_i \qquad (6.45)$$

$$= B_{ij} \epsilon_j^o + D_{ij} k_j \qquad (6.46)$$

Note the reappearance of the same coupling matrix here as that in Equation 6.37. We can now combine Equations 6.37 and 6.46.

$$\begin{aligned} N_i &= A_{ij} \epsilon_j^o + B_{ij} k_j \\ M_i &= B_{ij} \epsilon_j^o + D_{ij} k_j \end{aligned} \qquad (6.47)$$

These six equations represent the stress-strain relation in terms of modulus of a general laminate. The modulus is a 6 X 6 matrix. These equations when expanded into longhand expressions are:

$$
\begin{aligned}
N_1 &= A_{11}\epsilon_1^o + A_{12}\epsilon_2^o + A_{16}\epsilon_6^o + B_{11}k_1 + B_{12}k_2 + B_{16}k_6 \\
N_2 &= A_{21}\epsilon_1^o + A_{22}\epsilon_2^o + A_{26}\epsilon_6^o + B_{21}k_1 + B_{22}k_2 + B_{26}k_6 \\
N_6 &= A_{61}\epsilon_1^o + A_{62}\epsilon_2^o + A_{66}\epsilon_6^o + B_{61}k_1 + B_{62}k_2 + B_{66}k_6 \\
M_1 &= B_{11}\epsilon_1^o + B_{12}\epsilon_2^o + B_{16}\epsilon_6^o + D_{11}k_1 + D_{12}k_2 + D_{16}k_6 \\
M_2 &= B_{21}\epsilon_1^o + B_{22}\epsilon_2^o + B_{26}\epsilon_6^o + D_{21}k_1 + D_{22}k_2 + D_{26}k_6 \\
M_6 &= B_{61}\epsilon_1^o + B_{62}\epsilon_2^o + B_{66}\epsilon_6^o + D_{61}k_1 + D_{62}k_2 + D_{66}k_6
\end{aligned}
\tag{6.48}
$$

Or, in matrix multiplication table we have in Table 6.1 the stiffness and its inverse, the compliance, of a general laminate.

table 6.1
stiffness and compliance of a general laminate

	ϵ_1^o	ϵ_2^o	ϵ_6^o	k_1	k_2	k_6
N_1	A_{11}	A_{12}	A_{16}	B_{11}	B_{12}	B_{16}
N_2	A_{21}	A_{22}	A_{26}	B_{21}	B_{22}	B_{26}
N_6	A_{61}	A_{62}	A_{66}	B_{61}	B_{62}	B_{66}
M_1	B_{11}	B_{12}	B_{16}	D_{11}	D_{12}	D_{16}
M_2	B_{21}	B_{22}	B_{26}	D_{21}	D_{22}	D_{26}
M_6	B_{61}	B_{62}	B_{66}	D_{61}	D_{62}	D_{66}

	N_1	N_2	N_6	M_1	M_2	M_6
ϵ_1^o	α_{11}	α_{12}	α_{16}	β_{11}	β_{12}	β_{16}
ϵ_2^o	α_{21}	α_{22}	α_{26}	β_{21}	β_{22}	β_{26}
ϵ_6^o	α_{61}	α_{62}	α_{66}	β_{61}	β_{62}	β_{66}
k_1	β_{11}	β_{21}	β_{61}	δ_{11}	δ_{12}	δ_{16}
k_2	β_{12}	β_{22}	β_{62}	δ_{21}	δ_{22}	δ_{26}
k_6	β_{16}	β_{26}	β_{66}	δ_{61}	δ_{62}	δ_{66}

Both 6 X 6 matrices are symmetric. This requires that one coupling compliance matrix is the transpose of the other.

As a comparison we show in Tables 6.2(*a*) and (*b*) the modulus and compliance of symmetric and homogeneous anisotropic laminates, respectively. In the homogeneous laminate, the flexural components are directly related to the in-plane components.

able 6.2(a)

tiffness and compliance of a symmetric anisotropic laminate

	ϵ_1^0	ϵ_2^0	ϵ_6^0	k_1	k_2	k_6
N_1	A_{11}	A_{12}	A_{16}			
N_2	A_{21}	A_{22}	A_{26}			
N_6	A_{61}	A_{62}	A_{66}			
M_1				D_{11}	D_{12}	D_{16}
M_2				D_{21}	D_{22}	D_{26}
M_6				D_{61}	D_{62}	D_{66}

	N_1	N_2	N_6	M_1	M_2	M_6
ϵ_1^0	a_{11}	a_{12}	a_{16}			
ϵ_2^0	a_{21}	a_{22}	a_{26}			
ϵ_6^0	a_{61}	a_{62}	a_{66}			
k_1				d_{11}	d_{12}	d_{16}
k_2				d_{21}	d_{22}	d_{26}
k_6				d_{61}	d_{62}	d_{66}

able 6.2(b)

tiffness and compliance of a homogeneous anisotropic laminate

	ϵ_1^0	ϵ_2^0	ϵ_6^0	k_1	k_2	k_6
N_1	A_{11}	A_{12}	A_{16}			
N_2	A_{21}	A_{22}	A_{26}			
N_6	A_{61}	A_{62}	A_{66}			
M_1						
M_2					$\frac{h^2}{12} A_{ij}$	
M_6						

	N_1	N_2	N_6	M_1	M_2	M_6
ϵ_1^0	a_{11}	a_{12}	a_{16}			
ϵ_2^0	a_{21}	a_{22}	a_{26}			
ϵ_6^0	a_{61}	a_{62}	a_{66}			
k_1						
k_2					$\frac{12}{h^2} a_{ij}$	
k_6						

Or, in matrix notation,

$$N = A\, \epsilon^o + B\, k \tag{6.49}$$

$$M = B\, \epsilon^o + D\, k \tag{6.50}$$

or in terms of a matrix multiplication table we have Table 6.3.

table 6.3

generalized stress-strain relations in terms of stiffness modulus

	ϵ^o	k
N	A	B
M	B	D

	ϵ^o	k
N	$\frac{N}{m}$	N
M	N	Nm

The matrix quantities above are column or row, and square matrices a follows, in curly, and square brackets, respectively:

$$N = N_i = \begin{Bmatrix} N_1 \\ N_2 \\ N_6 \end{Bmatrix}, \quad M = M_i = \begin{Bmatrix} M_1 \\ M_2 \\ M_6 \end{Bmatrix} \tag{6.51}$$

$$\epsilon^o = \epsilon_i^o = \begin{Bmatrix} \epsilon_1^o \\ \epsilon_2^o \\ \epsilon_6^o \end{Bmatrix}, \quad k = k_i = \begin{Bmatrix} k_1 \\ k_2 \\ k_6 \end{Bmatrix} \tag{6.52}$$

$$A = A_{ij} = \begin{bmatrix} A_{11} & A_{12} & A_{16} \\ A_{21} & A_{22} & A_{26} \\ A_{61} & A_{62} & A_{66} \end{bmatrix} \tag{6.53}$$

$$B = B_{ij} = \begin{bmatrix} B_{11} & B_{12} & B_{16} \\ B_{21} & B_{22} & B_{26} \\ B_{61} & B_{62} & B_{66} \end{bmatrix} \tag{6.54}$$

$$D = D_{ij} = \begin{bmatrix} D_{11} & D_{12} & D_{16} \\ D_{21} & D_{22} & D_{26} \\ D_{61} & D_{62} & D_{66} \end{bmatrix} \tag{6.55}$$

where $i,j = 1,2,6$.

We need also to define the inverse of matrices. We have

$$\boldsymbol{a} = \boldsymbol{A}^{-1}\text{, and } \boldsymbol{d} = \boldsymbol{D}^{-1} \tag{6.56}$$

These inverse relationships are implied in Tables 4.1 and 4.2 for the in-plane behavior; and in Tables 5.1 and 5.2 for the flexural behavior. We now will find the generalized stress-strain relation in terms of compliance, which is the inverse of the 6 X 6 modulus matrix shown in Table 6.1.

Premultiplying Equation 6.49 by the inverse of $\boldsymbol{A}$ or simply $\boldsymbol{a}$,

$$\boldsymbol{aN} = \boldsymbol{aA}\epsilon^o + \boldsymbol{aBk} \tag{6.57}$$

Since

$$\boldsymbol{aA} = \mathbf{1}\text{, where } \mathbf{1} \text{ is unity matrix} \tag{6.58}$$

We have

$$\epsilon^o = \boldsymbol{aN} - \boldsymbol{aBk} \tag{6.59}$$

Substitute this into Equation 6.50, we have

$$\boldsymbol{M} = \boldsymbol{BaN} - \boldsymbol{BaBk} + \boldsymbol{Dk} \tag{6.60}$$

$$= \boldsymbol{BaN} + (\boldsymbol{D} - \boldsymbol{BaB})\boldsymbol{k} \tag{6.61}$$

We can show the last two equations in a tabular form in Table 6.4. They are useful for structures with fixed cylindrical cross sections such as tubes and pressure vessels.

table 6.4
generalized stress-strain relation in terms of *N* and *k*

	N	k
ϵ^o	a	$-aB$
M	Ba	$D-BaB$

The relationship in Table 6.4 is a partial inversion of that in Table 6.3. We now will derive the complete inversion.

Continuing with our matrix algebra, we can premultiply Equation 6.61 by $(D\text{-}BaB)^{-1}$, then after transposing we have

$$k = -(D - BaB)^{-1} BaN + (D - BaB)^{-1} M \tag{6.62}$$

Substituting this into Equation 6.59, we have

$$\epsilon^o = aN - aBk$$

$$= [a + aB(D - BaB)^{-1} Ba] N - aB(D - BaB)^{-1} M \tag{6.63}$$

$$= \alpha N - \beta M \tag{6.64}$$

We can now show the last two equations in Table 6.5(*a*) which is now the complete inversion of Table 6.1 or 6.3; i.e., the independent variables are N and M. The material coefficients in this table are the components of compliance.

table 6.5(*a*)
generalized stress-strain relation in terms of compliance

	N	M
ϵ^o	$a + aB(D-BaB)^{-1}Ba$	$-aB(D-BaB)^{-1}$
k	$-(D-BaB)^{-1}Ba$	$(D-BaB)^{-1}$

The two coupling matrices are transposed of each other. They need not be symmetric.

t is useful to designate the compliance in Table 6.5(*a*) in Greek letters n the same matrix multiplication format.

table 6.5(*b*)
generalized stress-strain relation in terms of compliance in greek

	N	M
ϵ^o	α	β
k	β^T	δ

	N	M
ϵ^o	$\frac{m}{N}$	$\frac{1}{N}$
k	$\frac{1}{N}$	$\frac{1}{Nm}$

unit

'his table is the inverse of Table 6.3. This can be expanded like Equaion 6.48 for the modulus. This is shown as follows:

$$
\begin{aligned}
\epsilon_1^o &= \alpha_{11}N_1 + \alpha_{12}N_2 + \alpha_{16}N_6 + \beta_{11}M_1 + \beta_{12}M_2 + \beta_{16}M_6 \\
\epsilon_2^o &= \alpha_{21}N_1 + \alpha_{22}N_2 + \alpha_{26}N_6 + \beta_{21}M_1 + \beta_{22}M_2 + \beta_{26}M_6 \\
\epsilon_6^o &= \alpha_{61}N_1 + \alpha_{62}N_2 + \alpha_{66}N_6 + \beta_{61}M_1 + \beta_{62}M_2 + \beta_{66}M_6 \\
k_1 &= \beta_{11}N_1 + \beta_{21}N_2 + \beta_{61}N_6 + \delta_{11}M_1 + \delta_{12}M_2 + \delta_{16}M_6 \\
k_2 &= \beta_{12}N_1 + \beta_{22}N_2 + \beta_{62}N_6 + \delta_{21}M_1 + \delta_{22}M_2 + \delta_{26}M_6 \\
k_3 &= \beta_{16}N_1 + \beta_{26}N_2 + \beta_{66}N_6 + \delta_{61}M_1 + \delta_{62}M_2 + \delta_{66}M_6
\end{aligned}
\tag{6.65}
$$

'his equation is also shown in Table 6.1 together with its inverse, the nodulus. Note that in the absence of coupling,

$$B = 0$$

$$\beta = 0$$

he equations in Tables 6.5(*a*) and (*b*) reduce to those of the uncoupled r symmetric laminates of previous chapters, and Table 6.2(*a*).

$$\epsilon^o = aN$$

$$\alpha = a$$

$$k = dM \tag{6.66}$$

$$\delta = d$$

These relations are valid only for symmetric laminates. They are no valid for general laminates, i.e.,

$$\alpha \neq a$$

$$\delta \neq d \tag{6.67}$$

3. evaluation of components of modulus

We will now evaluate all the components of modulus listed in Equation 6.48 and Table 6.1. These components of modulus are evaluated by performing the following integrations:

$$[A_{ij}, B_{ij}\ D_{ij}] = \int_{-h/2}^{h/2} Q_{ij}[1,z,z^2]\,dz \tag{6.68}$$

This equation is written with the implied convention that each term in the bracketed quantity on the left-hand side of the equation has corresponding term in the bracket on the right-hand side of the equation. For general laminates, the limits of integration is from $-h/2$ to $h/2$. This is different from the limits for the symmetric laminates from the midplane ($z = 0$) to the top face of the plate where $z = h/2$. If each ply consists of homogeneous materials, with even number of plies we can replace the integration by summation as follows:

$$A_{ij} = h_o \sum_{t=1-n/2}^{n/2} Q_{ij}^{(t)}\ [t-(t-1)] \tag{6.69}$$

(continues)

$$B_{ij} = \frac{h_o^2}{2} \sum_{t=1-n/2}^{n/2} Q_{ij}^{(t)} \left[t^2 - (t-1)^2 \right]$$

$$D_{ij} = \frac{h_o^3}{3} \sum_{t=1-n/2}^{n/2} Q_{ij}^{(t)} \left[t^3 - (t-1)^3 \right] \qquad (6.69) \text{ (concluded)}$$

he modulus of general laminates is proportional to the modulus of ach ply multiplied by the weighting factor that appears in the bracket f the equation above. A similar weighting factor was used for the valuation of flexural modulus in Equation 5.49. The terms used in quation 6.69 are defined in Figure 6.2. Index t is the ordinal number or individual plies as they go from $1-n/2$ to $n/2$, where n is even and qual to the total number of plies. It is further assumed that all plies ave the same thickness. With the assumptions above, we can show the umerical values of the weighting factors in Equation 6.69 for a general aminate up to 16 plies thick. This is shown in Table 6.6.

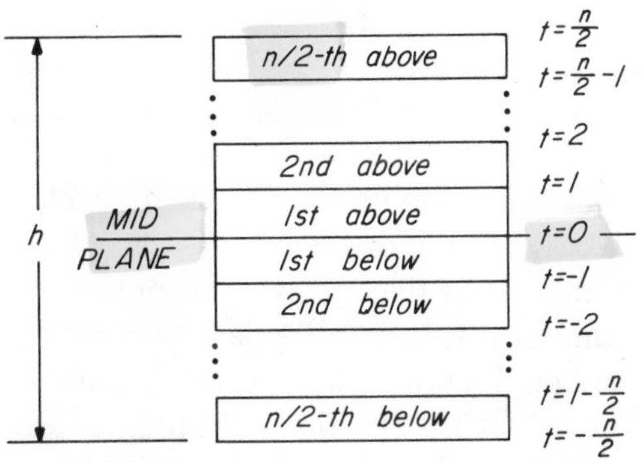

Figure 6.2 Nomenclatures for general laminates with even plies. Summation in Equation 6.69 is indexed from top surface of each ply; i.e., from $t = 1-n/2$ to $t = n/2$.

table 6.6
weighting factors for general laminates

	Ply Order	t	$t-(t-1)$ (A_{ij})	$t^2-(t-1)^2$ (B_{ij})	$t^3-(t-1)^3$ (D_{ij})	
	8th above	8	1	15	169	
	7th above	7	1	13	127	
	6th above	6	1	11	91	
	5th above	5	1	9	61	
	4th above	4	1	7	37	
	3rd above	3	1	5	19	
	2nd above	2	1	3	7	
mid-	1st above	1	1	1	1	mid
plane	1st below	0	1	−1	1	plan[e]
	2nd below	−1	1	−3	7	
	3rd below	−2	1	−5	19	
	4th below	−3	1	−7	37	
	5th below	−4	1	−9	61	
	6th below	−5	1	−11	91	
	7th below	−6	1	−13	127	
	8th below	−7	1	−15	169	

Note that the numerical values for the in-plane modulus is unity; th[e] in-plane modulus is independent of the stacking sequence. This was tru[e] for the symmetric laminates; this is also true for general laminates. Th[e] in-plane modulus is a function of the volume fraction of the constituer[nt] plies. The positions of the plies do not affect the in-plane modulus.

This, of course, is not true for the coupling and flexural moduli. Th[e] weighting factors increase as they go away from the mid-plane. This ca[n] be seen by the numerical values in the last two columns of Table 6.6[.] The values for the coupling modulus are antisymmetric with respect t[o] the mid-plane. The values for the flexural modulus are symmetric wit[h] respect to the mid-plane. These weighting factors as functions of th[e] ply ordinal number t are plotted in Figure 6.3.

The evaluation of the modulus of general laminates can also b[e] achieved by using either the direct summation in Equation 6.69, or th[e] geometric factors, the V's, as was done for the in-plane and flexur[al]

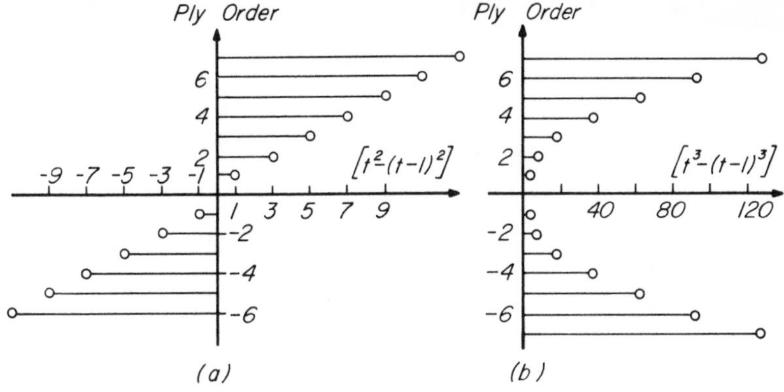

Figure 6.3 Weighting factors for coupling and flexural moduli. They are shown in (*a*) and (*b*) respectively. The factors are antisymmetric for coupling modulus; symmetric, for flexural modulus.

odulus earlier. The advantages of using the V's are:

- Close relationship between the formulas for the modulus and the multiple-angle transformation equations.
- The geometric parameters of a laminate such as the stacking sequence and unit ply thickness are embodies in the V's. If a different ply material is used or the same ply material changes properties (due to temperature, for example) the V's are not affected.
- Four of the V's are also used for the calculation of the non-mechanical stress resultants and moments due to temperature change and moisture absorption. The V's again reflect the geometric parameters of a laminate.

he limitation of the use of V's is that the material of the ply must main the same. This, of course, would not be the case of hybrid mposites. In place of the summation in Equation 6.69, we can use e V's in the following equations for the evaluation of the modulus of neral laminates. Taking a typical component of this stiffness modulus r symmetrically located core:

$$[A_{11},B_{11},D_{11}] = \int_{-h/2}^{h/2} Q_{11}[1,z^1,z^2]\,dz \qquad (6.7$$

$$= \int_{-h/2}^{h/2} [U_1 + U_2\cos 2\theta + U_3\cos 4\theta][1,z,z^2]\,dz \qquad (6.7$$

$$= U_1\int_{-h/2}^{h/2} [1,z,z^2]\,dz + U_2\int_{-h/2}^{h/2} \cos 2\theta[1,z,z^2]\,dz$$

$$+ U_3\int_{-h/2}^{h/2} \cos 4\theta[1,z,z^2]\,dz \qquad (6.7$$

$$= U_1\left[h,0,\frac{h^3}{12}\right] + U_2[V_{1A},V_{1B},V_{1D}] + U_3[V_{2A},V_{2B},V_{2D}] \qquad (6.7$$

Similarly we can obtain the other components of the modulus follows:

$$[A_{22},B_{22},D_{22}] = U_1\left[h,0,\frac{h^3}{12}\right] - U_2[V_{1A},V_{1B},V_{1D}] + U_3[V_{2A},V_{2B},V_2 \qquad (6.74$$

$$[A_{12},B_{12},D_{12}] = U_4\left[h,0,\frac{h^3}{12}\right] - U_3[V_{2A},V_{2B},V_{2D}] \qquad (6.75$$

$$[A_{66},B_{66},D_{66}] = U_5\left[h,0,\frac{h^3}{12}\right] - U_3[V_{2A},V_{2B},V_{2D}] \qquad (6.76$$

$$[A_{16},B_{16},D_{16}] = \frac{1}{2}U_2[V_{3A},V_{3B},V_{3D}] + U_3[V_{4A},V_{4B},V_{4D}] \qquad (6.7$$

$$[A_{26},B_{26},D_{26}] = \frac{1}{2}U_2[V_{3A},V_{3B},V_{3D}] - U_3[V_{4A},V_{4B},V_{4D}] \qquad (6.78$$

here the geometric factors are:

$$ {}_{1A} = \int_{-h/2}^{h/2} \cos 2\theta \, dz, \quad V_{1B} = \int_{-h/2}^{h/2} \cos 2\theta \, z dz, \quad V_{1D} = \int_{-h/2}^{h/2} \cos 2\theta \, z^2 dz \tag{6.79} $$

$$ {}_{2A} = \int_{-h/2}^{h/2} \cos 4\theta \, dz, \quad V_{2B} = \int_{-h/2}^{h/2} \cos 4\theta \, z dz, \quad V_{2D} = \int_{-h/2}^{h/2} \cos 4\theta \, z^2 dz \tag{6.80} $$

$$ {}_{3A} = \int_{-h/2}^{h/2} \sin 2\theta \, dz, \quad V_{3B} = \int_{-h/2}^{h/2} \sin 2\theta \, z dz, \quad V_{3D} = \int_{-h/2}^{h/2} \sin 2\theta \, z^2 dz \tag{6.81} $$

$$ {}_{4A} = \int_{-h/2}^{h/2} \sin 4\theta \, dz, \quad V_{4B} = \int_{-h/2}^{h/2} \sin 4\theta \, z dz, \quad V_{4D} = \int_{-h/2}^{h/2} \sin 4\theta \, z^2 dz \tag{6.82} $$

e formulas for the in-plane and flexural modulus are exactly the me as those in Tables 4.3 and 5.3. We only need to change the limits the integrals in Equation 5.40. The location of the core may not be mmetric. The simple core/thickness correction such as h^* does not ways exist. We present the formulas for the coupling modulus in uation 6.73 et al. in matrix multiplication in Table 6.7. The lack of variants is a key feature. All components of this coupling modulus n change sign.

table 6.7

formulas for coupling modulus of general laminates with symmetric core

	U_2	U_3
B_{11}	V_{1B}	V_{2B}
B_{22}	$-V_{1B}$	V_{2B}
B_{12}		$-V_{2B}$
B_{66}		$-V_{2B}$
B_{16}	$\frac{1}{2}V_{3B}$	V_{4B}
B_{26}	$\frac{1}{2}V_{3B}$	$-V_{4B}$

Only the volume fraction of ply orientations, not the lack of mi plane symmetry, affect the in-plane modulus of a general laminate. Th graphic solutions in Figure 4.11 and the other formulas in Chapter remain valid. The in-plane compliance, however, is sensitive to th stacking sequence of the laminate. For a general laminate the in-plar compliance is given in Table 6.5b and repeated here

$$\alpha = a + aB(D - BaB)^{-1}\, Ba \tag{6.8}$$

Note the presence of the coupling matrix. Only in the case of a sym metric laminate

$$\alpha = a \tag{6.84}$$

Again we can show the integration of the V's from Equation 6.79 t 6.82 can be replaced by summation if each ply consists of hom geneous material. This is done for a typical value of V as follows:

$$V_{1D} = \frac{1}{3} \sum_{i=1}^{m} \cos 2\theta_i [z_i^3 - z_{i-1}^3] \tag{6.8}$$

Again the quantity in the bracket can be expressed in terms of th ordinal number of the plies and this is done as before and the result

$$V_{1D} = \frac{h_o^3}{3} \sum_{t=1-n/2}^{n/2} \cos 2\theta_t [t^3 - (t-1)^3] \tag{6.86}$$

where index t is defined in Figure 6.2, and the value in the bracket given in Table 6.6.

The V's for the coupling modulus can also be replaced first by summation then by the ordinal numbers of the plies. This is done : follows:

$$V_{1B} = \int_{-h/2}^{h/2} \cos 2\theta z dz \tag{6.87}$$

$$= \frac{1}{2} \sum_{i=1}^{m} \cos 2\theta_i [z_i^2 - z_{i-1}^2] \tag{6.88}$$

$$= \frac{h_o^2}{2} \sum_{t=1-n/2}^{n/2} \cos 2\theta_t [t^2 - (t-1)^2] \tag{6.89}$$

Note that the weighting factors in the equations above are shown in Table 6.6 and Figure 6.3.

. unsymmetric cross-ply laminates

We will calculate the stiffness and compliance of a class of cross-ply laminates made of T300/5208. We will first examine a 16-ply laminate. The minimum number of ply groups is two, the laminate will have eight 0-degree plies in the lower half of the laminate and eight 90-degree plies in the upper half; see Figure 6.4(*a*). We can increase the ply groups to four which will have four plies at zero followed by four plies at 90 in the lower half of the laminate, then four plies of zero and four plies of 90 in the upper half; see Figure 6.4(*b*). As the number of ply groups increase, we will examine the effect of this on the modulus and compliance of the cross-ply laminate. In the limit as the number of ply groups increase we should recover the quasi-homogeneous laminate.

The modulus of cross-ply laminates can be calculated using Equation 6.69 and the numerical values in Table 6.6. There is no need to calculate the in-plane modulus of laminates because the same rule of mixtures equation for the symmetric laminates is applicable to the general laminates. We would therefore concentrate on the coupling modulus and the flexural modulus. There are two ways of calculating the coupling and flexural moduli from Table 6.6. The first method is to follow Equation 6.69 precisely, and the laminate shown in Figure 6.4(*a*) is

$$[0_8/90_8]_T \tag{6.90}$$

$$B_{11} = \frac{h_o^2}{2}[Q_{11}(-15-13-11-9-7-5-3-1) + Q_{22}(15+13+11+9+7+5+3+1)]$$

$$= 64\frac{h_o^2}{2}[-Q_{xx} + Q_{yy}] \tag{6.91}$$

Let

$$h_o = 125 \times 10^{-6}\,\text{m}$$

$$8h_o = 10^{-3}\,\text{m}$$

$$Q_{xx} = 181.8\ \text{GPa}$$

$$Q_{yy} = 10.3\ \text{GPa} \tag{6.92}$$

$$B_{11} = \frac{10^{-3}}{2}(-181.8 + 10.3) = -85.7\ \text{kN} \tag{6.93}$$

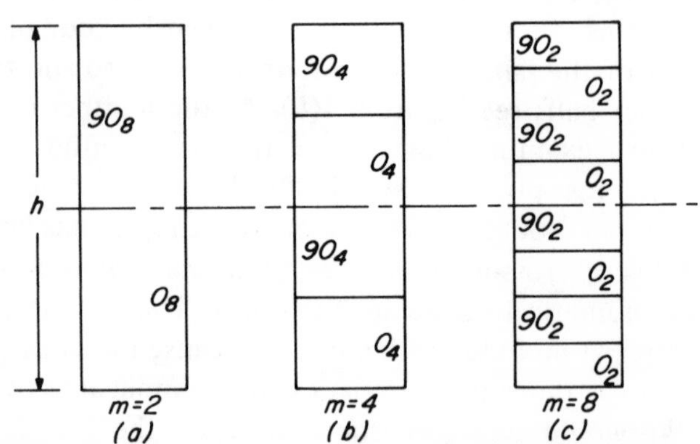

Figure 6.4 Unsymmetric cross-ply laminates with different ply groupings. This particular change in grouping is done by subdividing each ply group into two sub-groups.

There is another way of using Table 6.6 for the same cross-ply minate. This is done by treating each ply group as a new unit ply. hen the effective unit ply thickness in Equation 6.69 in this case is creased eight fold. The weighting factor now will be −1 for the degree ply and +1 for the 90-degree ply, as compared to −64 and 64 respectively. With this method, we can arrive at Equation 6.91 irectly.

If the ply group is doubled, the laminate now shown in Figure .4(*b*) is

$$[0_4/90_4/0_4/90_4]_T \text{ or } [0_4/90_4]_{2T} \tag{6.94}$$

sing the first method:

$$B_{11} = \frac{h_o^2}{2}[Q_{xx}(-15-13-11-9)+Q_{yy}(-7-5-3-1)$$

$$+Q_{xx}(1+3+5+7)+Q_{yy}(9+11+13+15)]$$

$$= \frac{32h_o^2}{2}[-Q_{xx}+Q_{yy}] \tag{6.95}$$

$$= -42.8 \text{ kN} \tag{6.96}$$

his is one half of the value in Equation 6.93. Using the second ethod:

$$\text{The equivalent unit ply thickness} = 4h_o \tag{6.97}$$

$$B_{11} = \frac{(4h_o)^2}{2}[-3Q_{xx} - Q_{yy} + Q_{xx} + 3Q_{yy}]$$

$$= \frac{32h_o^2}{2}[-Q_{xx}+Q_{yy}] \tag{6.98}$$

his is the same as Equation 6.95. There is a fundamental difference etween the coupling modulus and the in-plane and flexural moduli in a general laminate. The principal components of the coupling modul can be negative. This does not violate materials stability because a components of the coupling modulus are off-diagonal terms in th 6 X 6 matrix in Table 6.1.

Since the Poisson and shear components are equal, i.e.,

$$Q_{xy}^{(0)} = Q_{xy}^{(90)}$$
$$Q_{ss}^{(0)} = Q_{ss}^{(90)} \qquad (6.9$$

Then

$$B_{12} = B_{66} = 0 \qquad (6.10$$

for all cross-ply laminates. Similarly, since shear coupling is zero f on-axis plies,

$$B_{16} = B_{26} = 0 \qquad (6.10$$

For flexural modulus, it is easier to use the second method (th equivalent ply thickness method). From Equations 6.69 and 6.90

$$D_{11} = \frac{(8h_o)^3}{3}\,[Q_{xx} + Q_{yy}] = \frac{1}{3}(181.8 + 10.3) \qquad (6.10$$

$$= 64.0 \text{ Nm} \qquad (6.10$$

From Equation 6.94 for the 4-ply group laminate:

$$D_{11} = \frac{(4h_o)^3}{3}\,[7Q_{xx} + Q_{yy} + Q_{xx} + 7Q_{yy}] \qquad (6.10$$

$$= 64.0 \text{ Nm} \qquad (6.10$$

Note that flexural modulus is not affected by the number of pl groups. This is because the weighting factor is symmetric (see Table 6. and Figure 6.3) and the stiffness of unidirectional plies is also sym metric with respect to the material axes. The resulting stiffness c ısymmetric cross-ply laminates is shown in Figure 6.5. The ply group ımber m has effect on the coupling modulus $\boldsymbol{B}$ only, and m does not ›pear at all in $\boldsymbol{A}$ and $\boldsymbol{D}$.

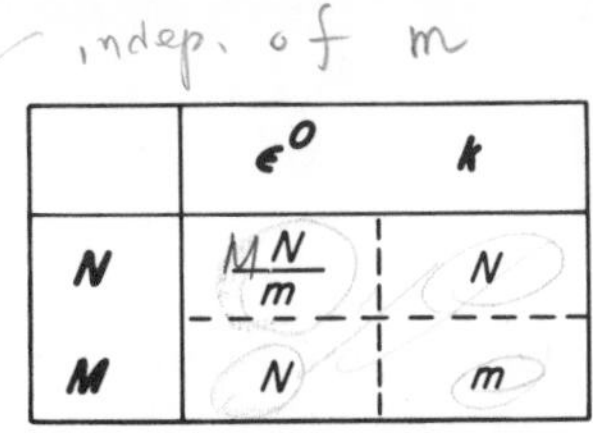

	N	M
ϵ^0	$\frac{m}{GN}$	$\frac{1}{MN}$
k	$\frac{1}{MN}$	$\frac{1}{KNm}$

	ϵ_1^0	ϵ_2^0	ϵ_6^0	k_1	k_2	k_6
N_1	192	5.7		$-\frac{171}{m}$		
N_2	5.7	192			$\frac{171}{m}$	
N_6			14.3			
M_1	$-\frac{171}{m}$			64	1.9	
M_2		$\frac{171}{m}$		1.9	64	
M_6						4.7

	N_1	N_2	N_6	M_1	M_2	M_6
ϵ_1^0	12.9	−0.39		17.3		
ϵ_2^0	−0.39	12.9			−17.3	
ϵ_6^0			69.7			
k_1	17.3			38.8	−1.1	
k_2		−17.3		−1.1	38.8	
k_6						209

igure 6.5 Stiffness and compliance of T300/5208 unsymmetrical 16-ply cross-ply ıminates. The stiffness is for all m values although m affects only the coupling iffness. The compliance is for the case of $m = 2$; see Figure 6.4(a). The physical imensions are also shown.

We can compare the $\boldsymbol{A}$ matrix in Figure 6.5 with the in-plane :iffness of the laminate $[0/90]_S$ in Equation 4.51, we see they are lentical. There is no influence on this stiffness by the number of ply roup, or with or without the midplane symmetry.

We can also see that the $\boldsymbol{D}$ in Figure 6.5 is equal to the flexural ıodulus of a quasi-homogeneous [0/90] laminate shown in Equations .69 and 5.74. They are identical because for this particular laminate he weighting factor in Table 6.6 for $\boldsymbol{D}$ is symmetric with respect to the ıidplane. The 0 and 90 degrees are so located in Figure 6.4, their ontribution to the flexural modulus is not sensitive to their position, e., $[0/90]_T$ or $[90/0]_T$, or to the number of ply groups if they change ccording to the pattern in Figure 6.4. If we have a laminate such as

$[0_4/90_8/0_4]_T$ the flexural modulus will be highly sensitive to the stacking sequence, although the in-plane modulus will remain constant.

Since the flexural modulus is equal to that of a quasi-homogeneous laminate, we see that Equation 5.109 is valid

$$D_{ij} = \frac{h^2}{12} A_{ij} \tag{6.106}$$

where $h = 2 \times 10^{-3}$ m for our unsymmetric laminate in Figure 6.5. This equation is the necessary but not sufficient condition for a quasi-homogeneous laminate. The other condition is for $\boldsymbol{B} = 0$; i.e. no in-plane and flexural coupling. We may call this general laminate pseudo-homogeneous.

Being an unsymmetric laminate, the compliance in Figure 6.5 will be affected by the presence of the $\boldsymbol{B}$ matrix. The inequalities in Equation 6.67 are valid. We can see this if we compare the following:

$$\alpha_{11} = 12.9\ (\text{GN/m})^{-1}$$

$$a_{11} = 5.205\ (\text{GN/m})^{-1}$$

The latter is taken from Equation 4.51. Similarly,

$$\delta_{11} = 38.8\ (\text{kNm})^{-1}$$

$$d_{11} = 15.62\ (\text{kNm})^{-1}$$

The latter is taken from Equation 5.74. It is interesting that the compliance in Figure 6.5 is also pseudo-homogeneous because

$$\delta_{ij} = \frac{12}{h^2} \alpha_{ij} \tag{6.107}$$

and $\boldsymbol{\beta}$ is not zero. Equation 6.107 is analogous to Equation 5.110 for truly or quasi-homogeneous laminates.

Typical results of the selected components of stiffness and compliance of our unsymmetric cross-ply laminates as ply groups change are shown in Table 6.8.

table 6.8

selected stiffness and compliance components T300/5208 unsymmetrical cross-ply laminates

m	$B_{11} = -B_{22}$	$\alpha_{11} = \alpha_{22}$	$\beta_{11} = -\beta_{22}$	$\delta_{11} = \delta_{22}$
	kN	$(\text{GN/m})^{-1}$	$(\text{MN})^{-1}$	$(\text{kNm})^{-1}$
2	−85.7	12.9	17.3	38.8
4	−42.8	6.12	4.09	18.3
8	−21.4	5.41	1.81	16.2
∞	0	5.20	0	15.6

With these constants, it is possible to show the strain distribution in a minate subjected to uniaxial stress resultant or bending moment.

- For $N_1 \neq 0$

  From Equations 6.26 and 6.65

$$\begin{aligned} \epsilon_1 &= \epsilon_1^o + zk_1 \\ &= [\alpha_{11} + z\beta_{11}]N_1 \end{aligned} \tag{6.108}$$

$$\begin{aligned} z_{max} &= h/2 = 8h_o = 8 \times 125 \times 10^{-6}\,\text{m} \\ &= 10^{-3}\,\text{m} \end{aligned} \tag{6.109}$$

For $m = 2$

$$\epsilon_1/N_1 = (12.9 \pm 17.3)10^{-9} \tag{6.110}$$

$$= 30.2, -4.4\ (\text{GN/m})^{-1} \tag{6.111}$$

For $m = \infty$

$$\epsilon_1/N_1 = 5.20\ (\text{GN/m})^{-1} \tag{6.112}$$

- For $M_1 \neq 0$

$$\epsilon_1 = [\beta_{11} + z\delta_{11}]M_1 \tag{6.113}$$

For $m = 2$

$$\epsilon_1/M_1 = (17.3 \pm 38.8)10^{-6} \tag{6.114}$$

$$= 56.1, -21.5\ (\mathrm{MN})^{-1} \tag{6.115}$$

For $m = \infty$

$$\epsilon_1/M_1 = \pm 15.6\ (\mathrm{MN})^{-1} \tag{6.116}$$

The strain distribution due to simple tension or bending is shown in Figure 6.6. Under tensile load, unsymmetric cross-ply laminate will warp. The upper half will stretch more than the lower half. The 90-degree plies are in the upper half for the $m = 2$ case. The particular coupling in this laminate induces greater strains which may be undesirable. With the stiffness and compliance of the laminate known, a systematic study on the effect of various lamination parameters becomes possible. With the exception of the 6 X 6 matrix inversion general laminates are as simple as symmetric laminates.

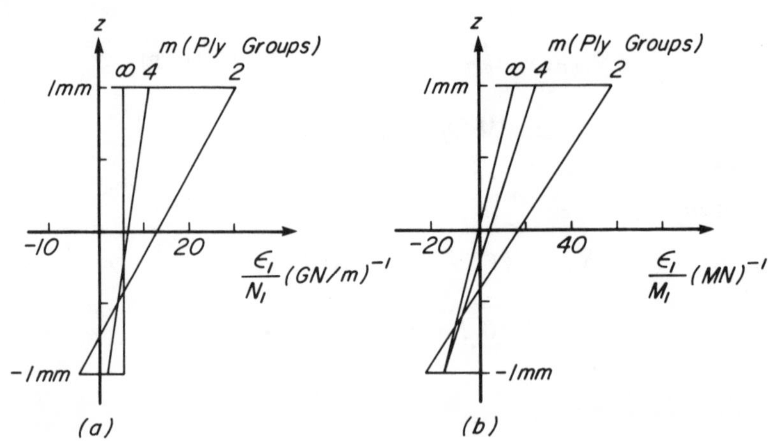

Figure 6.6 Strain distribution of unsymmetric cross-ply laminates. As number of ply groups varies from 2 to infinity, the strain changes. Figure on the left is uniaxial tension; on the right, simple bending. The material is T300/5208.

In conclusion, unsymmetric cross-ply laminates are different from ·mmetric laminates because of two nonzero components of the ›upling moduli. Their values, as shown in Equation 6.93, 6.96 et al., ıd in Figure 6.5 are:

$$B_{11} = -B_{22} = -[Q_{xx} - Q_{yy}]\frac{h^2}{m} \tag{6.117}$$

$$= -\frac{2h^2}{m}U_2 \tag{6.118}$$

here

h = total laminate thickness

m = number of ply groups

$U_2 = \frac{1}{2}[Q_{xx} - Q_{yy}]$

hus the coupling components are proportional to the difference etween the principal modulus, and inversely proportional to m. As the umber of ply groups increase, the coupling modulus vanishes. In the mit when m becomes infinity, bending and stretching are uncoupled. s shown in Figure 6.6, strain is constant in (a), and is antisymmetric ı (b).

A simple illustration of the coupling term of this laminate is seen by ıe induced moment needed to prevent warpage. This is shown in igure 6.7. In this laminate the lower half is 0 degree; the upper half, 0 degrees; i.e., $[0/90]_T$.

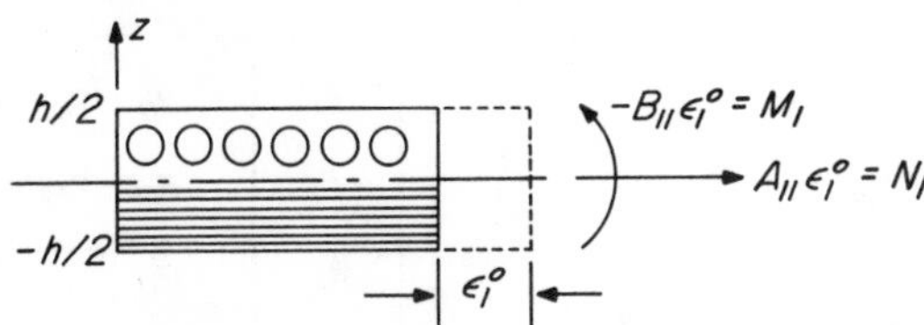

Figure 6.7 Stress resultant and induced moment of an unsymmetric cross-ply laminate under uniaxial extension. The induced moment is in the direction that prevents warpage.

We have seen the range of variation of the stiffness and compliance of an unsymmetric cross-ply laminate as functions of the number of ply groups. The features of this class of laminates can be summarized by the generalized stress-strain relation in Table 6.9 where matrix multiplication is implied.

table 6.9
generalized stress-strain relations of unsymmetric cross-ply laminates

	ϵ_1^0	ϵ_2^0	ϵ_6^0	k_1	k_2	k_6
N_1	A_{11}	A_{12}		B_{11}		
N_2	A_{21}	A_{22}			B_{22}	
N_6			A_{66}			
M_1	B_{11}			D_{11}	D_{12}	
M_2		B_{22}		D_{21}	D_{22}	
M_6						D_{66}

	N_1	N_2	N_6	M_1	M_2	M_6
ϵ_1^0	a_{11}	a_{12}		β_{11}		
ϵ_2^0	a_{21}	a_{22}			β_{22}	
ϵ_6^0			a_{66}			
k_1	β_{11}			δ_{11}	δ_{12}	
k_2		β_{22}		δ_{21}	δ_{22}	
k_6						δ_{66}

But the table above can be rearranged to show that the shear and twisting components are uncoupled. This is shown in Table 6.10 which is merely a repackaged arrangement of Table 6.9.

table 6.10
repackaged stress-strain relations of unsymmetric cross-ply laminates

	ϵ_1^0	ϵ_2^0	k_1	k_2	ϵ_6^0	k_6
N_1	A_{11}	A_{12}	B_{11}			
N_2	A_{21}	A_{22}		B_{22}		
M_1	B_{11}		D_{11}	D_{12}		
M_2		B_{22}	D_{21}	D_{22}		
N_6					A_{66}	
M_6						D_{66}

	N_1	N_2	M_1	M_2	N_6	M_6
ϵ_1^0	a_{11}	a_{12}	β_{11}			
ϵ_2^0	a_{21}	a_{22}		β_{22}		
k_1	β_{11}		δ_{11}	δ_{12}		
k_2		β_{22}	δ_{21}	δ_{22}		
ϵ_6^0					a_{66}	
k_6						δ_{66}

e only need to invert a 4 X 4 matrix instead of a 6 X 6. Secondly, the ear and twisting behavior of our unsymmetric cross-ply laminate haves exactly like a homogeneous plate. The ply grouping and mid-ane symmetry have no effect on the shear and twisting properties. his fact can be utilized in the design of a part requiring the minimum ge. It is not necessary to use symmetric laminates.

antisymmetric laminates

his class of laminates is neither symmetric nor unsymmetric. It is also lled quasi-symmetric. Instead of a symmetry with respect to the mid-ane that

$$\theta(z) = \theta(-z) \tag{6.119}$$

e have an antisymmetry that

$$\theta(z) = -\theta(-z) \tag{6.120}$$

he ply orientations are odd functions with respect to the midplane. A vo-ply angle-ply laminate such as

$$[-\phi/\phi]_T \tag{6.121}$$

antisymmetric. This class of laminates has values beyond academic iriosity.

With antisymmetry, the following components of stiffness remain ymmetric or even;

$$\left.\begin{aligned} Q_{11}(z) &= Q_{11}(-z) \\ Q_{22}(z) &= Q_{22}(-z) \\ Q_{12}(z) &= Q_{12}(-z) \\ Q_{66}(z) &= Q_{66}(-z) \end{aligned}\right\} \tag{6.122}$$

he shear or normal coupling components, however, are antisymmetric r odd;

$$\left.\begin{aligned} Q_{16}(z) &= -Q_{16}(-z) \\ Q_{26}(z) &= -Q_{26}(-z) \end{aligned}\right\} \tag{6.123}$$

The weighting factors for the integration of Equation 6.68 are anti-symmetric or odd for the coupling modulus; and symmetric or even, for the flexural as well as the in-plane modulus. These factors are shown in Figure 6.3 and Table 6.6. Because of the interaction between odd and even functions, the following components of modulus will vanish for antisymmetric laminates:

$$B_{11} = B_{22} = B_{12} = B_{66} = 0 \tag{6.124}$$

$$D_{16} = D_{26} = A_{16} = A_{26} = 0 \tag{6.125}$$

We will now examine a special antisymmetric laminate made of T300/5208 material with the following stacking sequence for $m = 2$, 4, 8, and 16, respectively.

$$[-45_8/45_8]_T,\ [-45_4/45_4]_{2T},\ [-45_2/45_2]_{4T},\ [-45/45]_{8T} \tag{6.126}$$

These laminates are shown in Figure 6.8.

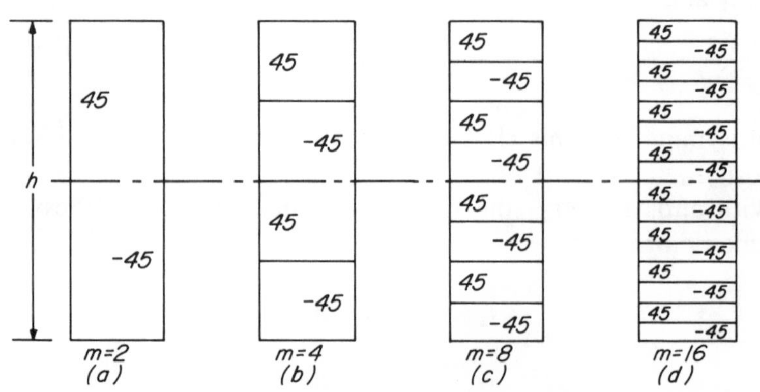

Figure 6.8 Special antisymmetric laminates with different number of ply groups.

For ply orientation of ±45 degrees, the coupling terms of a unidirectional composite are:

$$Q_{16} = Q_{26} = \pm\frac{1}{2}U_2 \qquad (6.12$$

$$= \pm\frac{1}{4}[Q_{xx} - Q_{yy}] \qquad (6.12$$

om Equation 6.69 and Table 6.6 we have for $m = 2$:

$$B_{16} = B_{26} = \frac{[8h_o]^2}{2} U_2 = \frac{10^{-6}}{2} 85.7 \tag{6.129}$$

$$= 42.8 \text{ kN} \tag{6.130}$$

s $m = 4$, we can readily show that

$$B_{16} = B_{26} = 21.4 \text{ kN} \tag{6.131}$$

ıus a general relation that

$$B_{16} = B_{26} = \frac{h^2}{m} U_2 \tag{6.132}$$

ems to be valid.

The flexural modulus for this laminate is orthotropic because the ear coupling terms are cancelled. The remaining components will have like those of a pseudo-homogeneous laminate, for which the llowing relation applies:

$$D_{ij} = \frac{h^3}{12} Q_{ij} \tag{6.133}$$

sing the unidirectional stiffness for $\theta = 45$ degrees from Chapter 3 able 3.5).

$$Q_{11} = Q_{22} = 56.6 \text{ GPa}$$
$$Q_{12} = 42.32 \text{ GPa} \tag{6.134}$$
$$Q_{66} = 46.59 \text{ GPa}$$

he shear and normal coupling terms of this modulus are not needed. or our 16-ply laminate

$$\frac{h^3}{12} = \frac{(16 \times 125 \times 10^{-6})^3}{12} = 666 \times 10^{-12} \text{m}^3 \tag{6.135}$$

Combining the equations above,

$$D_{11} = D_{22} = 56.6 \times 666 = 37.7 \text{ Nm} \tag{6.136}$$

$$D_{12} = 42.32 \times 666 = 28.2 \text{ Nm} \tag{6.137}$$

$$D_{66} = 46.59 \times 666 = 31.0 \text{ Nm} \tag{6.138}$$

The flexural modulus for this laminate, like that for the unsymmetri cross-ply, will be unaffected by the number of ply groups. In fact, the are the same as those for symmetric laminates, in Equation 5.114 with out the shear coupling terms. This laminate is pseudo-homogeneous.

The resulting stiffness and compliance of an antisymmetric laminat is shown in Table 6.11. Before we invert the 6 X 6 matrix to determin the compliance, we can save ourselves much work if we repackage th modulus matrix in Table 6.11 into two 3 X 3 matrices which we ca invert by hand if necessary. This rearrangement was done for the cross ply laminate in Table 6.10. For the antisymmetric case, the repackage matrices are shown in Table 6.12. In addition to the simplification i matrix algebra, the couplings in antisymmetric laminates are ver specific and special. It is a challenge to the designer to capitalize on th special quality and opportunity provided by antisymmetric laminate

table 6.11
stiffness and compliance of an antisymmetric laminate

	ϵ_1^0	ϵ_2^0	ϵ_6^0	k_1	k_2	k_6
N_1	A_{11}	A_{12}				B_{16}
N_2	A_{21}	A_{22}				B_{26}
N_6			A_{66}	B_{61}	B_{62}	
M_1			B_{16}	D_{11}	D_{12}	
M_2			B_{26}	D_{21}	D_{22}	
M_6	B_{61}	B_{62}				D_{66}

	N_1	N_2	N_6	M_1	M_2	M_6
ϵ_1^0	a_{11}	a_{12}				β_{16}
ϵ_2^0	a_{21}	a_{22}				β_{26}
ϵ_6^0			a_{66}	β_{61}	β_{62}	
k_1			β_{61}	δ_{11}	δ_{12}	
k_2			β_{62}	δ_{21}	δ_{22}	
k_6	β_{16}	β_{26}				δ_{66}

Table 6.12
Packaged stiffness and compliance of the same antisymmetric laminate in Table 6.11

	ϵ_1^o	ϵ_2^o	k_6	k_1	k_2	ϵ_6^o
N_1	A_{11}	A_{12}	B_{16}			
N_2	A_{21}	A_{22}	B_{26}			
M_6	B_{61}	B_{62}	D_{66}			
M_1				D_{11}	D_{12}	B_{16}
M_2				D_{21}	D_{22}	B_{26}
N_6				B_{61}	B_{62}	A_{66}

	N_1	N_2	M_6	M_1	M_2	N_6
ϵ_1^o	a_{11}	a_{12}	β_{16}			
ϵ_2^o	a_{21}	a_{22}	β_{26}			
k_6	β_{16}	β_{26}	δ_{66}			
k_1				δ_{11}	δ_{12}	β_{61}
k_2				δ_{21}	δ_{22}	β_{62}
ϵ_6^o				β_{61}	β_{62}	a_{66}

Numerical data of $[-45/45]_T$ laminates of T300/5208 are shown in Figure 6.9. The modulus is for all values of m, the number of ply groups. This correction factor appears in the denominator of the coupling components only. The compliance shown in Figure 6.9 is for the case of m=2 only.

	ϵ^o	k
N	$\frac{N}{m}$	N
M	N	Nm

	N	M
ϵ^o	$\frac{m}{N}$	$\frac{1}{N}$
k	$\frac{1}{N}$	$\frac{1}{Nm}$

	ϵ_1^o	ϵ_2^o	ϵ_6^o	k_1	k_2	k_6
N_1	113	84				$\frac{85.7}{m}$
N_2	84	113				$\frac{85.7}{m}$
N_6			93	$\frac{85.7}{m}$	$\frac{85.7}{m}$	
M_1			$\frac{85.7}{m}$	37.7	28.2	
M_2			$\frac{85.7}{m}$	28.2	37.7	
M_6	$\frac{85.7}{m}$	$\frac{85.7}{m}$				31.0

	N_1	N_2	N_6	M_1	M_2	M_6
ϵ_1^o	23.7	−11.1				−17.3
ϵ_2^o	−11.1	23.7				−17.3
ϵ_6^o			26.7	−17.3	−17.3	
k_1			−17.3	71.1	−33.4	
k_2			−17.3	−33.4	71.1	
k_6	−17.3	−17.3				80.0

Figure 6.9 Stiffness and compliance of T300/5208 antisymmetric laminates [−45/45]. The modulus is for all values of m; the compliance, for m = 2 only.

Similar to the comments made to the unsymmetric cross-ply laminate, we can compare the $\boldsymbol{A}$ in Figure 6.9 with the symmetric angle-ply in Table 4.9. Since the 16-ply thickness is 2×10^{-3}m, the comparable components are identical; i.e.,

$$A_{11} = 56.6 \times 2 \times 10^{-3} = 113 \text{ MN/m} \tag{6.139}$$

Again, compare the $\boldsymbol{D}$ in Figure 6.9 with that for the symmetric angle-ply laminate in Equation 5.114, we have identical components if we use the case m = infinity; i.e., the quasi-homogeneous, square symmetric laminate. The relation of Equation 6.106 is valid. So this antisymmetric laminate is pseudo-homogeneous. We can see that the compliance of this laminate is also pseudo-homogeneous.

As the number of ply groups increase, the compliance components will tend toward the value for quasi-homogeneous laminate, like the symmetric angle-ply. For comparison between the antisymmetric and symmetric angle-ply laminates, we show the change in the compliance components as functions of m in Table 6.13. The compliance for symmetric laminates is taken from Equation 5.115 et al.

table 6.13

selected compliance components of T300/5208 symmetric and antisymmetric angle-ply laminates

m	$d_{11} = d_{22}$	d_{66}	$d_{16} = d_{26}$	$\alpha_{11} = \alpha_{22}$	$\beta_{16} = \beta_{26}$	$\delta_{11} = \delta_{22}$
	$(\text{kNm})^{-1}$	$(\text{kNm})^{-1}$	$(\text{kNm})^{-1}$	$(\text{GN/m})^{-1}$	$(\text{MN})^{-1}$	$(\text{kNm})^{-1}$
2*	–	–	–	23.71	−17.33	71.13
4	66.03	58.35	−18.95	20.40	− 4.09	61.21
8	60.83	36.25	− 5.88	20.05	− 1.81	60.17
16	60.09	33.12	− 2.68	19.98	− 0.87	59.95
∞	59.85	32.19	0	19.95	0	59.85

*The $m = 2$ case is not possible for the symmetric angle-ply laminate.

Only the 16 and 26 components reduce rapidly as m increases, all other compliance components vary only modestly. With these compliance components, we can determine the strain distribution of an antisymmetric laminate subjected to a simple stress resultant or moment.

• $N_1 \neq 0$

$$\begin{aligned} \epsilon_1^o &= \alpha_{11} N_1 \\ \epsilon_2^o &= \alpha_{21} N_1 \\ k_6 &= \beta_{16} N_1 \end{aligned} \tag{6.140}$$

• $M_6 \neq 0$

$$\begin{aligned} \epsilon_1^o &= \beta_{16} M_6 \\ \epsilon_2^o &= \beta_{26} M_6 \\ k_6 &= \delta_{66} M_6 \end{aligned} \tag{6.141}$$

ecause of the special coupling of this laminate, the same strains (in-ane and curvature components) are induced by totally different pplied stresses.

An opportunity for reducing or eliminating twisting curvature is ssible when both stress resultant and twisting movement are present. his can occur in a rotor or a fan blade when centrifugal stress and rodynamic twisting occur simultaneous. The conditions for zero visting curvature is:

$$\begin{aligned} k_6 &= \beta_{16} N_1 + \delta_{66} M_6 = 0 \\ \frac{N_1}{M_6} &= -\frac{\delta_{66}}{\beta_{16}} \end{aligned} \tag{6.142}$$

The compliance components in Equation 6.142 can be manipulated ithin certain limits set by the properties of the constituent ply or plies the case of hybrids. Since the sign of the coupling compliance is ontrollable, while the flexural compliance is not, the curvature in quation 6.142 is controllable. The case of zero curvature is one special ase. This is a unique characteristic of composite materials. It is also nportant to realize that it is the ratio, not the absolute values, that an eliminate the curvature.

- $M_1 \neq 0$

$$\begin{aligned} k_1 &= \delta_{11} M_1 \\ k_2 &= \delta_{21} M_1 \\ \epsilon_6^o &= \beta_{61} M_1 \end{aligned} \tag{6.143}$$

- $N_6 \neq 0$

$$\begin{aligned} k_1 &= \beta_{61} N_6 \\ k_2 &= \beta_{62} N_6 \\ \epsilon_6^o &= \alpha_{66} N_6 \end{aligned} \tag{6.144}$$

Again, an opportunity existing for reduction or elimination of in-plane shear strain is possible, such that

$$\begin{aligned} \epsilon_6^o &= \alpha_{66} N_6 + \beta_{61} M_1 = 0 \\ \frac{N_6}{M_1} &= -\frac{\beta_{61}}{\alpha_{66}} \end{aligned} \tag{6.145}$$

All these highly coupled relations for antisymmetric laminates offer unique opportunities in the design of laminates. Under combined stresses, certain mode of deformation can be reduced or controlled. Strain distribution can also be altered for more favorable conditions. Composite materials are more than a highly competitive replacement of conventional materials. Composite materials can perform functions not possible with conventional materials. The opportunities available for new functions abound and need to be fully exploited.

6. the parallel axis theorem

Up to this point, the midplane of a laminate is the $z = 0$ plane. The parallel axis theorem deals with the general case where the midplane of a laminate or a ply is not on the $z = 0$ plane. This theorem is analogous to that for the moment of inertia of a rigid body.

In Figure 6.10 we show the relation between the laminate with spect to the $z = 0$ plane. The relation between z and z' is:

$$
\begin{aligned}
z &= z' - d, \quad \text{or} \\
z' &= z + d
\end{aligned}
\tag{6.146}
$$

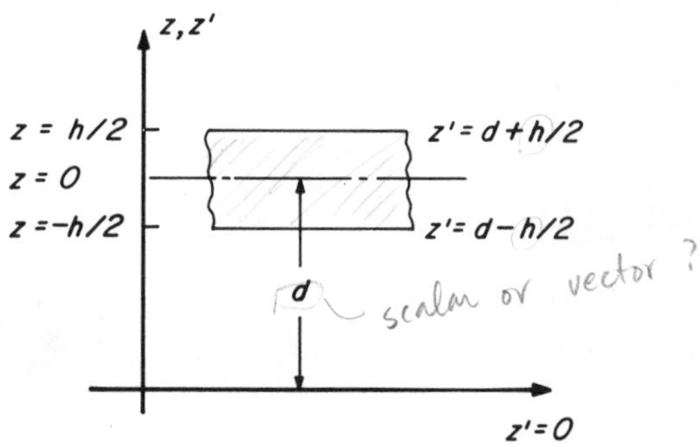

Figure 6.10 Relation between laminate midplane and the transferred plane.

The stress resultant and moments in the new, transferred plane can defined:

$$
\begin{aligned}
N_i' &= \int_{d-h/2}^{d+h/2} \sigma_i' dz' \\
&= \int_{-h/2}^{h/2} \sigma_i dz \\
&= N_i
\end{aligned}
\tag{6.147}
$$

om Equation 6.146 we can say

$$dz' = dz$$

hen

$$
\begin{aligned}
z' &= d \pm h/2 \\
z &= \pm h/2
\end{aligned}
\tag{6.148}
$$

$$M'_i = \int_{d-h/2}^{d+h/2} \sigma'_i z' dz' \tag{6.149}$$

$$= \int_{-h/2}^{h/2} \sigma_i (z+d) dz$$

$$= \int \sigma_i z dz + d \int \sigma_i dz$$

$$= M_i + d N_i \tag{6.150}$$

Note that the moment must be corrected by the transfer distance d. We can now derive the transfer of stress-strain relation of a general laminate.

$$N'_i = \int_{d-h/2}^{d+h/2} \sigma'_i dz' \tag{6.151}$$

$$= \int_{d-h/2}^{d+h/2} Q_{ij}(\epsilon_j^{o'} + z' k'_j) dz'$$

$$= \left[\int_{-h/2}^{h/2} Q_{ij} dz \right] \epsilon_j^{o'} + \left[\int_{-h/2}^{h/2} Q_{ij}(z + d) dz \right] k'_j \tag{6.152}$$

$$N'_i = A_{ij} \epsilon_j^{o'} + [B_{ij} + d A_{ij}] k'_j \tag{6.153}$$

Similarly

$$M'_i = \int_{d-h/2}^{d+h/2} \sigma'_i z' dz' \tag{6.154}$$

$$= \int_{d-h/2}^{d+h/2} Q_{ij}(\epsilon_j^{o'} + z' k'_j) z' dz' \quad \text{(continued)}$$

$$= \left[\int_{-h/2}^{h/2} Q_{ij}(z+d)dz\right]\epsilon_j^{o'} + \left[\int_{-h/2}^{h/2} Q_{ij}(z+d)^2 dz\right]k_j'$$

$$= [B_{ij} + dA_{ij}]\epsilon_j^{o'} + [D_{ij} + 2dB_{ij} + d^2 A_{ij}]k_j' \qquad (6.155)$$

(concluded)

Therefore

$$\begin{cases} A_{ij}' = A_{ij} \\ B_{ij}' = B_{ij} + dA_{ij} \\ D_{ij}' = D_{ij} + 2dB_{ij} + d^2 A_{ij} \end{cases} \qquad (6.156)$$

This is the parallel axis theorem. The stiffness of a general laminate can be transferred by a distance d along the z-axis.

First, the theorem can be used to generate the stiffness of general laminates. This approach is particularly suited for odd number of plies. Let the in-plane and flexural modulus of each ply or ply group be the basic building block. The coupling modulus is zero for homogeneous ply. The transfer distance d will be the location of the midplane of each ply.

$$\begin{aligned} A_{ij}' &= \sum_{b=-p}^{p} A_{ij}^{(b)} \\ B_{ij}' &= \sum_{b=-p}^{p} d_b A_{ij}^{(b)} \\ D_{ij}' &= \sum_{b=-p}^{p} D_{ij}^{(b)} + \sum_{b=-p}^{p} d_b^2 A_{ij}^{(b)} \end{aligned} \qquad (6.157)$$

where $p = \frac{1}{2}(n-1)$

n = odd

= total number of plies

b = the ordinal number of plies, $-p \leqslant b \leqslant p$

If all plies in a laminate have the same ply thickness, the transfer distance can be replaced by the ordinal number of the plies. Let

$$
\begin{aligned}
b &= d_b/h_o \\
B'_{ij} &= h_o \sum_{b=-p}^{p} bA_{ij}^{(b)} \\
D'_{ij} &= \sum_{b=-p}^{p} D_{ij}^{(b)} + h_o^2 \sum_{b=-p}^{p} b^2 A_{ij}^{(b)}
\end{aligned}
\tag{6.158}
$$

We can modify Figure 6.2 to show the new ordinal number b in Figure 6.11. The formulation by the parallel axis theorem is best suited for odd number of plies. In Figure 6.11 the ordinal number t is for even number of plies; that for b, odd number. With the new ordinal number we can establish a table analogous to that for the old ordinal number t in Tables 5.4 and 6.6. The result of the new ordinal number for odd number of plies is shown in Table 6.14.

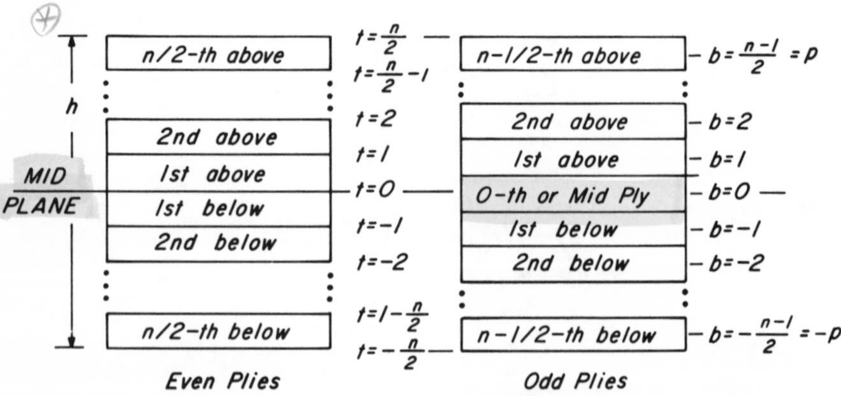

Figure 6.11 Ordinal numbers for laminates with even number of plies on the left; odd number, on the right.

table 6.14
numerical values of b's for calculation of stiffness of general laminates

	Ply order	b	Σb	b^2	Σb^2	
	8th above	8	36	64	204	
	7th above	7	28	49	140	
	6th above	6	21	36	91	
	5th above	5	15	25	55	
	4th above	4	10	16	30	
	3rd above	3	6	9	14	
	2nd above	2	3	4	5	
	1st above	1	1	1	1	
mid-plane	0th or mid ply	0	0	0	0	mid-plane
	1st below	−1	− 1	1	1	
	2nd below	−2	− 3	4	5	
	3rd below	−3	− 6	9	14	
	4th below	−4	−10	16	30	
	5th below	−5	−15	25	55	
	6th below	−6	−21	36	91	
	7th below	−7	−28	49	140	
	8th below	−8	−36	64	204	

We will now show how general laminates with odd plies can be calculated. For each ply we have

$$
\begin{aligned}
A_{ij} &= h_o Q_{ij} \\
B_{ij} &= 0 \\
D_{ij} &= \frac{h_o^3}{12} Q_{ij}
\end{aligned}
\tag{6.159}
$$

The stiffness of a general laminate with the same ply material and ply thickness is merely the sum of the contribution of each ply using the parallel axis theorem. We have

$$A'_{ij} = \sum_{b=-p}^{p} A_{ij}^{(b)}$$

$$= h_o \sum_{b=-p}^{p} Q_{ij}^{(b)} \tag{6.160}$$

$$B'_{ij} = \sum_{b=-p}^{p} B_{ij}^{(b)}$$

$$= h_o \sum bA_{ij}^{(b)}$$

$$= h_o^2 \sum_{b=-p}^{p} bQ_{ij}^{(b)} \tag{6.161}$$

$$D'_{ij} = \sum_{b=-p}^{p} D_{ij}^{(b)}$$

$$= \frac{h_o^3}{12} \sum Q_{ij}^{(b)} + h_o^3 \sum b^2 Q_{ij}^{(b)}$$

$$= \frac{h_o^3}{12} \sum_{b=-p}^{p} [1+12b^2]\, Q_{ij}^{(b)} \tag{6.162}$$

Let us first compute a one-ply laminate using Equations .150–6.157. First, from Table 6.14 for 0-th ply,

$$b = 0$$

We have

$$A'_{ij} = h_o Q_{ij}$$

$$B'_{ij} = 0$$

$$D'_{ij} = \frac{h_o^3}{12} Q_{ij}$$

Now if we have a three-ply laminate ($b = -1,0,1$) of T300/5208 with stacking sequence of $[0/90/0]_T$, we have from Table 6.14,

$$\begin{aligned} A'_{11} &= h_o[Q_{xx} + Q_{yy} + Q_{xx}] = 46.73 \text{ MN/m} \\ B'_{11} &= h_o^2[-Q_{xx} + 0 + Q_{xx}] = 0 \\ D'_{11} &= \frac{h_o^3}{12}[13Q_{xx} + Q_{yy} + 13Q_{xx}] = 771 \times 10^{-3} \text{ Nm} \end{aligned} \tag{6.163}$$

vhere the coefficient 13 comes from $[1+12b^2]$ for $b = \pm 1$.

If we have a 24-ply laminate with the same stacking sequence; i.e.,

$$[0_8/90_8/0_8]_T$$

ve can immediately compute the stiffness by the equivalent thickness f $8h_o$ and apply the appropriate scaling to the values in Equation .163, we will have:

$$\begin{aligned} A'_{11} &= 8(46.73) = 373.8 \text{ MN/m} \\ B'_{11} &= 0 \\ D'_{11} &= 8^3(771 \times 10^{-3}) = 394.7 \text{ Nm} \end{aligned} \tag{6.164}$$

If the ply materials are different from ply to ply, due to either ply orientations or use of multiple materials (hybrids), the proper modulus should be substituted into Equation 6.158 for each modulus designated by superscript *b*.

The use of the parallex axis theorem for the calculation of the stiffness of a general laminate is convenient when:

- **There are odd number of plies,**
- **All plies have the same thickness, and**
- **Plies can be of different materials (hybrids).**

If any one condition is not satisfied, the simple factors in Table 6.14 cannot be used without modifications.

The parallel axis theorem is also useful in determining the relative importance of stacking sequence in a laminate in a built-up structure, such as the cap or flange of a "T" section or the skin of a fuselage. Repeating Equation 6.156 here:

$$\begin{aligned} A'_{ij} &= A_{ij} \\ B'_{ij} &= B_{ij} + dA_{ij} \\ D'_{ij} &= D_{ij} + 2dB_{ij} + d^2 A_{ij} \end{aligned} \tag{6.165}$$

If we introduce the following stiffness components:

$$\begin{aligned} A^*_{ij} &= A_{ij}/h \\ B^*_{ij} &= 2B_{ij}/h^2 \\ D^*_{ij} &= 12D_{ij}/h^3 \end{aligned} \tag{6.166}$$

All the normalized components of stiffness will have the same physical unit as the stiffness of a homogeneous material; i.e., Pa. The parallel axis theorem in Equation 6.165 can be expressed in terms of the normalized components.

$$\begin{aligned} A^{*\prime}_{ij} &= A^*_{ij} \\ B^{*\prime}_{ij} &= B^*_{ij} + 2A^*_{ij}\frac{d}{h} \\ D^{*\prime}_{ij} &= D^*_{ij} + 12B^*_{ij}\frac{d}{h} + 12A^*_{ij}\left[\frac{d}{h}\right]^2 \end{aligned} \tag{6.167}$$

Let us examine a cross-ply laminate of T300/5208, in Figure 5.11.

$$[0_4/90_4]_S,\ v_o = v_{90} = \frac{1}{2}$$

From Table 4.6, and Equation 4.48:

$$\begin{aligned} A^*_{11} &= A_{11}/h = U_1 + U_3 = 76.37 + 19.71 \\ &= 96.08 \text{ GPa} \end{aligned} \tag{6.168}$$

From Table 5.7, and Equation 5.64 for $m = 4$

$$\begin{aligned} D^*_{11} &= U_1 + \frac{3}{m} U_2 + U_3 \\ &= 160.37 \text{ GPa} \end{aligned} \tag{6.169}$$

Because of symmetry

$$B^*_{11} = 0 \tag{6.170}$$

Substituting these values into Equation 6.167 for the normalized flexural modulus, we have

$$\begin{aligned} D^{*\prime}_{11} &= 160 + 12 \times 96 \left[\frac{d}{h}\right]^2 \\ &= 160 + 1152 \left[\frac{d}{h}\right]^2 \end{aligned} \tag{6.171}$$

Note the numerical values as a function of the normalized transfer distance. We have seen here that the sensitivity of the transfer distance to the flexural modulus. If the number of ply groups change, as we have seen in Equation 5.66 et al., only the first term of Equation 6.171 is affected. So the effect of ply groups or stacking sequence in general will be negligible when the transfer distance is increased beyond the laminate thickness. Similar effect of a sandwich core can be expected. The second term in Equation 6.171 becomes dominant. It determines the

table 6.15
normalized flexural modulus as function of transfer distance for T300/5208 cross-ply laminates

$\frac{d}{h}$	$D^{*\prime}_{11}$	$\frac{D^{*\prime}_{11}}{12A^*_{11}\left[\frac{d}{h}\right]^2}$
0	160	00
0.1	171	14.88
1	1,312	1.138
10	115,360	1.001,38
100	11,520,160	1.000,013,8

stiffness of a thin shell. The first term is still important because it controls the local stability of the shell. Results are listed in Table 6.15.

Now let us apply the example above to an unsymmetric construction such as the cross-ply laminate in Equation 6.90 and Figure 6.4(*a*). The stiffness of this laminate is shown in Figure 6.5. We can compute the following normalized components where the 16-ply laminate with a thickness of 2 mm:

$$A^*_{11} = 192/2 \times 10^{-3} = 96 \text{ GPa}$$

$$B^*_{11} = -\frac{171}{m}\frac{2}{(2 \times 10^{-3})^2} = -43 \text{ GPa for } m = 2 \tag{6.172}$$

$$D^*_{11} = 64\frac{12}{(2 \times 10^{-3})^3} = 96 \text{ GPa}$$

Substituting these values into Equation 6.166, we have

$$A^{*\prime}_{11} = 96 \text{ GPa}$$

$$B^{*\prime}_{11} = -43 + 192\frac{d}{h} \tag{6.173}$$

$$D^{*\prime}_{11} = 96 - 513\frac{d}{h} + 1152\left[\frac{d}{h}\right]^2$$

When the transfer distance is ten times the laminate thickness, the coupling and flexural modulus components in the 1-direction become insensitive (within 5 percent variation) to the stacking sequence. For thin shells, where radius is at least ten times the thickness, the stacking sequence, sandwich core and midplane symmetry contribute to the local behavior of the shell only. They have little effect on the gross behavior of the shell.

The numerical results in Table 6.16 are plotted in Figure 6.12, where we show the normalized flexural modulus as a function of the normalized transfer distance. Both the symmetric and unsymmetric cross-ply laminates are shown. As the transfer distance increases beyond ten, the two curves merge into one; i.e., the last term in Equations 6.171 and 6.173 become dominant.

table 6.16
the normalized flexural modulus for unsymmetric laminates as functions of transfer distance

$\frac{d}{h}$	$B_{11}^{*\prime}$	$\dfrac{B_{11}^{*\prime}}{192\frac{d}{h}}$	$D_{11}^{*\prime}$ (Unsymm)	$\dfrac{D_{11}^{*\prime}}{1152\left[\frac{d}{h}\right]^2}$
0	−43	∞	96	∞
0.1	−23.8	−1.239	56.22	4.880
1	149	.776	753	.638
10	1877	.977	110,166	.9563
100	19157	.9977	11,468,796	.9955

The unsymmetric laminate curve falls below that of the symmetric laminate. Thus asymmetry reduces the flexural modulus. From Equation 6.167, we can derive the condition for eliminating the coupling matrix by proper transfer of the axis.

$$B_{ij}^{*\prime} = 0 = B_{ij}^{*} + 2A_{ij}^{*}\frac{d}{h}$$

$$\frac{d}{h} = -\frac{B_{11}^{*}}{2A_{11}^{*}} \qquad (6.174)$$

Substituting this into the flexural modulus,

$$D_{11}^{*\prime} = D_{11}^{*} - 3\frac{[B_{11}^{*}]^2}{A_{11}^{*}} \tag{6.175}$$

or substituting Equation 6.166 into this,

$$D_{11}' = D_{11} - \frac{B_{11}^2}{A_{11}} \tag{6.176}$$

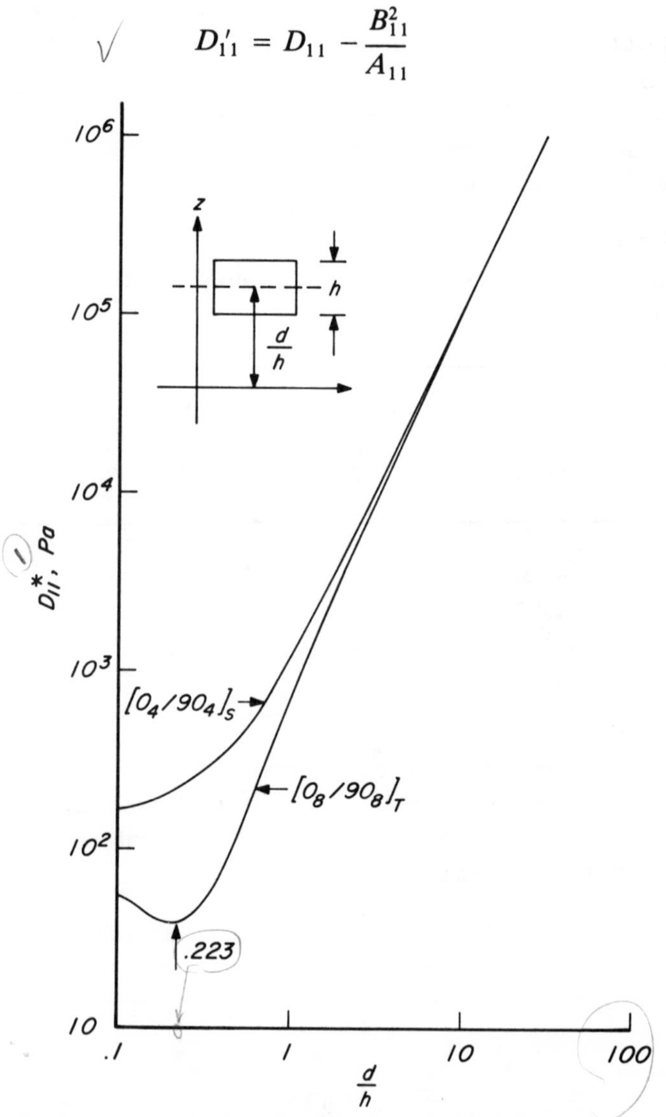

Figure 6.12 Normalized flexural modulus as a function of normalized transfer distance.

Note that the transferred stiffness is always less than the original. This equation only applies to one component at a time. For example, we have cylindrical bending along the 2-axis, instead of the 1-axis,

$$D'_{22} = D_{22} - \frac{B_{22}^2}{A_{22}} \tag{6.177}$$

Using the data in Equation 6.173,

$$\frac{d}{h} = \frac{43}{192} = .223 \tag{6.178}$$

$$D_{11}^{*\prime} = 38.89 \text{ GPa} \quad \text{or}$$

$$D'_{11} = 38.89 \frac{h^3}{12} = 25.92 \text{ Nm} \tag{6.179}$$

This is lower than original untransferred modulus of 37.7 Nm shown in Equation 6.136. In fact, at this transfer distance, the flexural modulus reaches a minimum. This is shown in Figure 6.12. Also at this transfer distance, the laminate behaves like a symmetric laminate with a reduced flexural modulus. The buckling equation for isotropic material can be directly applied for this unsymmetric laminate under cylindrical bending.

We have seen the use of the parallel axis theorem for a variety of problems. The theorem is helpful in separating the local from the global behavior of a built-up structure. It is also helpful in identifying the effects of stacking sequence, the midplane symmetry, and the contribution of a sandwich core. The theorem illustrates the difference between the absolute or unnormalized modulus and the normalized modulus. In the design and sizing of structures, the laminate modulus as a function of the number of plies, ply orientations, and transfer distance must be optimally selected. The absolute modulus is preferred. The parallel axis theorem is easy to use if plies are added to an existing laminate without performing the integration or summation from the bottom to the top plies. The marginal return of an additional ply can be quickly established.

7. transformation of the coupling stiffness and compliance

It is useful to establish the transformation of the coupling stiffness and compliance matrix. With such relations we can show the unsymmetric cross-ply laminate and the ±45 degrees antisymmetric laminate are related to each other through a rotation of 45 degrees. With the transformation relations we can calculate the stiffness and compliance of our unsymmetric laminate for any angle of rigid body rotation. For our cross-ply laminates listed in Table 6.5, we have for $[0/90]_T$ the following components:

$$
\begin{aligned}
B'_{11} &= -B'_{22} = -85.7 \text{ kN} \\
B'_{12} &= B'_{66} = B'_{16} = B'_{26} = 0
\end{aligned}
\tag{6.180}
$$

From the transformation equation in Table A.7, we can immediately write down the transformed coupling modulus for $[\theta/90+\theta]_T$ as follows:

$$
\begin{aligned}
B_{11} &= -B_{22} = (m^4-n^4)B'_{11} = B'_{11}\cos 2\theta = -85.7\cos 2\theta \\
B_{12} &= B_{66} = 0 \\
B_{16} &= B_{26} = \frac{1}{2}B'_{11}\sin 2\theta = -42.8\sin 2\theta
\end{aligned}
\tag{6.181}
$$

The transformation of the coupling modulus is very simple and follows the trigonometric functions of the double angle. When θ equals 0 degree, the 16 and 26 components are zero. When θ equals −45 degrees, the 11 and 22 components are zero while the 16 and 26 reach maximum. Components related to the 12 and 66 components are identically zero for all angles. The nonzero transformed coupling modulus are shown in Figure 6.13.

From Table A.7 we can write the transformation equation for the antisymmetric laminate [−45/45] will have the following components for the coupling modulus based on m=2 in Figure 6.9:

$$
\begin{aligned}
B'_{16} &= B'_{26} = +42.8 \text{ kN} \\
B'_{11} &= B'_{22} = B'_{12} = B'_{66} = 0
\end{aligned}
\tag{6.182}
$$

From this laminate orientation, the coupling modulus of a laminate $[\theta-45/\theta+45]_T$ can be obtained from the following nonzero components:

$$
\begin{aligned}
B_{11} &= -B_{22} = -4(m^3n+mn^3)B'_{16} \\
&= -2B'_{16}\sin 2\theta = -85.7\sin 2\theta \qquad (6.183) \\
B_{16} &= B_{26} = (m^4-n^4)B'_{16} = 42.8\cos 2\theta
\end{aligned}
$$

The result of this transformation is identical to that shown in Figure 6.13 except the origin for our laminate should be displaced to the left by 45 degrees. The transformation relation of this equation is more useful than that in Equation 6.181 because the starting point is from the symmetry axis of the antisymmetric laminates. The matrix inversion can be reduced from one 6 X 6 to two 3 X 3. Thus we can use the transformation equation to generate the modulus and compliance of a cross-ply laminate in Figure 6.5 by rotating the results of an antisymmetric laminate in Figure 6.9.

We need equations comparable to that of 6.183 for our coupling compliance matrix β. This can be derived from Table A.10 because β is symmetric for the case of a bidirectional laminate. The result is:

$$
\begin{aligned}
\beta'_{16} &= \beta'_{26} = -17.3\ (\text{MN})^{-1} \\
\beta_{11} &= -\beta_{22} = -2(m^3n+mn^3)\beta'_{16} \\
&= -\beta'_{16}\sin 2\theta = 17.3\sin 2\theta \\
\beta_{16} &= \beta_{26} = (m^4-n^4)\beta'_{16} \qquad (6.184) \\
&= \beta'_{16}\cos 2\theta = -17.3\cos 2\theta \\
\beta_{12} &= \beta_{66} = 0
\end{aligned}
$$

From the transformation relations above, we can see that a rotation of +45 degrees will result in the laminate $[0/90]_T$ for which the nonzero components are:

$$\beta_{11} = -\beta_{22} = 17.3\ (\text{MN})^{-1} \qquad (6.185)$$

This agrees with the result in Figure 6.5 for the unsymmetric cross-ply.

We have thus established an expedient method of generating the stiffness and compliance of antisymmetric laminates with arbitrary laminates orientations. The entire antisymmetric laminate can undergo arbitrary rigid body rotation. The process described above requires the inversion of two 3 X 3 instead of one 6 X 6 matrix provided β is symmetric.

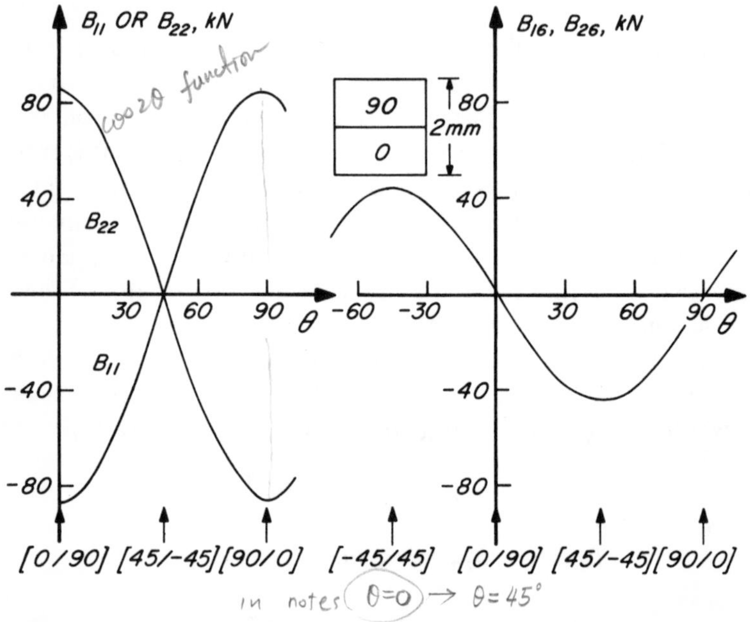

Figure 6.13 Transformation of the coupling modulus of a T300/5208 cross-ply laminate: $[0_8/90_8]_T$.

8. conclusions

Index notation is almost a prerequisite in the study of general laminates. A highly coupled behavior is available and provides opportunities for design and fabrication not possible with the conventional material. The governing stress-strain relations are as conceptually simple as the symmetric laminates. The stiffness of a general laminate can be easily manipulated to provide any degree of in-plane versus flexural coupling. In many cases, only the coupling matrix $\boldsymbol{B}$ is sensitive to the asymmetry of stacking sequence. Every component of the compliance on the other hand, is sensitive to the asymmetric stacking sequence of the laminate.

The parallel axis theorem is a useful tool for calculating the modulus of any laminate. Constructions of hybrid, built-up structures can be readily expressed in terms of its modulus. The sensitivity of stacking sequence, asymmetry, transfer distances and changes of materials or finite widths can all be assessed in a straight-forward manner.

A structure such as a wing can have symmetric but different laminates of the top and bottom covers. The wing will be asymmetric. With a properly chosen transfer distance like that is in Equation 6.174, the wing will bend like a simple beam with a reduced flexural stiffness. Numerous combinations of symmetric and asymmetric constructions, including hybrids, can be utilized to create novel responses of built-up structures. The local versus global stiffness is easy to differentiate. The use of prestress can shift failure modes to more advantageous combinations and locations. The essence of composite materials lies in the judicious choice of:

- Ply materials ($Q_{xx}^{(i)}, \ldots$)
- Local property (A_{ij}, B_{ij}, D_{ij})
- Global or structural property ($A'_{ij}, B'_{ij}, D'_{ij}$)

In addition, the process of curing and environmental effects (to be presented in Chapter 8) can be chosen to provide the most desirable prestress. We should not penalize composite materials by eliminating anisotropy and asymmetry. We should instead improve our analytical capability so we can do justice to the effective use of composite materials. We should not limit ourselves to ten constants and make the remaining 26 zero. We should try to take advantage of all 36 constants.

9. homework problems

a. Discuss the pros and cons of calculating the modulus of general laminates of the following approaches:

(1) Direct integration of Equation 6.68 or summation of Equation 6.69.

(2) Separation of geometric factors from material constants in Equation 6.79 et al.

(3) The parallel axis theorem in Section 6.

b. What is the consequence of not having the minus sign in the definition of curvature in Equation 5.9 on the modulus of a general laminate? What is the consequence if the factor of 2 for the twisting curvature is left out in Equation 5.9?

c. Calculate the components in Table 6.4 which is a partial inversion of the modulus of Table 6.3 for T300/5208 cross-ply laminates shown in Figure 6.4. Is the stiffness matrix symmetric, asymmetric or anti-symmetric?

d. What is the stiffness of a circular cylinder with unsymmetric cross-ply wall of the last problem subjected to uniaxial extension along its axis? How does it compare with a symmetric laminate of the same total thickness? Compare the ply stress in the cylindrical wall of the symmetric and unsymmetric construction.

e. Calculate the partially inverted modulus in Table 6.4 for a T300/5208 antisymmetric laminate shown in Figure 6.8. The result can be applied to a thin wall tube under an applied torque. What is the resulting ply stress and strain? How do they compare with those for a symmetric laminate of the same wall thickness?

f. Calculate unsymmetric ±30-degree angle-ply laminates of T300/5208 for various ply groups similar to those in Figure 6.8. How can we write down the results for the same laminates for ±60-degree?

g. Write a general relation for the stiffness of a built-up structure with piece-wise variable widths. How can hybrid (variable materials) be introduced?

h. How are engineering constants defined for general laminates? Show the bending and torsional stiffness of T300/5208 ±45-degree antisymmetric laminates as functions of ply groups. Compare the results with the symmetric laminates in Figure 5.17.

i. General laminates offer the widest choice of coupling between various behavioral variables. Centrifugal force (in-plane stress resultant), for example, can be used to reduce or eliminate the twisting curvature shown in Equation 6.142. Centrifugal force can also be used to reduce or eliminate bending curvature; see Table 6.9. What are the conditions for eliminating bending and twisting curvature simultaneously by centrifugal force? Can T300/5208 satisfy the conditions?

nomenclature

A_{ij}, B_{ij}, D_{ij}	= Stiffness of symmetric or unsymmetric laminates; units vary
$A^*_{ij}, B^*_{ij}, D^*_{ij}$	= Normalized stiffness; Pa
a_{ij}	= In-plane compliance of a symmetric laminate, in N^{-1}; i,j = 1,2,6
b	= Width of a beam or
	= The ordinal number of plies in a laminate with odd plies
d_{ij}	= Flexural compliance of a symmetric laminate
d	= Transfer distance from the reference axis in the parallel axis theorem
h	= Total thickness of laminate, in m
h_o	= Unit ply thickness, in m
k_i	= Curvature, in m^{-1}; $i = 1,2,6$
M_i	= Moment, in Nm^{-1}; $i = 1,2,6$
m	= Total number of ply groups
n	= Total number of plies in a laminate
N_i	= Stress resultant, in Nm^{-1}; $i = 1,2,6$
Q_{ij}	= Stiffness of a unidirectional composite; i,j = 1,2,6 or x,y,s
t	= The ordinal number of plies in a laminate with even plies
$V_{iA,B,D}$	= Geometric factors for the in-plane, coupling and flexural moduli of an unsymmetric laminate; i = 1 to 4
U_i	= Linear combinations of the stiffness of a unidirectional composite; i = 1 to 5
$\alpha_{ij}, \beta_{ij}, \delta_{ij}$	= Compliance of unsymmetric laminates; units vary
$\alpha^*_{ij}, \beta^*_{ij}, \delta^*_{ij}$	= Normalized compliance; Pa^{-1}
σ_i	= Stress components in a lamina
ϵ_i, ϵ_i^o	= Total and in-plane strain in a laminated composite

chapter 7

strength of composite materials

꞉ strength of unidirectional and multidirectional composites can be cribed by quadratic interaction failure criteria in stress and strain ce. The first ply failure envelope in stress space can determine the imum strength. This envelope in strain space can be approximated a right ellipsoid in the *p-q-r* strain space. This envelope becomes ependent of ply orientations. The design of composite laminates omes analogous to that for conventional materials.

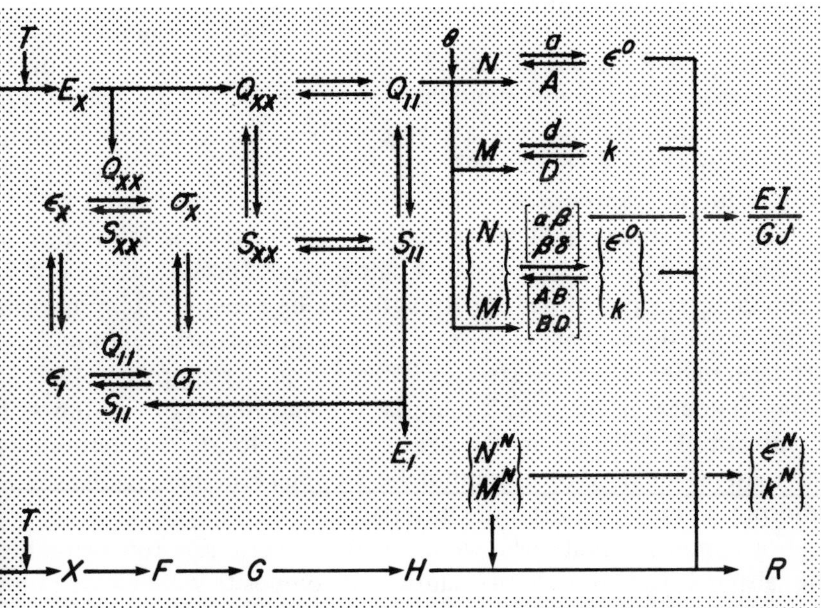

1. failure criteria

For the determination of strength of any material it is the usual practice to estimate the stress at the time and location when failure occurs. In the case of conventional materials we need only to determine the maximum tensile, compressive, or shear stress and can then make some observation about the failure and the failure mechanism. This process is relatively straightforward because isotropic materials have no preferential orientation and usually one strength constant will suffice. The isotropic material is essentially a one-dimensional or one-constant material. The Young's modulus for stiffness will suffice because Poisson's ratio is taken to be about 0.3, and the uniaxial tensile strength will also suffice because the shear strength is taken to be about 50 to 60 percent of the tensile.

For composite materials, however, the one-constant approach fo stiffness or for strength is no longer adequate. We saw earlier that fou elastic constants were needed for the stiffness. We will see later in thi chapter six constants for the strength of unidirectional composites ar needed. The number of constants however do not introduce conceptua difficulty. We know that unidirectional composites have highly direc tionally dependent strengths. The longitudinal strength can be twent times that of the transverse and shear strengths. So for any state o applied stress, all three stress components must be examined before judgment on the cause of failure can be made. We cannot say quickl the specific stress component that is responsible for the failure. Prob ably all three components are responsible. The effect of combine stresses must be systematically determined and can be regarded as way of life for composites.

The determination of strength using failure criteria is based on th assumption that the material is homogeneous (properties do not var from point to point) and its strength can be experimentally measure with simple tests. Failure criteria provide the analytic relation for th strength under combined stresses. There is another approach of strengt using fracture mechanics. A material is assumed to contain flaws. Th dominant flaw based on its size, shape and location determines th strength when its growth cannot be stopped. In this chapter we wi interpret strength using failure criteria.

For composite materials, we need a failure criterion for the unidire tional plies. The strength of a laminated composite will be based on th rength of the individual plies within a laminate. We would expect ccessive ply failures as the applied load to a laminate increases. We ill have the first ply failure (FPF) to be followed by other ply failures ntil the last ply failure which would be the ultimate failure of the minate. The ply stress and ply strain calculations for symmetric and neral laminates are intended for strength determination. This is the bject of this chapter.

There are two popular approaches for failure criteria of unidirec- onal composites. They are all based on the on-axis stress or strain as e basic variable with different tensile and compressive strengths.

the maximum stress and strain criteria

$$
\begin{aligned}
\sigma_x &\leqslant X \text{ or } X' \\
\sigma_y &\leqslant Y \text{ or } Y' \qquad (7.1) \\
\sigma_s &\leqslant S
\end{aligned}
$$

Failure occurs when one of the equalities is met. Using the linear relation we can express the equation above in the following maximum strain criterion:

$$
\begin{aligned}
\epsilon_x^* &\leqslant \frac{X}{E_x} \text{ or } \frac{X'}{E_x} \\
\epsilon_y^* &\leqslant \frac{Y}{E_y} \text{ or } \frac{Y'}{E_y} \qquad (7.2) \\
\epsilon_s^* &\leqslant \frac{S}{E_s}
\end{aligned}
$$

Failure occurs when one of the equalities is met. These two criteria are not the same. One of the homework problems in Chapter 1 showed the difference. Only when Poisson's ratio of the unidirectional material is zero, the criteria become identical. Conceptually they are similar. Each component of stress or strain has its own criterion and is not affected by the other components. There is no interaction.

b. the quadratic interaction criterion

$$F_{ij}\sigma_i\sigma_j + F_i\sigma_i = 1 \qquad (7.$$

This can be expressed in strain components.

$$G_{ij}\epsilon_i\epsilon_j + G_i\epsilon_i = 1 \qquad (7.$$

where the F's and G's are strength parameters analogous to t constants in Equations 7.1 and 7.2. Failure occurs when eith equation is met.

We will choose the quadratic criterion in this book. It is simp versatile and analytic. Established rules on transformation, invarian and symmetry are applicable. It includes interaction among the stress strain components analogous to the von Mises criterion for isotro materials. Many have used variations of this failure criterion.

Failure criteria serve important functions in the design and sizing composite laminates. They should provide a convenient framework model for mathematical operations. The framework should remain t same for different definitions of failures, such as the ultimate strengt the proportional limit, yielding, endurance limit, or a working stre based on design or reliability considerations. The criteria are not i tended to explain the mechanisms of failure. Failures in composi materials involve many modes; viz., fiber failures, matrix failures, int facial failures, delamination, and buckling. Furthermore, the vario modes interact and can occur concurrently and sequentially. Failu analysis based on some post-mortem examination without due conside ation of the dynamic process of failure can be misleading.

2. quadratic failure criterion

Equation 7.3 can be expanded for the case of two-dimensional stress, $i,j = 1,2,6$.

$$\begin{aligned} F_{11}\sigma_1^2 &+ 2F_{12}\sigma_1\sigma_2 + F_{22}\sigma_2^2 + F_{66}\sigma_6^2 \\ &+ 2F_{16}\sigma_1\sigma_6 + 2F_{26}\sigma_2\sigma_6 \qquad (7. \\ &+ F_1\sigma_1 + F_2\sigma_2 + F_6\sigma_6 = 1 \end{aligned}$$

ince our unidirectional composite is in its orthotropic axes, as shown ı Figure 7.1, the strength should be unaffected by the direction or sign f the shear stress component. If shear stress is reversed, the strength hould remain the same. Sign reversal for the normal stress components, ay from tensile to compressive, is expected to have a significant effect n the strength of our composite. Thus, all terms in Equation 7.5 that ontain linear or first-degree shear stress must be deleted from the quation. There are three such terms:

$$F_{xs}\sigma_x\sigma_s, \quad F_{ys}\sigma_y\sigma_s, \quad F_s\sigma_s \tag{7.6}$$

here $i,j = x,y,s$ is applied to Equation 7.3.

Since the stress components are in general not zero, the only way to nsure that the terms above vanish is for

$$F_{xs} = F_{ys} = F_s = 0 \tag{7.7}$$

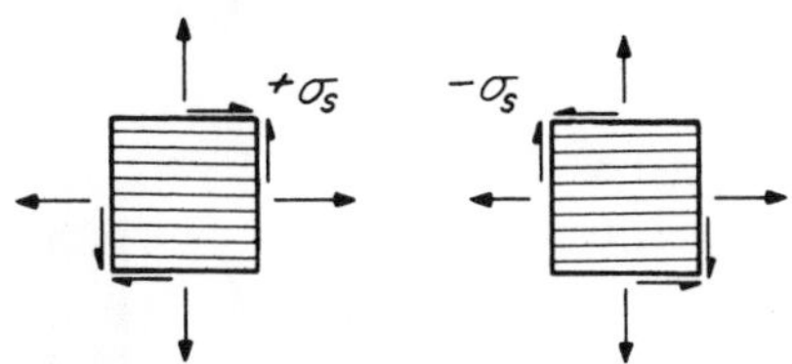

Figure 7.1 On-axis positive and negative shears. They should have no effect on the strength of unidirectional composites. Coupling between shear and normal components cannot exist in this orthotropic orientation.

/ith the removal of the three terms, Equation 7.5 can be simplified.

$$\begin{aligned} F_{xx}\sigma_x^2 + 2F_{xy}\sigma_x\sigma_y + F_{yy}\sigma_y^2 + F_{ss}\sigma_s^2 \\ + F_x\sigma_x + F_y\sigma_y = 1 \end{aligned} \tag{7.8}$$

here are four quadratic strength parameters analogous to the four ndependent components of modulus. There are two linear strength arameters as a result of the difference in tensile and compressive

strengths. There is no counterpart of this in the modulus because tensil
and compressive moduli are assumed to be equal.

Of the six material constants or strength parameters, five can b
measured by performing simple tests.

- Longitudinal Tensile and Compressive Tests

  Let X = Longitudinal tensile strength
  X' = Longitudinal compressive strength

These strengths are measured by uniaxial tests shown in Figure 7.2.
Substituting the measured strength into Equation 7.8,

If $\sigma_x = X$,

$$F_{xx}X^2 + F_x X = 1 \tag{7.9}$$

If $\sigma_x = -X'$,

$$F_{xx}X'^2 - F_x X' = 1 \tag{7.10}$$

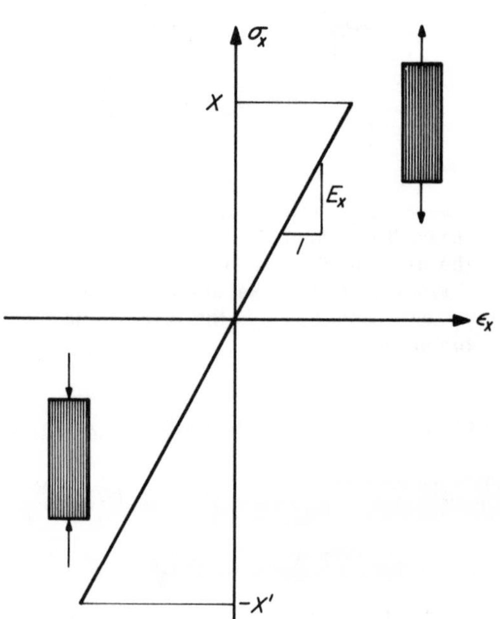

Figure 7.2 Uniaxial longitudinal tensile and compressive tests.

We have two equations for two unknowns from which we can get

$$F_{xx} = \frac{1}{XX'}$$

$$F_x = \frac{1}{X} - \frac{1}{X'} \tag{7.11}$$

- Transverse Tensile and Compressive Tests

  Let Y = Transverse tensile strength
  Y' = Transverse compressive strength

  Using the same approach as the longitudinal tests and by reason of symmetry, we know

$$F_{yy} = \frac{1}{YY'}$$

$$F_y = \frac{1}{Y} - \frac{1}{Y'} \tag{7.12}$$

- Longitudinal Shear Test

  Let S = Longitudinal shear strength

  Substituting this value into the shear stress in Equation 7.8,

$$F_{ss} = \frac{1}{S^2} \tag{7.13}$$

We have obtained five of the six coefficients in our failure criterion of Equation 7.8. The one remaining term is related to the interaction between the two normal stress components. The only way that this coefficient can be measured is for both normal stress components to be nonzero; this requires a combined stress or biaxial test. This experimental task unfortunately is not as easy to perform as the simple uniaxial or shear test.

Although the exact value for the interaction term is indeterminate at this time, there are upper and lower bounds imposed on this value based on a geometric consideration. A conic section or quadratic curves can go from an ellipse to parallel lines, and to hyperbola depending on the value of the interaction term, the coefficient of the product of two normal stress components in Equation 7.8. The criterion that dictates which branch of the quadratic curve it belongs to is based on the value of the following discriminant:

$$\text{Discriminant} = F_{xx}F_{yy} - F_{xy}^2 \begin{cases} > 0 \text{ for ellipse} \\ = 0 \text{ for parallel lines} \\ < 0 \text{ for hyperbola} \end{cases} \tag{7.14}$$

In order to insure that the failure criterion represents a closed curve in the plane of the normal stress components, this discriminant is constrained by the value shown for the ellipse in the equation above. The curve has to be closed in order to avoid infinite strength.

If we introduce a dimensionless or normalized interaction term,

$$F_{xy}^* = F_{xy} \Big/ \sqrt{F_{xx}F_{yy}} \tag{7.15}$$

The range of values of the discriminant in Equation 7.14 can now be expressed by the range of values of the normalized interaction term:

$$\begin{aligned} -1 < F_{xy}^* < 1 &\text{ for ellipse} \\ F_{xy}^* = \pm 1 &\text{ for parallel lines} \\ F_{xy}^* < -1, 1 < F_{xy}^* &\text{ for hyperbola} \end{aligned} \tag{7.16}$$

We can rearrange Equation 7.8 in terms of the following dimensionless parameters:

$$\begin{aligned} x &= \sqrt{F_{xx}}\,\sigma_x \\ y &= \sqrt{F_{yy}}\,\sigma_y \\ z &= \sqrt{F_{ss}}\,\sigma_s \end{aligned} \tag{7.17}$$

n addition, we also need the following dimensionless parameters:

$$F_x^* = \frac{F_x}{\sqrt{F_{xx}}} = \frac{X' - X}{\sqrt{XX'}}$$
$$F_y^* = \frac{F_y}{\sqrt{F_{yy}}} = \frac{Y' - Y}{\sqrt{YY'}} \tag{7.18}$$

vith these parameters, Equation 7.8 becomes:

$$x^2 + 2F_{xy}^* xy + y^2 + z^2 + F_x^* x + F_y^* y = 1 \tag{7.19}$$

This equation represents a family of ellipses. In the $z = 0$ plane the loci for all materials can be described by ellipses with the following features:

- We can find the x-axis intercepts by letting $y = 0$, in Equation 7.19, we will have

$$x^2 + F_x^* x - 1 = 0$$

$$x = -\frac{F_x^*}{2} \pm \sqrt{\left(\frac{F_x^*}{2}\right)^2 + 1}$$

From Equation 7.18, we can show by direct substitution

$$x = \sqrt{\frac{X}{X'}}, \quad -\sqrt{\frac{X'}{X}} \tag{7.20}$$

- Similarly, we can show the y-axis intercepts as

$$y = \sqrt{\frac{Y}{Y'}}, \quad -\sqrt{\frac{Y'}{Y}} \tag{7.21}$$

- F_{xy}^* will govern both the slenderness ratio and the inclination of the major axis; i.e., +45 degrees for negative F_{xy}^*, and −45 degrees for positive F^*_{xy}. In place of the biaxial stress test to determine th sixth strength parameter, we assume that the orthotropic failur criterion in Equation 7.19 is a generalization of the von Mise criterion

$$F^*_{xy} = -\frac{1}{2} \quad (7.22$$

where the von Mises criterion can be expressed as follows:

$$x^2 - xy + y^2 = 1 \quad (7.23$$

- The linear terms will determine the displacements of the center.

The stress failure criterion in Equation 7.8 for unidirectional com posites can be expressed in strain-space. This is often more convenien than that in stress-space because strain distribution across the thicknes of a laminate is idealized as constant or at most a linear function of th z-axis. Thus strain at any ply in a laminate can be readily determine from which the failure criterion in strain-space can be applied directly This is the motivation for expressing the criterion in strain rather tha stress-space. Since our material is assumed to be linearly elasti up to failure, the one-to-one correspondence between strain and stres is always valid. For each stress there is one and only one correspondin strain.

In order to derive Equation 7.4 from 7.3, we need only to substitut the stress components by strain components using the on-axis stress strain equation listed in Table 1.6, thus

$$F_{ij}Q_{ik}Q_{jf}\epsilon_k\epsilon_f + F_iQ_{ij}\epsilon_j = 1 \quad (7.24$$

We can define

$$\begin{aligned} G_{kf} &= F_{ij}Q_{ik}Q_{jf} \\ G_j &= F_iQ_{ij} \end{aligned} \quad (7.25$$

So that the failure criterion in strain space is

$$G_{kj}\epsilon_k\epsilon_j + G_k\epsilon_k = 1 \quad (7.26$$

We can expand this equation and invoke symmetry as we did in Equation 7.7, to have:

$$G_{xx}\epsilon_x^2 + 2G_{xy}\epsilon_x\epsilon_y + G_{yy}\epsilon_y^2 + G_{ss}\epsilon_s^2 + G_x\epsilon_x + G_y\epsilon_y = 1 \tag{7.27}$$

where

$$G_{xx} = F_{xx}Q_{xx}^2 + 2F_{xy}Q_{xx}Q_{xy} + F_{yy}Q_{xy}^2$$

$$G_{yy} = F_{xx}Q_{xy}^2 + 2F_{xy}Q_{xy}Q_{yy} + F_{yy}Q_{yy}^2$$

$$G_{xy} = F_{xx}Q_{xx}Q_{xy} + F_{xy}[Q_{xx}Q_{yy} + Q_{xy}^2] + F_{yy}Q_{xy}Q_{yy} \tag{7.28}$$

$$G_{ss} = F_{ss}Q_{ss}^2 = [Q_{ss}/S]^2$$

$$G_x = F_xQ_{xx} + F_yQ_{xy}$$

$$G_y = F_xQ_{xy} + F_yQ_{yy}$$

This equation is already dimensionless. Such representation has many advantages, which include the generality of the equation in all physical dimensions. These material constants have the same values in SI and English units.

The strength of a unidirectional composite for a given state of strain can be obtained directly from solving the quadratic equation of Equation 7.27. As was the case of the strength in Equation 7.8, there will be two roots: one for the given strain components; the other for the same strain components but with the signs reversed.

3. sample strength data

We will use unidirectional T300/5208 composite as an example for the strength calculation. The measured strength data of this material are:

$$\begin{aligned}
X &= \text{Longitudinal tensile} = 1500 \text{ MPa} \\
X' &= \text{Longitudinal compressive} = 1500 \text{ MPa} \\
Y &= \text{Transverse tensile} = 40 \text{ MPa} \\
Y' &= \text{Transverse compressive} = 246 \text{ MPa} \\
S &= \text{Longitudinal shear} = 68 \text{ MPa}
\end{aligned} \tag{7.29}$$

From Equations 7.11 et al., we can calculate the following:

$$
\begin{aligned}
F_{xx} &= 0.4444\ (\text{GPa})^{-2} \\
F_{yy} &= 101.6\ (\text{GPa})^{-2} \\
F_{ss} &= 216.2\ (\text{GPa})^{-2} \\
F_{x} &= 0 \\
F_{y} &= 20.93\ (\text{GPa})^{-1}
\end{aligned}
\tag{7.30}
$$

From Equations 7.15 and 7.22

$$
\begin{aligned}
F_{xy} &= \sqrt{F_{xx}F_{yy}}\,F^{*}_{xy} \\
&= -3.360\ (\text{GPa})^{-2}
\end{aligned}
\tag{7.31}
$$

In stress-space, the allowable strength curves for each material i[s] anchored by four points representing the four measured strengths. These points are the intercepts of the stress axes shown as solid dots in Figure 7.3. It is necessary that all failure envelopes must pass through these intercepts or focal points.

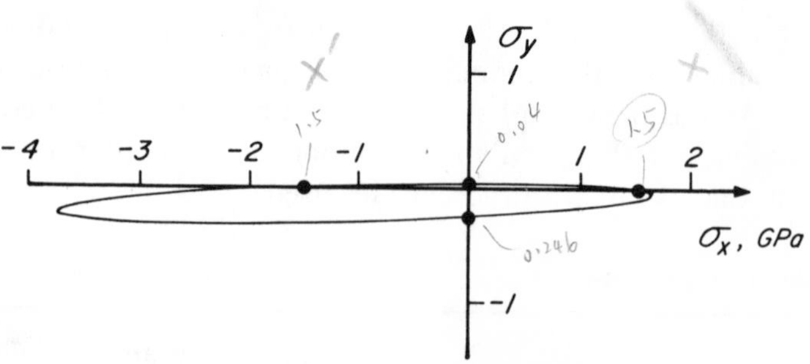

Figure 7.3 The four intercepts of the strength curve in zero shear stress plane for T300/5208. The assumed strength curve drawn through these points is based on the generalized von Mises criterion.

n allowable strength curve is drawn through these four points with an sumed interaction term of the generalized von Mises criterion in juation 7.22. Note the high degree of directionality in strength. The irve is highly elongated. As we have seen earlier, uniaxial stress duces biaxial strain because of the Poisson's effect. This is shown in gure 1.9. The four fixed points in strain-space that correspond to the tercepts in Figure 7.3 can be calculated from the on-axis stress-strain lation.

- when $\sigma_x = X$,

$$\epsilon_x^* = \epsilon_{x(a)} = X/E_x = 8.287 \times 10^{-3}$$

$$\epsilon_y^* = \epsilon_{y(a)} = -\nu_x \, \epsilon_x = -2.320 \times 10^{-3} \tag{7.32}$$

- when $\sigma_x = -X'$,

$$\epsilon_x^* = \epsilon_{x(a)} = X'/E_x = -8.287 \times 10^{-3}$$

$$\epsilon_y^* = \epsilon_{y(a)} = -\nu_x \, \epsilon_x = 2.320 \times 10^{-3} \tag{7.33}$$

- when $\sigma_y = Y$,

$$\epsilon_y^* = \epsilon_{y(a)} = Y/E_y = 3.883 \times 10^{-3}$$

$$\epsilon_x^* = \epsilon_{x(a)} = -\nu_y \, \epsilon_y = -0.0618 \times 10^{-3} \tag{7.34}$$

- when $\sigma_y = -Y'$,

$$\epsilon_y^* = \epsilon_{y(a)} = -Y'/E_y = -23.88 \times 10^{-3}$$

$$\epsilon_x^* = \epsilon_{x(a)} = -\nu_y \, \epsilon_y = 0.380 \times 10^{-3} \tag{7.35}$$

hese focal points are shown in Figure 7.4.

Now we will plot the allowable strength curves in strain-space for arious values of the normalized interaction term. The family of rength curves in strain-space is shown in Figure 7.5 using the four ocal points in Figure 7.4. The generalized von Mises criterion is equal

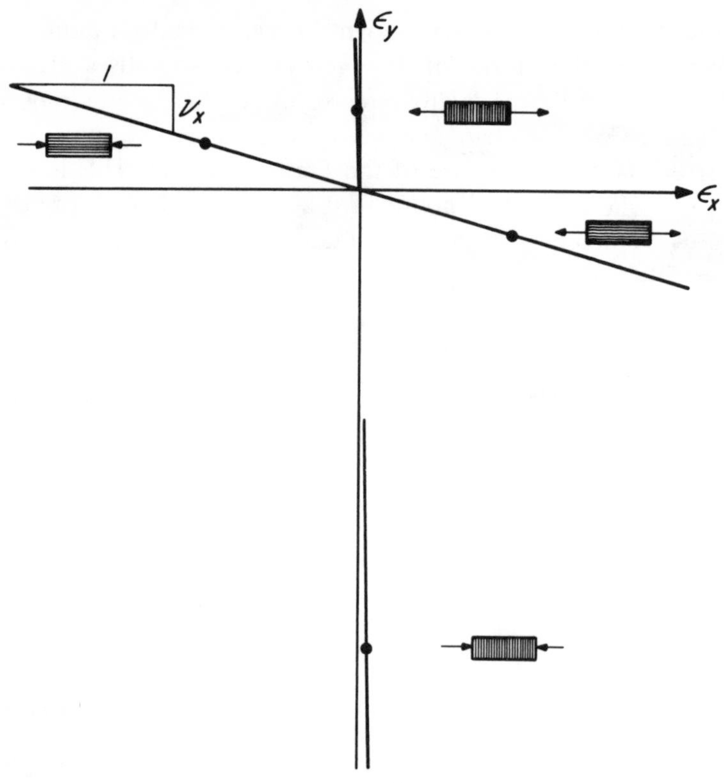

Figure 7.4 Four focal points of the allowable strength curve in strain-space, with zero shear strain. The material is T300/5208.

to the curve with $F^*_{xy} = -\frac{1}{2}$. With this assumed interaction term, th failure criteria in strain space (Equation 7.27) can be calculated wit Equation 7.28 and the modulus of the material. For T300/5208, w have

$$G_{xx} = .444 \times (181.8)^2 - 2 \times 3.36 \times 181.8 \times 2.89 + 101.6 \times (2.89)^2 \quad (7.36)$$

$$= 12004$$

Similarly

$$G_{yy} = 10680 \qquad G_{xy} = -3069 \qquad G_{ss} = 11117$$
$$G_x = 60.64 \qquad G_y = 216.5 \quad (7.37)$$

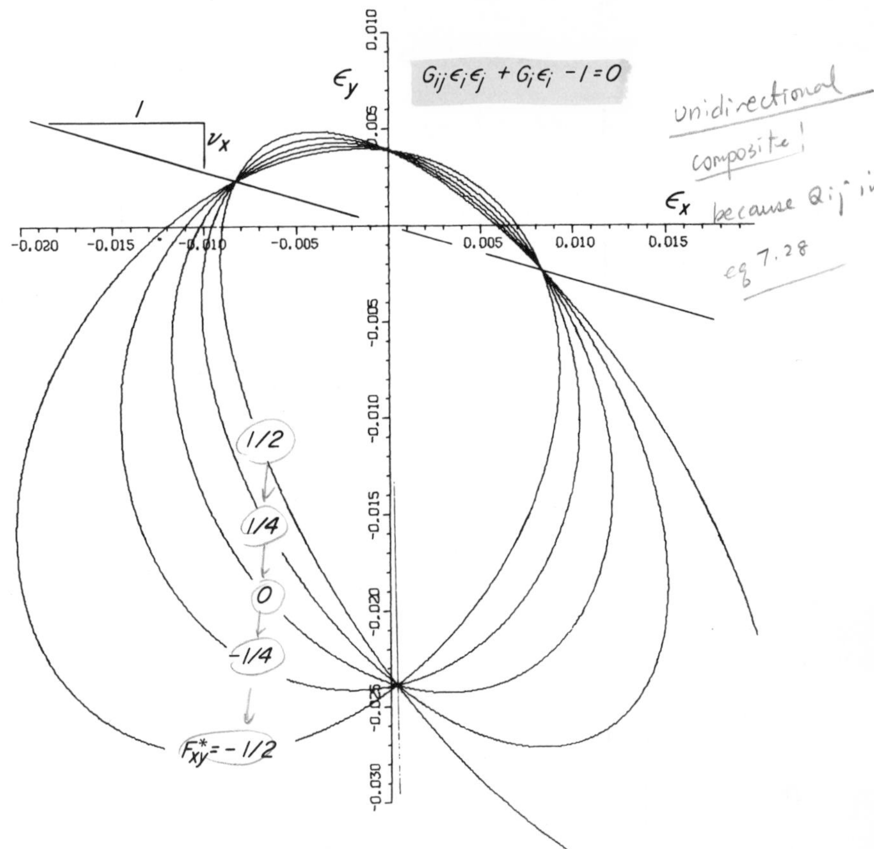

Figure 7.5 Allowable strength curves in strain-space for various values of the interaction term for T300/5208. These curves show much less directionality than comparable curves in stress-space.

Strength data for other unidirectional composites can be found in Table 7.1. The elastic constants of the same materials are listed in Chapter 1. The strength parameters in stress and strain space can be calculated following the example for T300/5208 in Equations 7.30 and 7.36. The normalized interaction term is assumed to be $-\frac{1}{2}$. We call the criterion with this value the generalized von Mises criterion. The strength parameters for the stress and strain space representations are listed in Tables 7.2 and 7.3, respectively. As a comparison a high strength aluminum is listed in the same tables. The stiffness of this metal is based on the Young's modulus of 69 GPa and Poisson's ratio of 0.3, same as those shown in Equation 1.33.

table 7.1
typical strengths of unidirectional composites in MPa

Type	Material	v_f	Long. tens. X	Long. comp. X'	Trans. tens. Y	Trans. comp. Y'	Shear S
T300/5208	Graphite /Epoxy	0.70	1500	1500	40	246	68
B(4)/5505	Boron /Epoxy	0.50	1260	2500	61	202	67
AS/3501	Graphite /Epoxy	0.66	1447	1447	51.7	206	93
Scotchply 1002	Glass /Epoxy	0.45	1062	610	31	118	72
Kevlar 49 /Epoxy	Aramid /Epoxy	0.60	1400	235	12	53	34
	Aluminum		400	400	400	400	230

table 7.2

strength parameters in stress space for unidirectional composites*

Type	Material	F_{xx} $(GPa)^{-2}$	F_{yy} $(GPa)^{-2}$	F_{xy} $(GPa)^{-2}$	F_{ss} $(GPa)^{-2}$	F_x $(GPa)^{-1}$	F_y $(GPa)^{-1}$
T300/5208	Graphite /Epoxy	.444	101.6	− 3.36	216.2	0	20.93
B(4)/5505	Boron /Epoxy	.317	81.15	− 2.53	222.7	.393	11.44
AS/3501	Graphite /Epoxy	.476	93.48	− 3.33	115.4	0	14.50
Scotchply 1002	Glass /Epoxy	1.543	273.3	−10.27	192.9	− .697	23.78
Kevlar 49 /Epoxy	Aramid /Epoxy	3.039	1572	−34.56	865.0	−3.541	64.46
	Aluminum	6.25	6.25	− 3.125	18.90	0	0

*Based on generalized von Mises criterion: $F^*_{xy} = -\frac{1}{2}$.

table 7.3
strength parameters in strain space for unidirectional composites (dimensionless)*

Type	Material	G_{xx}	G_{yy}	G_{xy}	G_{ss}	G_x	G_y
T300/5208	Graphite /Epoxy	12004	10680	−3069	11117	60.64	216.5
B(4)/5505	Boron /Epoxy	10374	27646	−2988	6961	129.6	214.3
AS/3501	Graphite /Epoxy	7375	7467	−1746	5828	39.17	130.57
Scotchply 1002	Glass /Epoxy	1913	18881	1712	3306	24.56	198.05
Kevlar 49 /Epoxy	Aramid /Epoxy	13453	47656	2068	4576	−149.8	350.8
	Aluminum	28387	28387	1976	13313	0	0

*Based on generalized von Mises criterion: $F^*_{xy} = -\frac{1}{2}$.

transformation equations for strength parameters

'e have dealt with the on-axis orientation of the failure criterion. We an establish the off-axis criterion by establishing the transformation quations. We have the following:

- On-axis: stress, and the quadratic and linear strength parameters:

$$\sigma_x, \sigma_y, \sigma_s; \quad F_{xx}, F_{yy}, F_{xy}, F_{ss}; \quad F_x, F_y$$

- Off-axis: stress, and the quadratic and linear strength parameters:

$$\sigma_1, \sigma_2, \sigma_6; \quad F_{11}, F_{22}, \ldots; \quad F_1, F_2, F_6$$

'e wish to express the off-axis strength in terms of the on-axis rength. This is precisely the same as the transformation of compliance Table 3.8.

We can derive this by substituting the stress transformation equation to the on-axis failure criterion in Equation 7.8. The stress transforma-on is used to replace the on-axis stress to an off-axis stress; i.e., able 2.1.

$$\begin{aligned}
&F_{xx}(m^2\sigma_1 + n^2\sigma_2 + 2mn\sigma_6)^2 \\
&\quad + 2F_{xy}(m^2\sigma_1 + n^2\sigma_2 + 2mn\sigma_6)(n^2\sigma_1 + m^2\sigma_2 - 2mn\sigma_6) \\
&\quad + F_{yy}(n^2\sigma_1 + m^2\sigma_2 - 2mn\sigma_6)^2 \\
&\quad + F_{ss}[-mn\sigma_1 + mn\sigma_2 + (m^2 - n^2)\sigma_6]^2 \\
&\quad + F_x(m^2\sigma_1 + n^2\sigma_2 + 2mn\sigma_6) \\
&\quad + F_y(n^2\sigma_1 + m^2\sigma_2 - 2mn\sigma_6) = 1
\end{aligned} \tag{7.38}$$

earranging the above, we have

$$\begin{aligned}
&[m^4F_{xx} + n^4F_{yy} + 2m^2n^2F_{xy} + m^2n^2F_{ss}]\sigma_1{}^2 \\
&\quad + \ldots + 2mn(F_x - F_y)\sigma_6 = 1
\end{aligned}$$

Matching this with Equation 7.5, we have

$$F_{11} = m^4 F_{xx} + n^4 F_{yy} + 2m^2 n^2 F_{xy} + m^2 n^2 F_{ss}$$

$$\cdots\cdots \qquad (7.39$$

$$F_6 = 2mn(F_x - F_y)$$

The quadratic strength parameter transforms exactly like the com pliance in Table 3.8; the linear, the same as strain in Table 2.5. Th only difference is that Table 2.5 transforms from the off-axis to th on-axis. The transformation of F's and those of all other material prop erties are from the on-axis to the off-axis. The sign of the sine function must change. The results are listed in Tables 7.4(a) and (b), wher matrix multiplication is implied.

table 7.4(a)

transformation of quadratic strength parameters in stress space in power functions

	F_{xx}	F_{yy}	F_{xy}	F_{ss}
F_{11}	m^4	n^4	$2m^2n^2$	m^2n^2
F_{22}	n^4	m^4	$2m^2n^2$	m^2n^2
F_{12}	m^2n^2	m^2n^2	m^4+n^4	$-m^2n^2$
F_{66}	$4m^2n^2$	$4m^2n^2$	$-8m^2n^2$	$(m^2-n^2)^2$
F_{16}	$2m^3n$	$-2mn^3$	$2(mn^3-m^3n)$	mn^3-m^3n
F_{26}	$2mn^3$	$-2m^3n$	$2(m^3n-mn^3)$	m^3n-mn^3

$m = \cos\theta,\ n = \sin\theta$

table 7.4(b)

transformation of linear strength parameters in stress space in power functions

	F_x	F_y
F_1	m^2	n^2
F_2	n^2	m^2
F_6	$2mn$	$-2mn$

$m = \cos\theta,\ n = \sin\theta$

milarly we can derive the same equations in terms of multiple angle nctions. These are shown in Tables 7.5(*a*) and (*b*).

table 7.5(*a*)
transformation of quadratic strength parameters in stress space in multiple angle functions

	1	U_2	U_3
F_{11}	U_1	$\cos 2\theta$	$\cos 4\theta$
F_{22}	U_1	$-\cos 2\theta$	$\cos 4\theta$
F_{12}	U_4		$-\cos 4\theta$
F_{66}	U_5		$-4\cos 4\theta$
F_{16}		$\sin 2\theta$	$2\sin 4\theta$
F_{26}		$\sin 2\theta$	$-2\sin 4\theta$

here the U's are defined like those for the compliance in Equation 56. The F_{ij} shall replace the S_{ij}.

table 7.5(*b*)
transformation of linear strength parameters in stress space in multiple angle functions

	p	q
F_1	1	$\cos 2\theta$
F_2	1	$-\cos 2\theta$
F_6		$2\sin 2\theta$

here $p = \frac{1}{2}(F_x + F_y)$, $q = \frac{1}{2}(F_x - F_y)$

We can derive the transformation equations for the strength param- ers in strain space by substituting the transformed strain components to the failure criterion in Equation 7.27. The results are shown in able 7.6 for the power functions formulation and Table 7.7 for the multiple angle functions. The quadratic parameters transform like th stiffness components; the linear parameters, the stress components.

table 7.6(*a*)

transformation of quadratic strength parameters in strain space in power functions

	G_{xx}	G_{yy}	G_{xy}	G_{ss}
G_{11}	m^4	n^4	$2m^2n^2$	$4m^2n^2$
G_{22}	n^4	m^4	$2m^2n^2$	$4m^2n^2$
G_{12}	m^2n^2	m^2n^2	m^4+n^4	$-4m^2n^2$
G_{66}	m^2n^2	m^2n^2	$-2m^2n^2$	$(m^2-n^2)^2$
G_{16}	m^3n	$-mn^3$	mn^3-m^3n	$2(mn^3-m^3n)$
G_{26}	mn^3	$-m^3n$	m^3n-mn^3	$2(m^3n-mn^3)$

$m=\cos\theta$, $n=\sin\theta$

table 7.6(*b*)

transformation of linear strength parameters in strain space in power functions

	G_x	G_y
G_1	m^2	n^2
G_2	n^2	m^2
G_6	mn	$-mn$

table 7.7(*a*)

transformation of quadratic strength parameters in strain space in multiple angle functions

	I	U_2	U_3
G_{11}	U_1	$\cos 2\theta$	$\cos 4\theta$
G_{22}	U_1	$-\cos 2\theta$	$\cos 4\theta$
G_{12}	U_4		$-\cos 4\theta$
G_{66}	U_5		$-\cos 4\theta$
G_{16}		$\frac{1}{2}\sin 2\theta$	$\sin 4\theta$
G_{26}		$\frac{1}{2}\sin 2\theta$	$-\sin 4\theta$

where the U's are similar to those for the stiffness in Equation 3.15.

table 7.7(b)

transformation of linear strength parameters in strain space in multiple angle functions

	p	q
G_1	1	$\cos 2\theta$
G_2	1	$-\cos 2\theta$
G_6		$\sin 2\theta$

here $p = \frac{1}{2}(G_x + G_y), q = \frac{1}{2}(G_x - G_y)$

We can easily determine the off-axis strength of unidirectional com-
osites by using the transformation relations of the strength param-
ers. The failure criterion in stress space is

$$F_{ij}\sigma_i\sigma_j + F_i\sigma_i = 1; \quad i,j = 1,2,6 \tag{7.40}$$

here the F's are of the off-axis orientation. The simplest case is the
iaxial tensile and compressive strengths of an off-axis unidirectional
mposite. The failure criterion above is reduced to:

$$F_{11}\sigma_1^2 + F_1\sigma_1 - 1 = 0 \tag{7.41}$$

sing the power function formulation of the transformation equation
Tables 7.4(a) and (b) respectively, we have

$$\begin{aligned} F_{11} &= m^4 F_{xx} + n^2 F_{yy} + 2m^2 n^2 F_{xy} + m^2 n^2 F_{ss} \\ F_1 &= m^2 F_x + n^2 F_y \end{aligned} \tag{7.42}$$

here $m = \cos\theta$, $n = \sin\theta$.

There are two roots in the solution of Equation 7.41 corresponding
the tensile and compressive strengths.

The off-axis shear strength is simply the roots of:

$$F_{66}\sigma_6^2 + F_6\sigma_6 - 1 = 0 \qquad (7.4$$

where $F_{66} = 4m^2m^2(F_{xx} + F_{yy}) - 8m^2n^2F_{xy} + (m^2 - n^2)^2 F_{ss}$

$F_6 = 2mn\,(F_x - F_y)$

Using the strength data in Table 7.2 we can predict the off-ax uniaxial strengths of AS/3501 when θ is 45 degrees

$$F_{11} = \frac{1}{4}(.476 + 93.48 - 2 \times 3.33 + 115.4)$$

$$= \frac{1}{4}(202.7) = 50.2\ (\text{GPa})^{-2} \qquad (7.4$$

$$F_1 = \frac{1}{2}(14.5) = 7.25\ (\text{GPa})^{-1} \qquad (7.4$$

Solving for the roots:

$$50.2 \times 10^{-18}\ \sigma_1^2 + 7.25 \times 10^{-9}\ \sigma_1 - 1 = 0 \qquad (7.4$$

$$\sigma_1 = -72.2 \pm \sqrt{(72.2)^2 + 19920} = -72.2 \pm 158$$

$$= 86, -230\ \text{MPa} \qquad (7.4$$

Repeating the same calculation for different ply orientation, we arriv at the solid lines in Figure 7.6. Available data are shown as circles.*

The normalized interaction term is taken to be −½ for the calcul tion above. This is an assumption. If we use the maximum and min mum value of 1 and −1, the limits imposed by ellipses, we hav respectively

$$F_{xy} = \pm 6.66\ (\text{GPa})^{-2}$$

$$F_{11} = 55.6,\ 49.0\ (\text{GPa})^{-2}$$

*Provided by R. Y. Kim, University of Dayton Research Institute

bstituting these values into Equation 7.46 in place of 50.2 and solv- ; the quadratic equations, we will have two sets of roots:

$$\text{For } F^*_{xy} = 1, \quad \sigma_1 = 83.9, \quad -214 \text{ MPa}$$

$$F^*_{xy} = -1, \quad \sigma_1 = 86.9, \quad -234 \text{ MPa}$$

e predicted off-axis uniaxial strengths are not sensitive to the value the interaction term. At 45 degrees the contribution of this term to transformed F is the greatest; see Equation 7.42. As the angle ves away from 45 degrees the interaction term has even less effect. nversely, the uniaxial off-axis test cannot be used to compare or idate failure criteria. Other combined stress state which the off-axis t cannot produce would be more discriminating.

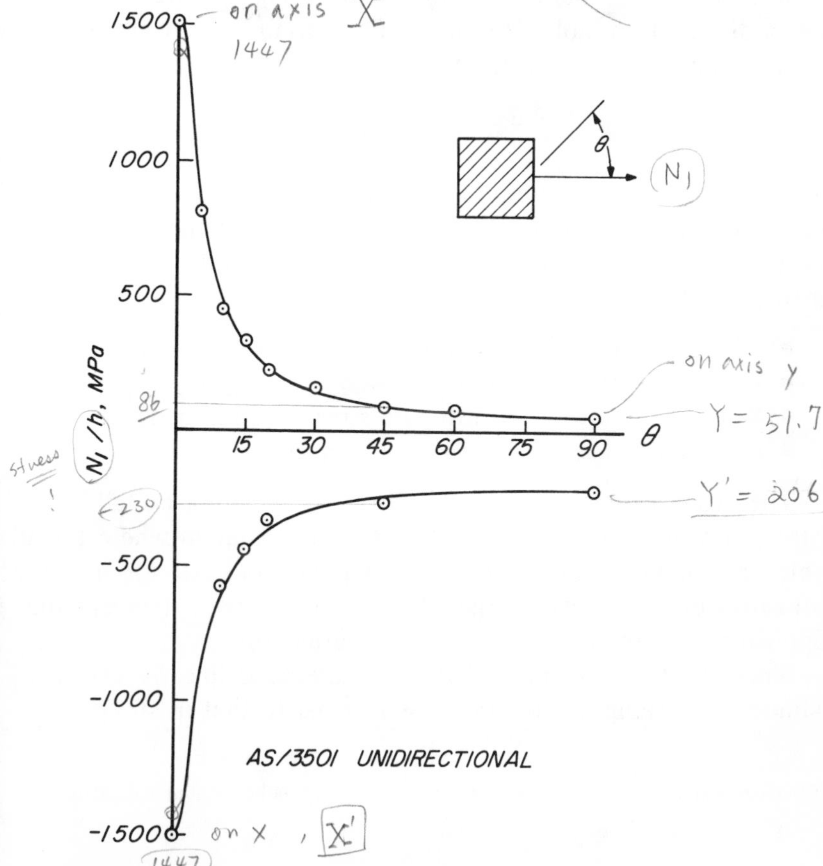

Figure 7.6 Uniaxial tensile and compressive strengths of AS/3501 graphite/epoxy composites as a function of fiber orientations.

5. strength/stress or strength ratios

Our failure criterion such as that in Equation 7.8 specifies the conditi of failure. The strength parameters expressed in F's are fixed for a giv material. The imposed stress components when substituted into t left-hand side of the failure criterion may produce any positive num ical value. When this value equals unity, the failure criterion is satisfie i.e., failure will occur under the given stress components. If the co ponents have greater values, the left-hand side of Equation 7.8 excee unity. This is not physically possible. The material cannot sustain su combination of stress components.

If the imposed stress is smaller, the left-hand side has a value le than unity. We conclude that failure has not occurred. Thus the failu criterion like that in Equation 7.8 provides only a go-or-no-go criteric

We can increase the information given by the failure criterion if use a different variable. We define this variable as the strength/stre ratio, or simply, strength ratio R:

$$\sigma_{i(a)} = R\,\sigma_i \qquad \epsilon_i^* = \epsilon_{i(a)} = R\,\epsilon_i \qquad (7.4$$

where stress or strain components without remarks are those appli or imposed; and subscript (a) or asterisk means the allowed or t ultimate stress or strain. Several features of this ratio should mentioned.*

- When applied stress or strain is zero, $R = \infty$.
- When the stress or strain is safe, $R > 1$.
- When the allowable or ultimate stress or strain is reached, $R =$
- R cannot be less than unity which has no physical reality.

The conventional failure criterion is a fail-or-no-fail criteric Strength ratios will not only define the upper bound where the allo able or ultimate exist ($R = 1$), but will also indicate the quantitati measures of the safety margin. If the ratio is two, it means that t applied stress can be doubled before failure occurs.

Since we have assumed that our material is linearly elastic up failure, the strength ratio in stress is equal to that in strain.

*The reciprocal of this is called stress ratio in some design handbooks of metallic structures.

We also assume proportional loading in Equation 7.48 that for each plied stress or strain, its unit vector remains fixed up to failure. This sumption is necessary to define a unique strength ratio from any arting point.

The starting point of stress or strain application need not be the rigin of the reference coordinate system. If our surface is a sphere, the arting point can be anywhere within the sphere. Moreover, there are any reasons for starting from points other than the center or origin. esides, initial stress or strain, different tensile and compressive rengths, and different longitudinal and transverse strengths will all ift the starting point of stress and strain application.

If the applied stress or strain is a unit vector, the resulting strength tio value becomes the allowable. This is a convenient feature of this tio from the standpoint of computation.

We will try to illustrate the meaning of strength ratios with simple xamples. Let us first use circles to represent the surfaces of constant rength ratios. The equation of this family of curves is:

$$x^2 + y^2 = 1/R^2, \text{ or} \tag{7.49}$$

$$R^2 = 1/(x^2 + y^2) \tag{7.50}$$

here x, y = stress or strain components.

rom the definition of strength ratios in Equation 7.48, we know

$$\begin{aligned} x_{(a)} &= Rx \\ y_{(a)} &= Ry \end{aligned} \tag{7.51}$$

$$R = x_{(a)}/x = y_{(a)}/y \tag{7.52}$$

ombining Equations 7.49 and 7.50, we have

$$x_{(a)}^2 + y_{(a)}^2 = 1 \tag{7.53}$$

he strength ratio becomes unity when the stress or strain is at its llowable level. The curves of Equation 7.49 are shown in Figure 7.7. e can make the following comments:

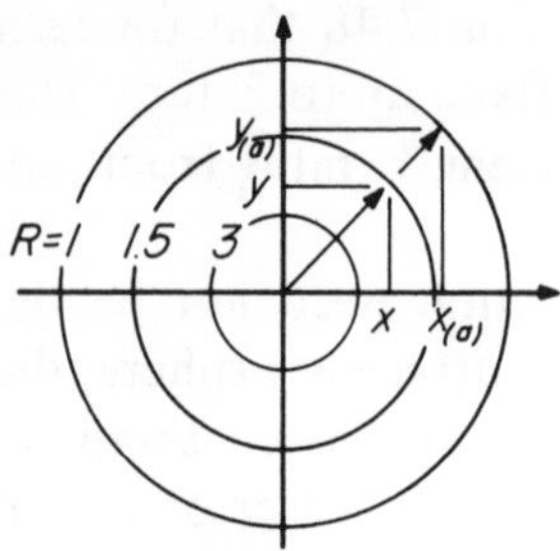

Figure 7.7 Curves for constant strength ratios for an idealized material.

- This material is isotropic with equ strengths in x and y directions. T tensile and compressive strengths a also equal.
- Proportional loading from the origin assumed. Each loading path follows radius. The strength ratio curves w change if proportional loading is n followed. This change will be d cussed later in this chapter.
- The combined stress effect can ce tainly be different from a circle. fact, the von Mises criterion for plasticity in isotropic materials nondimensional stress components is elliptical; see Equation 7.2

$$x^2 - xy + y^2 = 1/R^2, \quad \text{or} \qquad (7.5$$

$$R^2 = 1/(x^2 - xy + y^2) \qquad (7.5$$

Instead of concentric circles, we will have concentric ellipse However, the strength ratio concept is equally applicable. Th remarks on isotropy, equality in tension and compression, an proportional loading remain valid for the von Mises case.

If our material has different tensile and compressive strengths b remains isotropic with circular combined stress effect, Equation 7. must be modified as follows:

$$[x_{(a)} + d]^2 + [y_{(a)} + d]^2 = 1 \qquad (7.5$$

where d = one-half the difference between tensile and compressi strengths (positive d means higher compressive than tensile strength).

Substituting Equation 7.51 and rearranging, we have

$$\left[x + \frac{d}{R}\right]^2 + \left[y + \frac{d}{R}\right]^2 = 1/R^2, \quad \text{or} \qquad (7.5$$

$$(x^2 + y^2)R^2 + 2d(x + y)R - (1 - 2d^2) = 0 \qquad (7.5$$

his family of circles is shown in Figure 7.8. The following features can e noted.

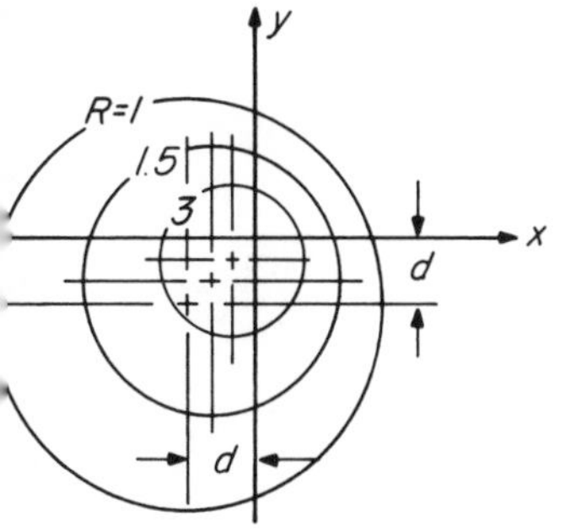

igure 7.8 Surfaces for constant rength ratios for an idealized otropic material having differ- t tensile and compressive rengths.

- The curves remain circular by the assumption made in Equation 7.56.
- The radius is inversely proportional to the square of the strength ratio.
- Proportional loading starts from the origin of the x-y coordinates.
- The center of each strength ratio is no longer fixed in one position.
- The circles are displaced by an amount inversely proportional to the strength ratio.

Another perturbation of the curves of constant strength ratio can come from initial stress.

Let us assume that initial tensile stress components exist. Then, analogous to Equation 7.57, we have

$$\left[x + \frac{x_o}{R}\right]^2 + \left[y + \frac{y_o}{R}\right]^2 = 1/R^2, \text{ or} \tag{7.59}$$

$$[x^2 + y^2]R^2 + 2[x_o x + y_o y]R - [1 - x_o^2 - y_o^2] = 0 \tag{7.60}$$

ıdging from this equation as compared with Equation 7.57, there is no ualitative difference between the two. The only difference is the dis- lacements of the centers of successive circles. The curves for Equation 59 will look essentially the same as those in Figure 7.8.

For our unidirectional composite, the failure criteria listed in Equa- ons 7.1 to 7.4 can be easily modified by introducing strength ratios efined in Equation 7.48.

$$\sigma_{i(a)} = R\sigma_i \qquad \epsilon_{i(a)} = R\epsilon_i \tag{7.61}$$

ailure occurs when the allowed stress or strain on the left-hand side of quation 7.61 is reached. Thus Equations 7.3 and 7.4 actually apply to the allowable quantities, or

$$F_{ij}\sigma_{i(a)}\sigma_{j(a)} + F_i\sigma_{i(a)} = 1$$

Substituting Equation 7.61,

$$[F_{ij}\sigma_i\sigma_j]R^2 + [F_i\sigma_i]R - 1 = 0$$

$$[G_{ij}\epsilon_i\epsilon_j]R^2 + [G_i\epsilon_i]R - 1 = 0 \quad \text{(7.6}$$

or

$$aR^2 + bR - 1 = 0 \quad \text{(7.6}$$

Instead of solving for the stress or strain, we solve for the strength rati There are to conjugate roots, R and R', corresponding to the appli stress/strain vector going in opposite directions; i.e.,

$$R, R' = \sqrt{(b/2a)^2 + (1/a)} \pm (b/2a) \quad \text{(7.6}$$

Usually only R is needed, R' is useful for bending.

6. in-plane strength of laminates

The in-plane strength of laminates is determined by examining t strength ratios of each ply orientation subjected to a given state stress resultants. The ply with the lowest strength ratio will fail fir The state of stress resultant when this ply failure occurs is call the first-ply-failure state. The plies with higher strength ratios w fail later, when the externally applied stress is increased. This successi ply failure progresses until the last ply or ultimate failure occurs. Tl ply-by-ply examination can be expressed in Figure 7.9 The relation b tween in-plane strain and the applied stress resultant is from Chapter

$$\epsilon_i^o = a_{ij}N_j \quad \text{(7.6}$$

Substituting this into Equation (7.62) for a ply with θ orientation:

$$[G_{ij}^{(\theta)}\,\epsilon_i^o\epsilon_j^o]R_{(\theta)}^2 + [G_i^{(\theta)}\,\epsilon_i^o]R_{(\theta)} - 1 = 0 \quad \text{(7.6}$$

$$[G_{ij}^{(\theta)}\,a_{ik}a_{jf}N_kN_f]R_{(\theta)}^2 + [G_i^{(\theta)}\,a_{ij}N_j]R_{(\theta)} - 1 = 0 \quad \text{(7.6}$$

$$[H_{kf}^{(\theta)}\,N_kN_f]R_{(\theta)}^2 + [H_j^{(\theta)}\,N_j]R_{(\theta)} - 1 = 0 \quad \text{(7.6}$$

here

$$H_{kf}^{(\theta)} = G_{ij}^{(\theta)}\, a_{ik} a_{jf}, \quad H_j^{(\theta)} = G_i^{(\theta)}\, a_{ij} \tag{7.69}$$

$R_{(\theta)}$ = Strength ratio of the θ ply orientation

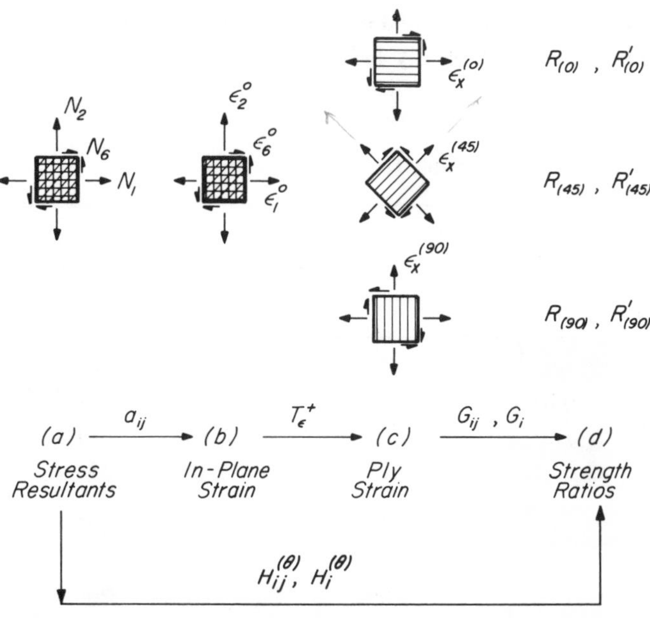

Figure 7.9 Ply-by-ply strength ratios of a laminate. For given laminate stress resultants or in-plane strains, ply strains can be calculated. Then the strength ratios can be readily determined. The lowest strength ratio ply is the first ply to fail.

The in-plane strength of a multidirectional laminate will have multi- e strength ratios; one set (R and R') for each ply orientation. The ply ith the lowest ratio will be the first to fail, the FPF. Two factors ontrol the ply failures in a laminate. First, the in-plane compliance $\boldsymbol{a}$. iis is a function of the ply volume fractions. The function is non- iear; see Figures 4.7 and 4.9. The specific ply orientation is the cond factor. The H functions in Equation 7.69 are vector products of e transformed strength parameters G's of the ply and the in-plane mpliance of the laminate. The H functions provide a direct link tween (a) and (d) in Figure 7.9 for each ply.

The ply failure stress levels for T300/5208 cross-ply laminates bas
on the normalized stress resultant (N/h) are listed in Table 7.8. F
each applied unit stress, the strength ratios are now equal to the ul
mate strengths. We list the strengths for the 0 and 90 degree plies. F
example, the first line in Table 7.8 shows the case of hydrostatic stre
applied to a [0/90] laminate. The tensile failure would occur at 3
MPa, and compressive failure at 1960 MPa. Both plies would fail simu
taneously. The first and last ply failures coincide.

table 7.8
selected ply failure stress of cross-ply laminates, in MPa

Laminates	Unit stress vector	0-degree ply $R_{(0)}$	$R'_{(0)}$	90-degree ply $R_{(90)}$	$R'_{(9}$
[0/90]	(1,1,0)	302	1960	302	19
	(1,0,0)	681	1107	373	22
	(0,1,0)	373	2268	681	11
	(1,−1,0)	856	351	351	8
[0_2/90]	(1,1,0)	240	1830	303	13
	(1,0,0)	892	1413	485	29
	(2,1,0)	208	1282	200	13
	(1,−1,0)	941	260	389	5
	(0,1,0)	262	1602	476	7
[0/90_2]	(1,1,0)	303	1334	240	18
	(1,0,0)	476	785	262	16
	(1,2,0)	200	1360	208	12
	(2,1,0)	192	543	126	9
[0_6/90]	(1,1,0)	133	951	207	5
	(1,0,0)	1159	1691	614	35
	(0,1,0)	135	829	244	4
	(1,−1,0)	675	135	300	2

The second line in Table 7.8 shows that under uniaxial tensile loa
the 90-degree ply would fail at 373 MPa, and the 0-degree at 681 MP
In compression, the first ply failure would be in the 0-degree ply

07 MPa, while the 90-degree will not fail until 2268 MPa.

The complete failure envelopes for various cross-ply laminates are own in stress space in Figure 7.10. The inner boundary or the over-ped area is the FPF locus; the outer boundary, the ultimate failure cus.

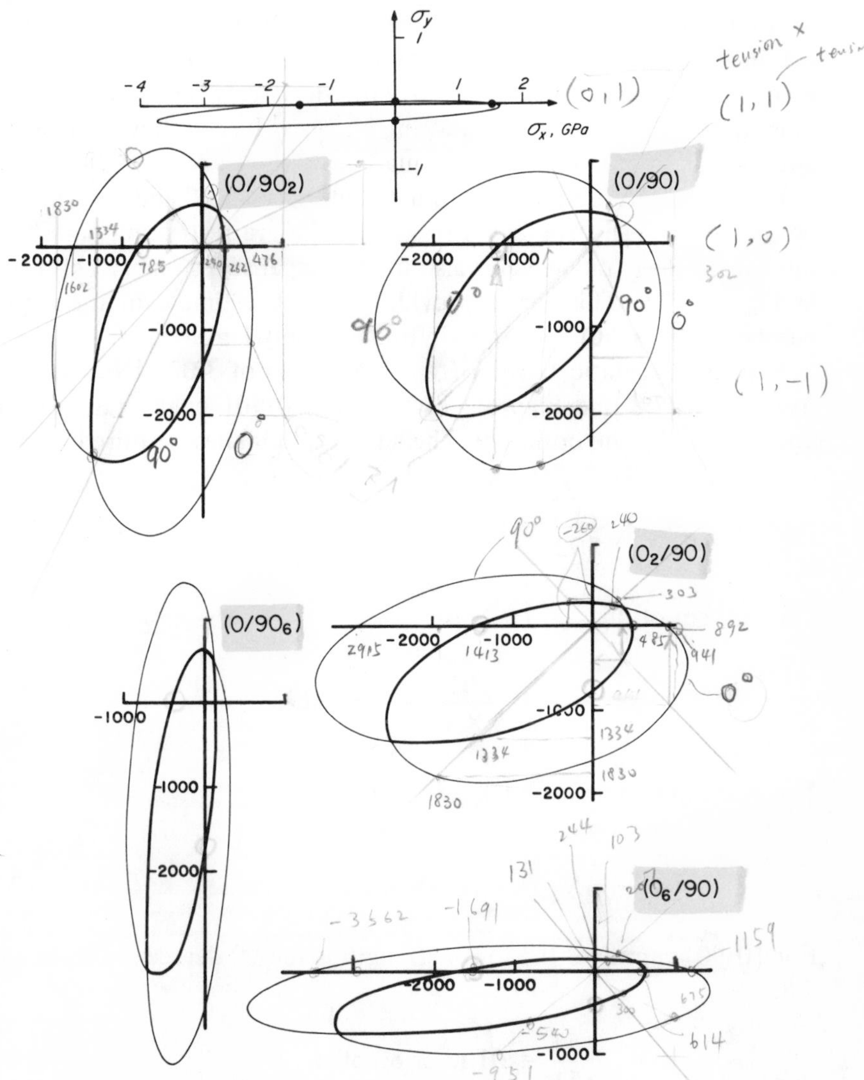

Figure 7.10 Failure envelopes in normal stress resultant space for T300/5208 cross-ply laminates. A unidirectional composite (Figure 7.3) is also shown to indicate the degree of change in ply failure envelopes within a laminate. The unit for the laminates is MPa.

If simultaneous failure of all plies is an optimum condition for laminate design, it is possible to achieve it only in the first and thir quadrants in the stress space. The envelopes do not coincide in th other quadrants. For hydrostatic stress, the [0/90] would be th optimum. This is intuitively obvious. If the stress ratio is 2:1 or the un vector is (2,1,0), we see in Table 7.8 that the $[0_2/90]$ laminate woul be approximately optimum. A more exact ratio should be 2:1.1. We ar within 10 percent of the optimum if we use 2:1. Thus a simple-minde model of matching the ply ratio with the stress ratios is fairly goo This approach is called the netting analysis. But netting analysis do not cover the second and fourth quadrants.

We can superpose all the laminates in Figure 7.10 and determine th optimum ply ratio for any stress ratio. We are restricting ourselves t the zero shear plane which is also the principal stress plane. This is don in Figure 7.11. This figure provides a quick estimate of the require number of plies and the ply ratio of a cross-ply laminate.

For a hydrostatic tension ($N_1 = N_2 = p$) of 3.02 MN/m we kno from Table 7.8 and Figure 7.11 that the [0/90] or ply ratio equal t unity would be the optimum. The laminate thickness required

$$h = \frac{3.02}{302} = .01 \text{ m} \tag{7.70}$$

The number of plies

$$n = \frac{h}{h_o} = \frac{.01}{125 \times 10^{-6}} = 80 \text{ plies} \tag{7.71}$$

The optimum laminate is $[0_{40}/90_{40}]$.

Suppose we have

$$N_1 = -N_2 = 3.51 \text{ MN/m} \tag{7.72}$$

For [0/90] laminate the lower strength is the 90 degree ply from Tabl 7.8.

$$h = \frac{3.51}{351} = .01 \text{ m or 80 plies} \tag{7.73}$$

ne laminate is the same as the above, or $[0_{40}/90_{40}]$. But, if we use $_2/90]$, the strength is now 389.

$$h = \frac{3.51}{389} = .00902 \text{ m or approximately 72 plies} \qquad (7.74)$$

ne optimum laminate should be $[0_{48}/90_{24}]$. Note the significant fference between the two laminates.

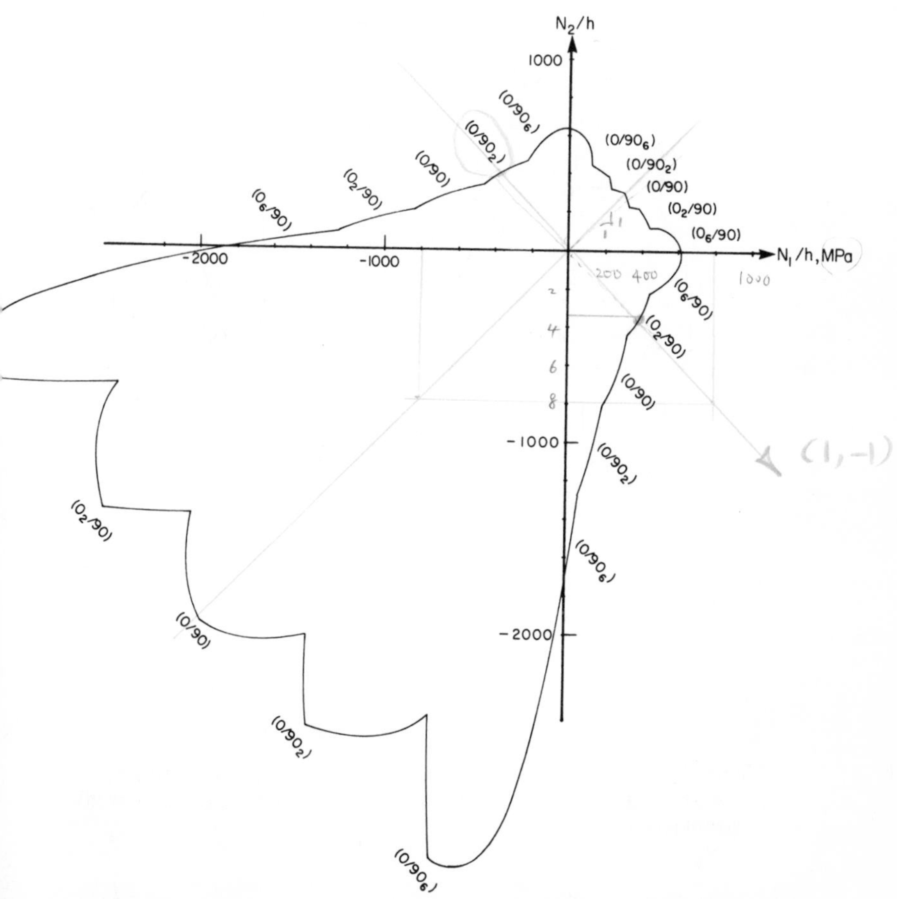

re 7.11 Maximum first ply failure envelope of T300/5208 cross-ply laminates. 0 and all 90-degree unidirectional composites are not included.

The failure envelopes of multidirectional laminates can also be sho in strain space. This is obtained by using the strength parameters strain space listed in Equation 7.37 for the T300/5208. Or we can u the parameters in Table 7.3 if we are interested in AS/3501. Subs tuting the data in Equation 7.66 for the multidirectional lamina we obtain the failure envelopes in strain space shown in Figure 7.12.

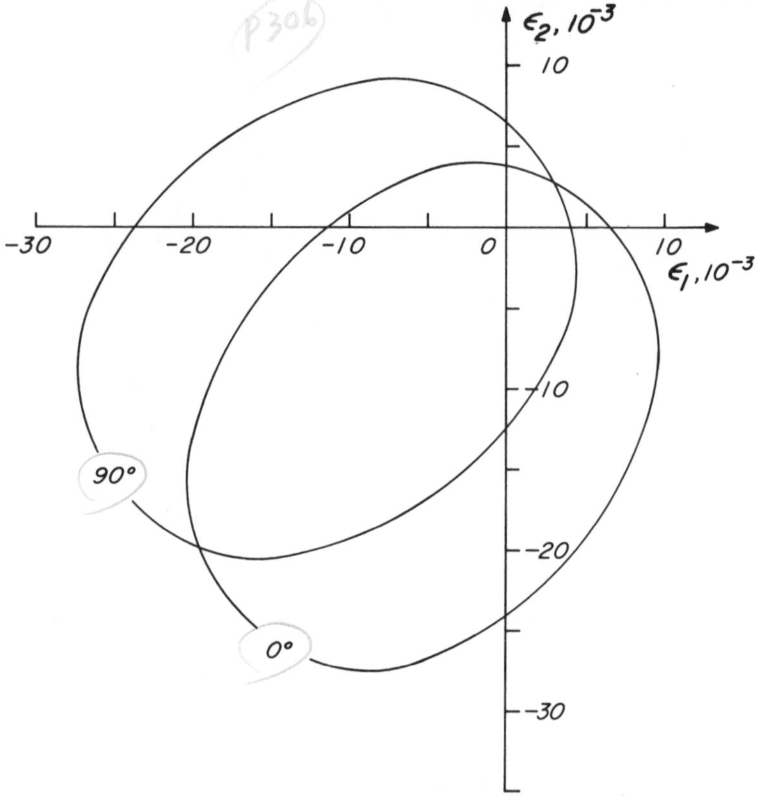

Figure 7.12 Failure envelopes of 0 and 90 degree plies in cross-ply laminates of T300/5208. Ply ratios do not change the failure envelopes in strain space.

The key feature of these envelopes is that they remain fixed for ea ply orientation, independent of the ply ratios. This is true because t laminate compliance does not appear in the failure criterion in Equ tion 7.66. For each laminate, the loading path in strain space will va This was not the case for the stress space version of the failure criteri

For different ply ratios, the externally applied loads will follow dif-erent loading paths in strain space. For uniaxial tension or compressive tresses, the resulting loading paths have slopes equal to their Poisson's atios. This was shown in Figure 7.4 for unidirectional composites 300/5208. For cross-ply laminates of various ply ratios, the loading aths as dictated by Poisson's ratios are shown in Figure 7.13.

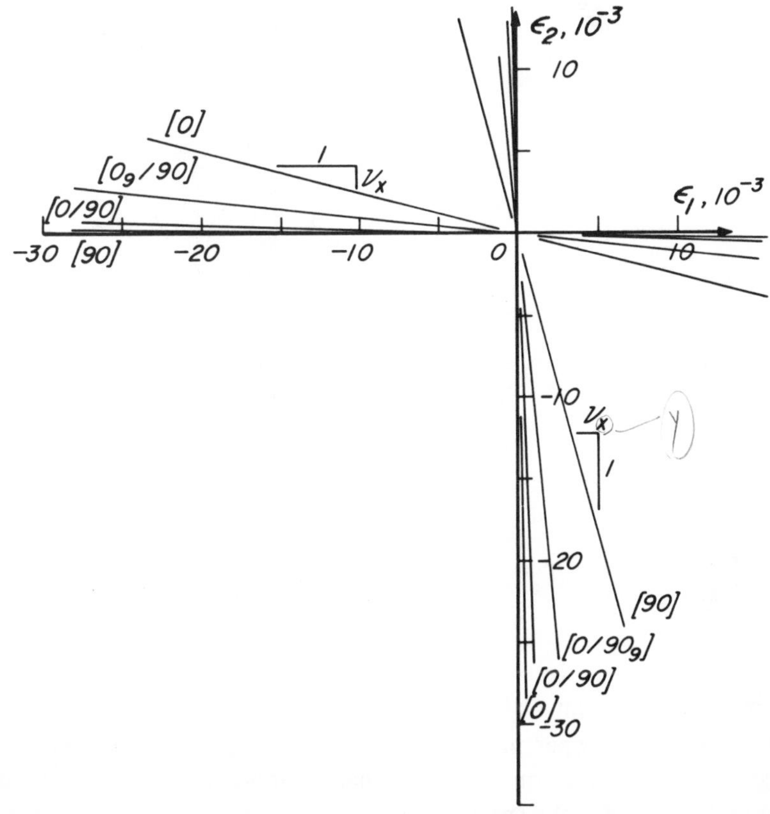

Figure 7.13 Loading paths for various ply ratios in cross-ply laminates of T300/5208. Loading is limited to uniaxial stresses. The slopes of the lines are therefore equal to the Poisson's ratio of particular laminates.

We can then combine the loading paths in Figure 7.13 with the ailure surfaces in Figure 7.12. This is done in Figure 7.14. For each oading vector, there is also the reversed (unloading) vector which hanges the sign for all the strain components. The strength ratio based

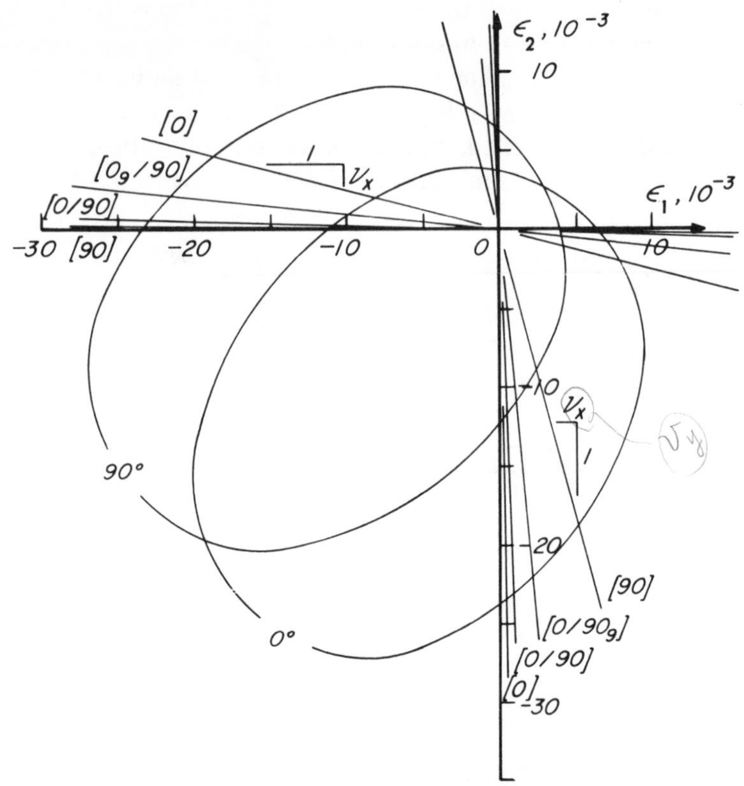

Figure 7.14 Superposed failure envelopes of plies in cross-ply laminates of T300/5208. This figure is the sum of Figures 7.12 and 7.13.

on strains will provide the numerical margin of safety; i.e., how mu increase the strain can sustain before failure.

The failure envelopes in strain or in-plane strain space in Figure 7. are governed by Equation 7.66. Laminate compliance is not included the equation. Only ply orientations appear through the strain para eters, the G's. For a bidirectional laminate, there are only two p envelopes and their shape and position remain fixed independent of t ply ratios. The loading path in strain space will change as we change t ply ratio. Several common loading paths are shown in Figure 7.13.

The failure envelopes of multidirectional laminates can be illustrat in a number of ways. Each representation has its own advantages a reveals one or more aspects of the interaction within a laminate. On with a good understanding of the lamina-laminate relation can we u composite materials effectively.

The laminate will have the first ply failure in the ply with the lowest rength ratio. After the first ply failure, the laminate may be able to ›ntinue until all plies have failed. The calculation of the laminate ›havior after the first ply failure is not easy to perform. The laminate ›mpliance is increased, or laminate modulus is decreased. Internal ımage is induced by the first ply failure. An iterative process is re-ıired to assess the successive ply failures. This process is not well ›fined and will not be covered here. If we assume that ply failures do ›t affect the laminate compliance, the ultimate strength of the lam-ate can be determined from the highest strength ratio among the ies. This approximation is not unreasonable if all ply failures up to ıe ultimate strength of the laminate are limited to matrix failures and › not involve fiber failures. The loss of transverse stiffness of many ıidirectional plies may not significantly affect the in-plane stiffness of ıe laminate. Figure 3.8 shows the change in the off-axis modulus of 300/5208 when transverse stiffness goes to zero.

It is reasonable to assume that damage initiation and accumulation › not occur if a laminate is kept below the first ply failure level. This vel is equivalent to the yield stress of conventional materials. It is a ›nservative criterion to design a laminate based on the FPF stress or rain. In the next section the FPF envelope can be approximated by mple geometric bodies in the strain space. A direct comparison of the rength capability between a multidirectional composite and other aterials becomes possible.

approximate first ply failure envelopes

he failure envelopes in strain space are independent of the ply orienta-ons. We showed the T300/5208 $[0_n/90_m]$ laminates in Figures 7.12 ıd 7.14. The loading paths in strain space, however, are sensitive to ıe ply ratios or stacking sequence. The laminate compliance enters to the determination of loading paths. Uniaxial loadings have slopes qual to the Poisson's ratio of the laminate. Several loading paths are ıown in Figures 7.13 and 7.14.

We show the failure envelopes of other ply orientations in Figure 15. There is a first ply failure domain common to all ply orientations, dependent of the stacking sequence of the laminate. In this plane $_1 - \epsilon_2$) the envelopes change shape and location as the ply orienta-on changes. They do not move like rigid bodies.

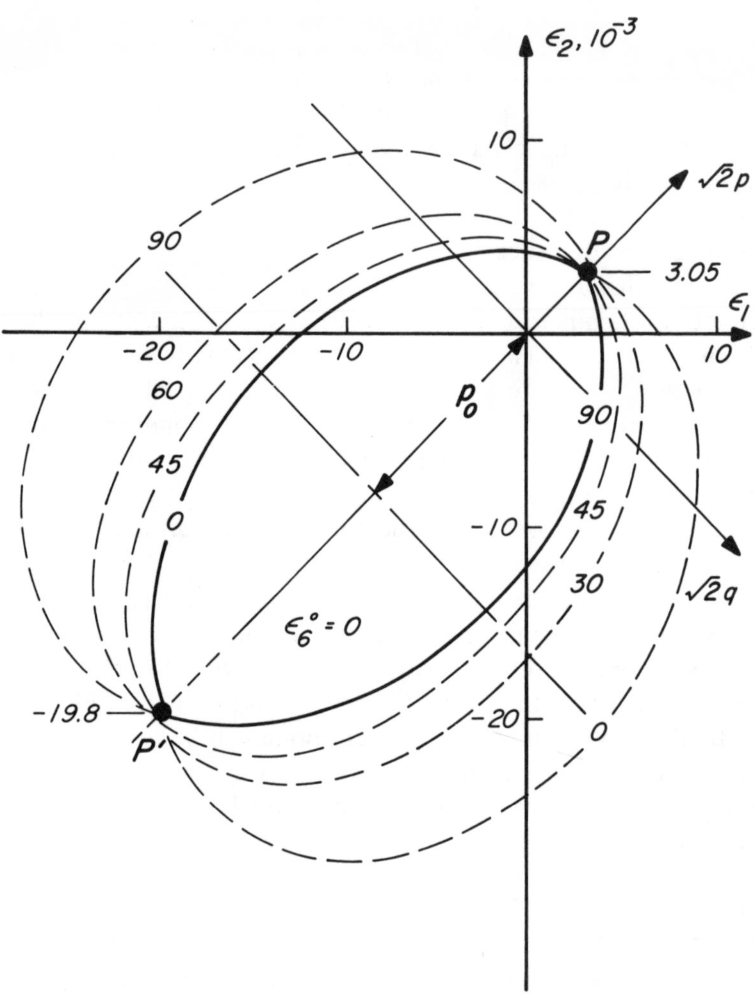

Figure 7.15 Failure envelopes of T300/5208 off-axis plies in the normal strain space.

We can repeat the representation of failure envelopes in the $\sqrt{2}q$ - $\sqrt{2}r$ space, where the square root of two comes from the coordinat transformation equations in Equation 2.38. We have previousl defined:

$$q = \frac{1}{2}(\epsilon_1 - \epsilon_2)$$

$$r = \frac{1}{2}\epsilon_6 \qquad (7.75$$

or the case of

$$p = \frac{1}{2}(\epsilon_1 + \epsilon_2) = 0 \tag{7.76}$$

ne failure envelope of a 0-degree ply is shown in Figure 7.16.

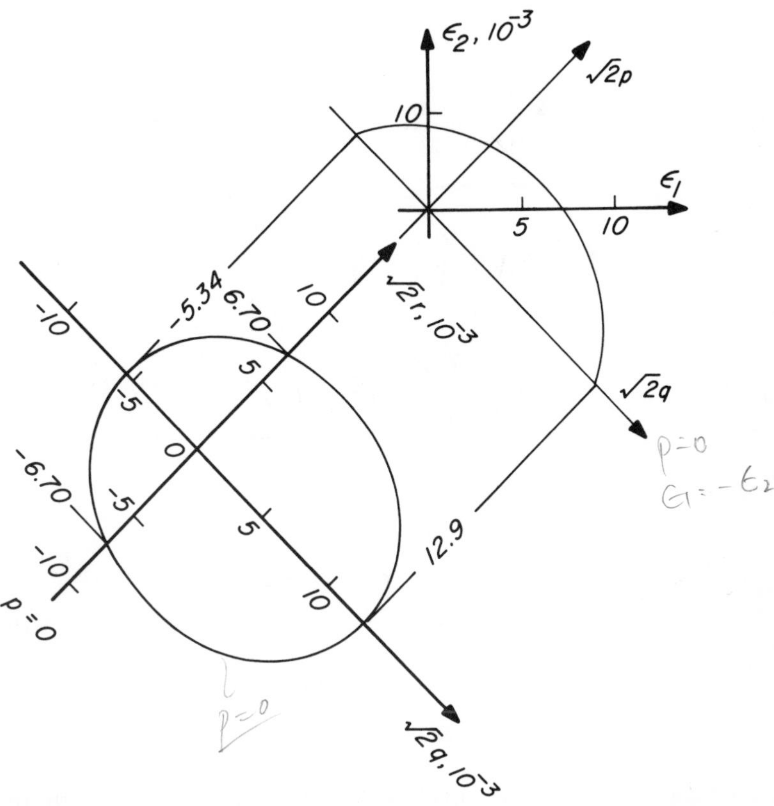

Figure 7.16 The failure envelope of a T300/5208 0-degree ply in the *q-r* strain space with zero *p*.

As ply orientation changes, the failure envelopes in the *q-r* space or the quivalent constant *p* space undergo rigid body rotations at an angular elocity of 2θ. The inner locus of the revolving failure envelope is the xact FPF envelope when we have infinite number of ply orientations. This envelope is conservative for finite number of orientations. For example, Figure 7.17 shows the margin of conservatism for a [0/30] aminate.

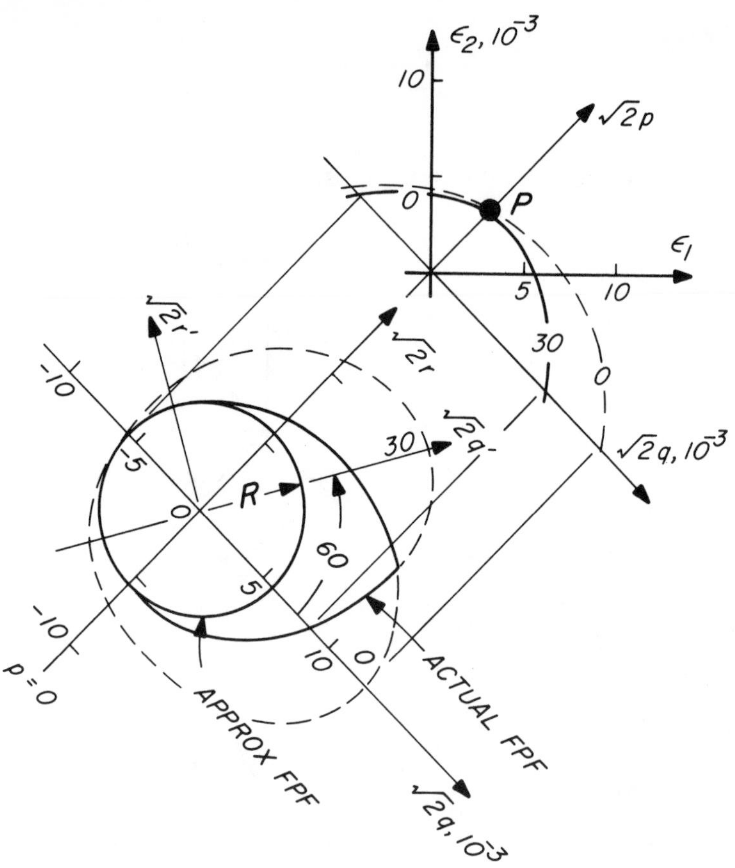

Figure 7.17 Difference between the actual and the approximate failure envelopes of a [0/30] laminate.

The approximate FPF envelope is compared directly with the actual FPF envelope of this laminate. For most other laminates, this margin becomes much smaller. Figure 7.18 shows the rapid convergence toward the approximate first ply failure envelope with radius ***R*** for a laminate of [0/30/90].

We can show the results above analytically by expressing the failure criterion in strain space in terms of p,q,r.* By direct substitution into the general (off-axis) failure criterion.

*See reference by H. T. Hahn and S. W. Tsai, "Graphical Determination of Stiffness and Strength of Composite Laminates," *Journal of Composite Materials*, Volume 8, pp. 160–177, 1974.

$$(G_{11} + 2G_{12} + G_{22})p^2 + (G_{11} - 2G_{12} + G_{22})q^2 + 4G_{66}r^2$$

$$2(G_{11} - G_{22})pq + 4(G_{16} + G_{26})pr + 4(G_{16} - G_{26})qr$$

$$+ (G_1 + G_2)p + (G_1 - G_2)q + 2G_6 r = 1 \tag{7.77}$$

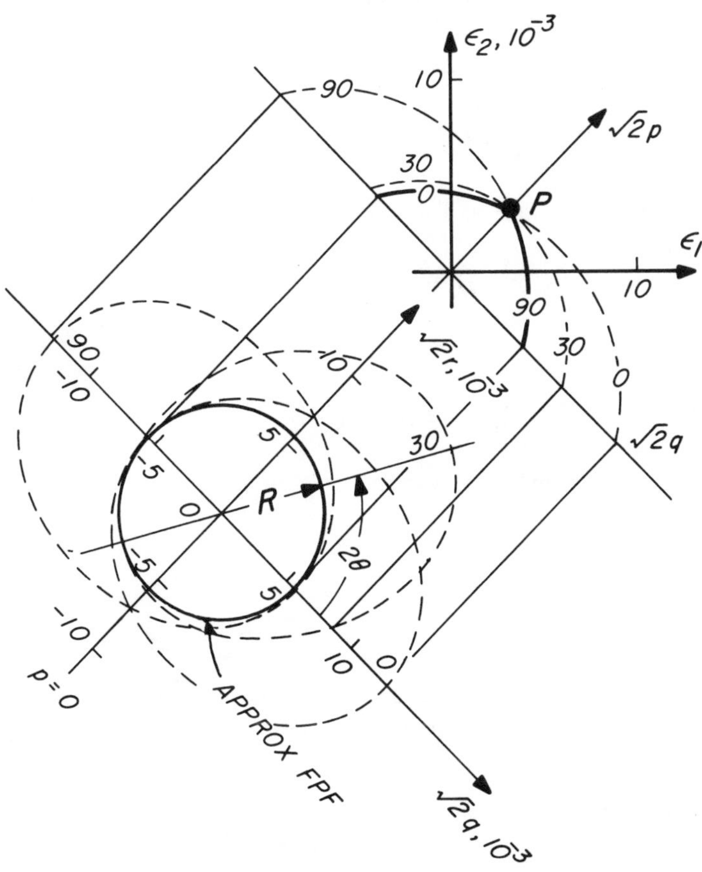

Figure 7.18 The approximate first ply failure envelope of a tridirectional T300/5208 laminate. The envelope is a circle in the q-r space which is convenient analytically.

Ve can substitute the transformation equations of the strength param-ters in Table 7.7, and move all invariant terms to the right-hand side of quation, we have

$$4(U_5 + U_3)q'^2 + 4(U_5 - U_3)r'^2 + (4U_2 p + 2q_G)q'$$

$$= 1 - 2[(U_1 + U_4)p^2 + p_G p] \tag{7.78}$$

where the U's are the linear combinations the G's (not the modulus o unidirectional composites). The transformed q and r are defined in Table 2.14. They correspond to the result of a rigid body rotation with angular displacement 2θ shown in Figures 7.16 to 7.18. Linear combinations of the linear strength parameters are defined in Table 7.7(b).

We can specialize the general equation of the failure envelope to special cases:

- Let q' and r' be zero.

$$2(U_1 + U_4)p^2 + 2p_G p - 1 = 0 \tag{7.79}$$

where for T300/5208 in Table 7.3

$$\begin{aligned} 2(U_1 + U_4) &= G_{xx} + G_{yy} + 2G_{xy} \\ &= 16546 \\ 2p_G &= G_x + G_y \\ &= 277 \end{aligned} \tag{7.80}$$

Solving for p,

$$p = 3.05 \times 10^{-3}, \; -19.79 \times 10^{-3} \tag{7.81}$$

These points are shown as $\boldsymbol{P}$ and $\boldsymbol{P'}$ in Figure 7.15. They represent the hydrostatic strain capability of the laminate. They are independent of the ply orientation.

- Let p and r' be zero.

$$4(U_5 + U_3)q'^2 + 2q_G q' - 1 = 0 \tag{7.82}$$

where

$$\begin{aligned} 4(U_5 + U_3) &= G_{xx} + G_{yy} - 2G_{xy} \\ &= 28822 \\ 2q_G &= G_x - G_y \\ &= -155.86 \end{aligned} \tag{7.83}$$

Solving for q',

$$q' = 9.185 \times 10^{-3}, \quad -3.77 \times 10^{-3}$$
$$\sqrt{2}q' = 12.98 \times 10^{-3}, \quad -5.34 \times 10^{-3} \qquad (7.84)$$

These points are the incepts of the q-axis shown in Figure 7.16.

- Let p and q' be zero.

$$4(U_5 - U_3)r'^2 = 1 \qquad (7.85)$$

where

$$4(U_5 - U_3) = 4G_{ss}$$
$$= 44471 \qquad (7.86)$$

$$r' = \pm 4.74 \times 10^{-3}$$
$$\sqrt{2}r' = \pm 6.70 \times 10^{-3} \qquad (7.87)$$

These points are the intercepts of the r-axis for the 0-degree ply. They are shown in Figure 7.16.

From the calculations above we know that the approximate first ply ailure envelope for T300/5208 is anchored by the two hydrostatic oints, $\boldsymbol{P}$ and $\boldsymbol{P'}$ in Figure 7.15 et al. The radius $\boldsymbol{R}$ in the q-r space from quation 7.84 is:

$$\boldsymbol{R} = \sqrt{2} \times 3.77 \times 10^{-3} = 5.34 \times 10^{-3} \qquad (7.88)$$

'his simple description of the failure envelope is accurate for laminates vith several ply orientations. This was shown in Figures 7.16 to 7.18. A irect comparison of a composite laminate with the conventional naterial is now possible. This is analogous to the quasi-isotropic contants of a composite materials which represent the minimum stiffness apability. Our invariant representation of the failure envelope here is lso the minimum capability in strength of our composite laminate.

This approximate FPF is conservative and is safe because damage initia tion and accumulation are not likely to occur.

In Figure 7.19 a direct comparison between the approximate FPF o T300/5208 and the high strength aluminum listed in Table 7.3 is pre sented. The aluminum envelope is approximate, and is based on th ultimate strength and on the von Mises failure criterion. The FPF envelope is closer to the conventional yielding than the ultimate. Th weight advantage of the graphite-epoxy composite is not included i the comparison in Figure 7.19. If yielding is used as the basis and th 40 percent weight advantage is claimed, the resulting circles for th high strength aluminum will be very small. It is fair to say that th strength advantage of composite materials is greater than the stiffnes advantage.

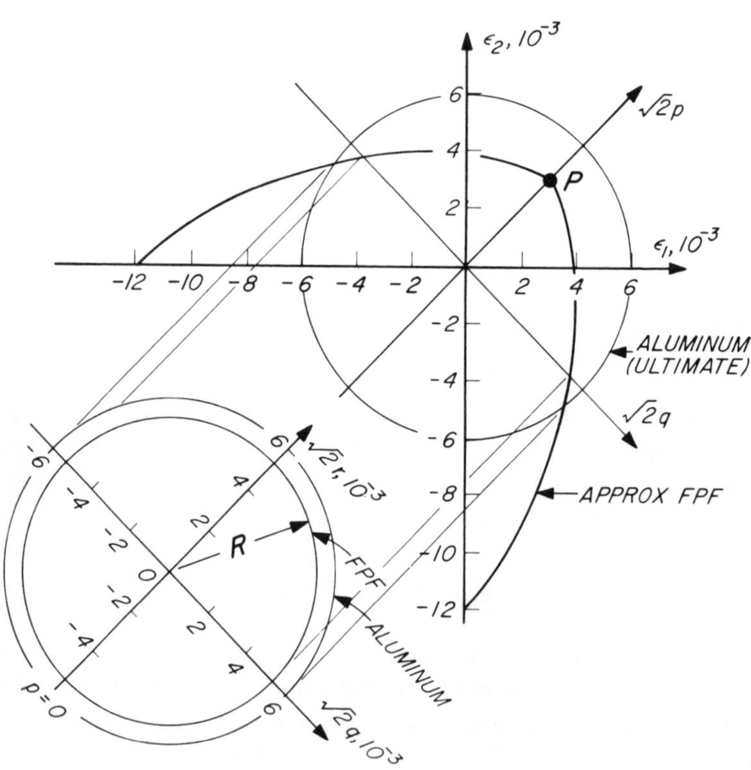

Figure 7.19 Comparison of failure envelopes of T300/5208 with a high strength aluminum. The approximate first ply failure envelope is used for the composite material. The aluminum failure is based on the ultimate and the von Mises criterion.

Further approximations of the FPF envelope can be made if more simplification is needed. For example, the approximate FPF in Figure 7.19 and 7.15 can be replaced by an ellipsoid with circular cross section in the q-r space. In place of the segmented FPF curves in the p-q plane, an elliptic cross section that inscribes or circumscribes the FPF curves can be found. This approximate FPF envelope will take the following form:

$$\frac{(p-p_o)^2}{a^2} + \frac{q^2 + r^2}{b^2} = ½ \tag{7.89}$$

where a, b are the semi-major axes of the ellipsoid, p_o the shift of the center of the ellipsoid from the origin. From Equations 7.79 et al., we can derive the following relations:

$$p_o = -\frac{p_G}{2(U_1 + U_4)} \tag{7.90}$$

$$a = \sqrt{p_o{}^2 + \frac{1}{2(U_1 + U_4)}} \tag{7.91}$$

$$b_{p=r=o} = \min\left\{ -\frac{q_G}{4(U_5 + U_3)} \pm \sqrt{\left[\frac{q_G}{4(U_5 + U_3)}\right]^2 + \frac{1}{4(U_5 + U_3)}} \right\} \tag{7.92}$$

This approximate envelope can also be expressed by:

$$\overline{G}_{ij}\, \epsilon_i\, \epsilon_j + \overline{G}_i\, \epsilon_i = 1 \tag{7.93}$$

where $\overline{G}_{ij}$, $\overline{G}_i$ are the strength parameters for the approximate FPF. Comparing Equation 7.89 and 7.93, we can show the following relations:

$$\overline{G}_{11} = \overline{G}_{22} = \frac{1}{4b^2}\,\frac{a^2 + b^2}{a^2 - p_o{}^2} \tag{7.94}$$

$$\bar{G}_{12} = -\frac{1}{4b^2}\frac{a^2 - b^2}{a^2 - p_o{}^2} \tag{7.95}$$

$$\bar{G}_{66} = \frac{1}{2}(\bar{G}_{11} - \bar{G}_{12}) = \frac{1}{4b^2}\frac{a^2}{a^2 - p_o{}^2} \tag{7.96}$$

$$\bar{G}_1 = \bar{G}_2 = -\frac{p_o^1}{a^2 - p_o{}^2} \tag{7.97}$$

$$\bar{G}_6 = \bar{G}_{26} = \bar{G}_{16} = 0 \tag{7.98}$$

There are only three independent strength parameters.

Other simplifications are possible, such as making

$$a = b = b_o \tag{7.99}$$

or

$$p_o = 0 \tag{7.100}$$

in Equation 7.89. Instead of ellipsoid for the approximate FPF, we have spheres.

When we have both conditions in Equations 7.99 and 7.100,

$$p^2 + q^2 + r^2 = b_o^2/2 \tag{7.101}$$

or

$$\epsilon_1{}^2 + \epsilon_2{}^2 + \frac{1}{2}\epsilon_6{}^2 = b_o^2 \tag{7.102}$$

where for T300/5208,

$$b_0 \cong 4 \times 10^{-3} \tag{7.103}$$

This simple approach is preferred over the maximum strain criterion because the analytic foundation is preserved in Equation 7.102.

The successive levels of approximation can be explicitly stated without changing internal consistency. The invariant nature and the ply orientation independence of various FPF envelopes are retained. The orientation dependency of the more complex relations in Equation

7.66 et al. can be replaced by approximate FPF envelopes. Direct comparisons among different composite materials and conventional materials are now possible. Laminate optimization and sizing can be carried out on a rational basis. The simplifications embodied in the approximate FPF envelopes can lead to straightforward and analytically consistent design procedure. To design for strength becomes as simple for composite materials as that for conventional materials.

8. conclusions

The quadratic interaction failure criterion in stress and strain space is recommended for unidirectional and multidirectional composite materials. This approach is easy to use because the coefficients of the failure criterion are components of tensors. Established transformation equations and invariants can be used. Depending on the accuracy desired, several levels of simplifications can be achieved. An important facet of this approach is the resulting rigid body rotation of the failure envelope in the q-r strain space. The angular displacement of the rotation is precisely twice the ply orientation. The inner locus of the rotated failure envelope is a circle. Taking advantage of the rigid body rotation, this approximate FPF is invariant, independent of the ply orientation. The ultimate failure of a laminate, on the other hand, cannot be readily reduced to an orientation independent representation. The approximate FPF envelope is conservative and provides a basis for simplified design and sizing procedure not possible with the exact FPF envelope. In the latter case, ply-by-ply orientation becomes necessary.

9. homework problems

a. What is the highest FPF stress of a T300/5208 cross-ply laminate under uniaxial compression? At what ply ratio does the maximum strength occur? For example, the FPF stress $[0_6/90]$ is 1691 MPa in Table 7.8. What is the physical explanation of this result above?

b. How do we calculate the exact intercept of approximate FPF on the positive ϵ_1 or ϵ_2 axis in Figure 7.19? Is it the same as the failure strain listed in Equation 7.32?

c. What can we say about the validity and usefulness of the maximum strain criterion in Equation 7.2?

d. What are the values of a, b and p_o in Equation 7.89 for T300/5208 assuming the ellipsoid would circumscribe the segmented FPF in Figure 7.15; i.e., passing through $\boldsymbol{P}$ and $\boldsymbol{P'}$? Is this envelope conservative? What are the values of $\bar{G}_{ij}$?

e. How can the approximate FPF in Problem *d* be represented in the stress-resultant space? Compare the result with all the exact FPF envelopes in Figure 7.10.

f. Show failure envelopes in stress and strain space of Kevlar 49 composite materials listed in Tables 7.2 and 7.3. What are the effects of the low longitudinal compressive strength on the approximate FPF envelopes?

omenclature

$_x, E_y, E_s$	=	Young's and shear moduli
$_j, F_i$	=	Strength parameters in stress space; $i,j = x,y,x$ or 1,2,6
$_{ij}, G_i$	=	Strength parameters in strain space
$_{ij}, \bar{G}_i$	=	First-ply-failure strength parameters in strain space
$^{(\theta)}_{ij}, H^{(\theta)}_i$	=	Strength parameters of the θ-degree ply in a laminate
q	=	Linear combinations of first rank strength parameters
$_{(\theta)}, R'_{(\theta)}$	=	Strength ratios of the θ-degree ply
	=	Longitudinal-transverse shear strength
	=	Linear combinations of second rank strength parameters
X'	=	Longitudinal tensile and compressive strengths
Y'	=	Transverse tensile and compressive strengths
y,s	=	Normalized stress components
	=	Stress components; $i = x,y,s$ or 1,2,6
	=	Strain components; $i = x,y,s$ or 1,2,6
*	=	Ultimate strains; $i = x, y, s$
	=	Poisson's ratios; $i = x,y$
F	=	First Ply Failure

chapter 8
hygrothermal behavior

formation is also possible upon change of temperature and upon orption of moisture. The matrix material is much more susceptible hygrothermal deformation than the fiber. The hygrothermal defor- tion of a unidirectional composite is therefore much higher in the nsverse direction than in the longitudinal direction. Such anisotropy deformation results in the presence of residual stresses in composite inates because the multidirectionality of fiber orientation prohibits e deformation. The temperature change and moisture absorption also nge mechanical properties. Therefore, hygrothermal behavior affects only dimensional stability but also safety of structures.

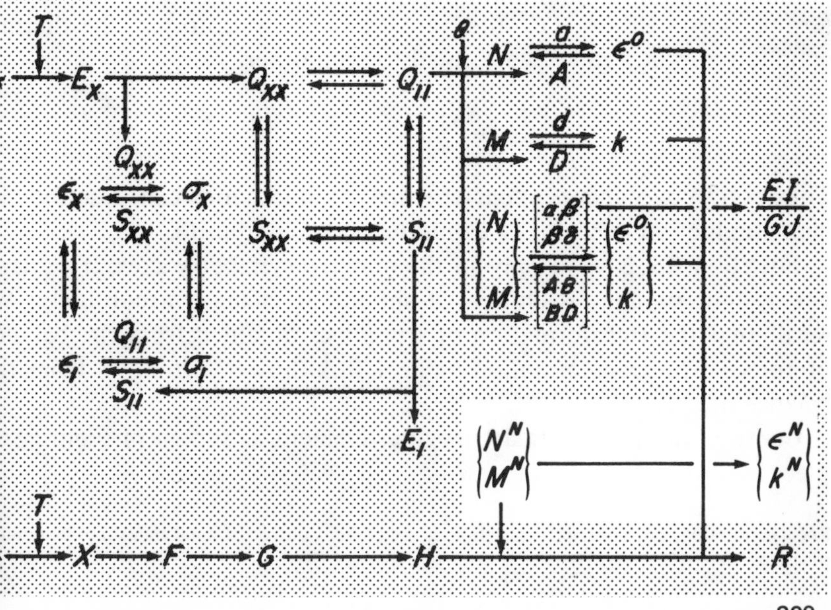

1. heat conduction and moisture diffusion

Whereas the mechanical behavior is described by the stiffness or the compliance matrix in Chapter 1, the heat conduction in the linear theory is described by the thermal conductivity matrix K_{ij}^T.* Specifically, the heat flux q_i^T per unit area per unit time in the x_i direction is related to the temperature gradient $T_{,j}$ in the x_j direction by

$$q_i^T = -K_{ij}^T \, T_{,j} \tag{8.1}$$

Note that T is the temperature and $T_{,j}$ is its partial derivative with respect to x_j.

In the material symmetry axes of unidirectional composite, the only heat flux possible due to the temperature gradient $T_{,i}$ is q_i^T. Therefore Equation 8.1 reduces to

$$\begin{aligned} q_x^T &= -K_x^T \, T_{,x} \\ q_y^T &= -K_y^T \, T_{,y} \\ q_z^T &= -K_z^T \, T_{,z} \end{aligned} \tag{8.2}$$

Furthermore, since unidirectional composites are isotropic in a plane normal to the fibers, i.e., they are transversely isotropic, the thermal conductivities K_y^T and K_z^T are equal to each other,

$$K_z^T = K_y^T \tag{8.3}$$

Thus, only two independent thermal conductivities can describe the heat conduction behavior of a unidirectional composite.

Equation 8.2 is valid only in the material symmetry axes. If a different system of axes is chosen, it should be changed accordingly. We recall that the transformation of stress was obtained from the balance of forces. The transformation of heat flux, however, follows from the balance of energy.

Consider an infinitesimal triangular element as shown in Figure 8.1

*Regular notation, not contracted notation, is used here. Contracted notation cannot be applied to vectors.

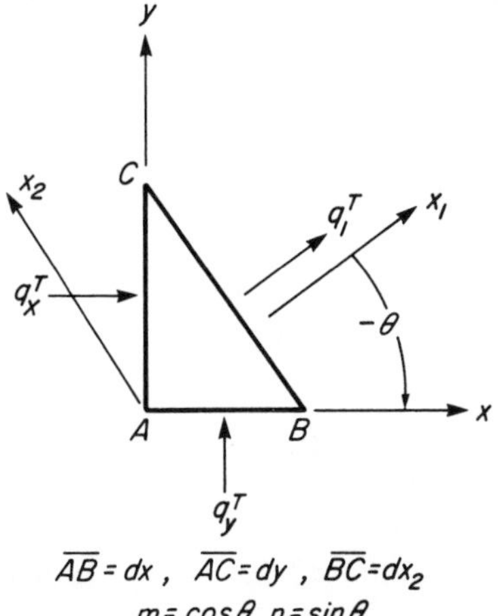

Figure 8.1 Energy balance for an infinitesimal triangular element. All heat flux components shown are positive. The x_1 axis is normal to $\overline{BC}$.

or convenience the thickness of the element is taken as unity. In the bsence of any heat source or sink within the element, the total heat ıflux must be equal to the total heat efflux. That is,

$$q_x^T\,dy + q_y^T\,dx = q_1^T\,dx_2 \tag{8.4}$$

ividing both sides by dx_2, and using $m = \cos\theta$, $n = \sin\theta$, we obtain

$$q_1^T = q_x^T\,m - q_y^T\,n \tag{8.5}$$

The equation for q_2^T follows similarly as

$$q_2^T = q_x^T\,n + q_y^T\,m \tag{8.6}$$

inally, since the rotation is around the z-axis, there is no change in q_z^T; e.,

$$q_3^T = q_z^T \tag{8.7}$$

The transformation of the temperature gradient can be deriv similarly to that of the strain, from the chain rule of differentiatic For example,

$$\frac{\partial T}{\partial x_1} = \frac{\partial T}{\partial x}\frac{\partial x}{\partial x_1} + \frac{\partial T}{\partial y}\frac{\partial y}{\partial x_1} \qquad (8.$$

Noting that the x-y axes have been rotated from the x_1-x_2 axes throu the angle $-\theta$, we rewrite Equation 8.8 as

$$\frac{\partial T}{\partial x_1} = \frac{\partial T}{\partial x} m - \frac{\partial T}{\partial y} n \qquad (8.$$

The equations for $\partial T/\partial x_2$ and $\partial T/\partial x_3$ are as follows:

$$\frac{\partial T}{\partial x_2} = \frac{\partial T}{\partial x} n + \frac{\partial T}{\partial y} m \qquad (8.1$$

$$\frac{\partial T}{\partial x_3} = \frac{\partial T}{\partial z} \qquad (8.1$$

With the transformation equations known for both heat flux a temperature gradient, we can now express the heat conductivities in t new coordinate system in terms of those in the material symmetry ax To this end we first substitute Equation 8.2 into Equation 8.5 to obta

$$q_1^T = -K_x^T\, T_{,x}\, m + K_y^T\, T_{,y}\, n \qquad (8.1$$

Solving Equation 8.9 and 8.10 for $T_{,x}$ and $T_{,y}$ and substituting t resulting equations into Equation 8.12 leads to

$$q_1^T = -(K_x^T\, m^2 + K_y^T\, n^2)\, T_{,1} - (K_x^T - K_y^T)\, mn\, T_{,2} \qquad (8.1$$

Therefore, the heat conductivities K_{11}^T and K_{12}^T in the new coordina system are given by

$$K_{11}^T = m^2\, K_x^T + n^2\, K_y^T$$

$$K_{12}^T = mn\, (K_x^T - K_y^T) \qquad (8.1$$

similar approach for q_2^T leads to K_{22}^T:

$$K_{22}^T = n^2\, K_x^T + m^2\, K_y^T \tag{8.15}$$

eedless to say, K_{33}^T is simply equal to K_z^T.

Equations 8.14 and 8.15 are the same as those for the stress if we quate K_{11}^T, K_{22}^T and K_{12}^T to σ_1, σ_2 and σ_6, respectively. Mathematally, the stress, the strain and the heat conductivity are the same cond-rank tensors although they represent physically different quanties. Thus, their transformation equations are the same.

The foregoing equations for heat conduction are equally applicable moisture diffusion. In the latter case, the heat flux q_i^T is replaced by e moisture flux q_i^H, and the temperature graduent $T_{,i}$ by the moisture ncentration gradient $H_{,i}$. Here the moisture concentration H is efined by

$$H = \lim_{\Delta V \to 0} \frac{\text{mass of moisture in } \Delta V}{\Delta V} \tag{8.16}$$

hus H is a measure of the amount of moisture at a point.

The relation between q_i^H and $H_{,j}$ is expressed in terms of the moisre diffusion coefficient K_{ij}^H so that

$$q_i^H = -K_{ij}^H\, H_{,j} \tag{8.17}$$

ecause of the material symmetry, Equation 8.17 for unidirectional omposites reduces to

$$\begin{aligned} q_x^H &= -K_x^H\, H_{,x} \\ q_y^H &= -K_y^H\, H_{,y} \\ q_z^H &= -K_z^H\, H_{,z} \end{aligned} \tag{8.18}$$

here

$$K_z^H = K_y^H \tag{8.19}$$

goes without saying that the same reduction was possible for q_i^T, see quation 8.2. Furthermore, the transformation equations for q_i^H, $H_{,i}$

and K_{ij}^H are the same as those for q_i^T, $T_{,i}$ and K_{ij}^T, respectively. Not that the transformation equation for q_i^H follows from the balance c mass whereas the balance of energy was used for q_i^T. The physic dimensions of the hygrothermal variables introduced so far are listed i Table 8.1.

table 8.1
units of hygrothermal properties

Temperature	K	Moisture concentration	g/m^3
Temperature gradient	K/m	Moisture concentration gradient	g/m^4
Heat flux	W/m^2	Moisture flux	$g/(m^2 \cdot s)$
Thermal conductivity	$W/(m \cdot K)$	Moisture diffusion coefficient	m^2/s
Thermal diffusivity	m^2/s	Specific moisture concentration	g/g
Specific heat	$J/(g \cdot K)$		

Of particular interest in studying the hygrothermal behavior of com posites is the one-dimensional diffusion through the thickness, i.e., i the z direction. This is the situation when a thin laminate is subjecte to a change in its environment, Figure 8.2. Suppose the initial temper ture and moisture concentration are uniform throughout the laminat and are denoted by T_o^* and H_o, respectively. The environmental chang is such that the temperature and moisture concentration at the surface are maintained at T_∞ and H_∞, respectively. It is then necessary t determine the distributions of T and H through the thickness as tim changes.

The governing equation for T is obtained from the balance of energy Consider an infinitesimal element dz, Figure 8.2(b). The heat influ through the unit area of the left face is q_z^T while the heat efflu through the right face is $q_z^T + \partial q_z^T/\partial z\, dz$. On the other hand, th increase in the energy stored per unit time within the element i $\rho C\, \partial T/\partial t\, dz$ where C is the specific heat. Since there is no heat sourc

*Here T_o is used to denote the initial temperature. However, in later sections T_o represents th stress-free temperature.

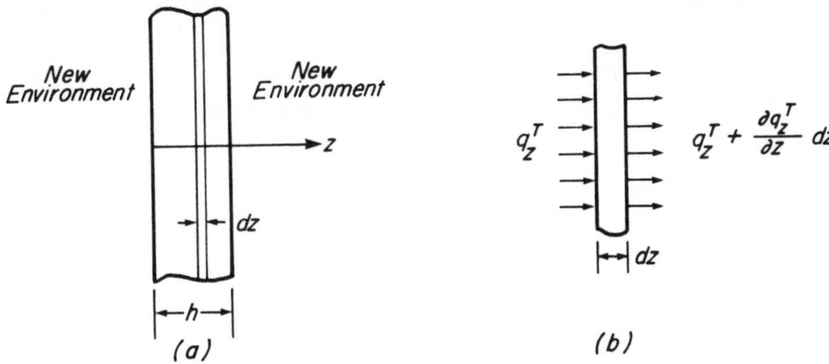

Figure 8.2 A thin laminate subjected to an environmental change.

sink, the balance of energy requires that

$$q_z^T - \left(q_z^T + \frac{\partial q_z^T}{\partial z} dz \right) = \rho C \frac{\partial T}{\partial t} dz, \tag{8.20}$$

,

$$-\frac{\partial q_z^T}{\partial z} = \rho C \frac{\partial T}{\partial t} \tag{8.21}$$

ere ρ is the mass density of the composite. Finally, substitution of uation 8.2 into Equation 8.21 leads to

$$\frac{\partial}{\partial z} \left(K_z^T \frac{\partial T}{\partial z} \right) = \rho C \frac{\partial T}{\partial t} \tag{8.22}$$

rthermore, if K_z^T and ρC are constant, Equation 8.22 can be duced to

$$\frac{K_z^T}{\rho C} \frac{\partial^2 T}{\partial z^2} = \frac{\partial T}{\partial t} \tag{8.23}$$

is is known as Fourier's equation.

The equation for moisture diffusion can be derived similarly frc the balance of mass as

$$\frac{\partial}{\partial z}\left(K_z^H \frac{\partial H}{\partial z}\right) = \frac{\partial H}{\partial t} \qquad (8.2$$

Again, if K_z^H is constant, Equation 8.24 can be reduced to Ficl equation

$$K_z^H \frac{\partial^2 H}{\partial z^2} = \frac{\partial H}{\partial t} \qquad (8.2$$

A comparison of Equations 8.23 and 8.25 reveals that these equ tions are identical in form to each other, indicating the similarity of t underlying processes. The thermal diffusivity $K_z^T/(\rho C)$ and the moistu diffusion coefficient K_z^H are a measure of the rate at which the tempe ature and the moisture concentration respectively change within t material. In general, these parameters depend on the temperature a moisture concentration. However, over the range of temperature a moisture concentration that prevails in typical applications of co posites, the thermal diffusivity is about 10^6 times greater than t moisture diffusion coefficient. Thus, the thermal diffusion takes pla 10^6 times faster than the moisture diffusion. As a result, the tempe ture will reach equilibrium long before the moisture concentratio does. This observation allows one to solve Equation 8.25 separate from Equation 8.23.

In the study of the hygrothermal behavior of composites the speci moisture concentration defined by

$$c = H/\rho \qquad (8.2$$

is frequently used in lieu of H. Physically, c represents the amount moisture as a fraction of the dry mass of composite, i.e.

$$c = \lim_{\Delta V \to 0} \frac{\text{mass of moisture in } \Delta V}{\text{mass of dry composite of volume } \Delta V} \qquad (8.2$$

In terms of c Equation 8.25 becomes

$$K^H \frac{\partial^2 c}{\partial z^2} = \frac{\partial c}{\partial t} \tag{8.28}$$

here the subscript has been dropped off K_z^H for convenience. The ppropriate boundary conditions are

$$\begin{aligned} c &= c_o \text{ for } 0 < z < h \text{ at } t \leqslant 0 \\ c &= c_\infty \text{ for } z = 0 \text{ and } h \text{ at } t > 0 \end{aligned} \tag{8.29}$$

lere h is the thickness of the laminate in Figure 8.2. The solution to quations 8.28 and 8.29 is given by [1]

$$\frac{c - c_o}{c_\infty - c_o} = 1 - \frac{4}{\pi} \sum_{j=0}^{\infty} \frac{1}{2j+1} \sin \frac{(2j+1)\pi z}{h} \exp\left[-\frac{(2j+1)^2 \pi^2 K^H t}{h^2} \right] \tag{8.30}$$

Equation 8.30 is shown graphically in Figure 8.3 where the non-imensional time $K^H t/h^2$ has been used. Note that c eventually reaches $_\infty$ throughout the laminate. Therefore, c_∞ is also called the equilib-ium (specific) moisture concentration.

In a moisture absorption test the final moisture concentration c_∞ is lways greater than the initial one c_o. The converse is true in a moisture esorption test. However, Equation 8.30 is equally valid in either case.

In actual experiments the sample frequently is weighed to determine he moisture content which is the total mass of the absorbed moisture ivided by the dry weight of the sample. The moisture content is in fact he same as the average specific moisture concentration $\bar{c}$ defined by

$$\bar{c} = \frac{1}{h} \int_o^h c \, dz \tag{8.31}$$

ubstituting Equation 8.30 into Equation 8.31 and noting that

$$\begin{aligned} \bar{c} &= c_o \text{ at } t = 0 \\ \bar{c} &= c_\infty \text{ at } t = \infty \end{aligned} \tag{8.32}$$

we obtain

$$\frac{\bar{c}-c_o}{c_\infty - c_o} = 1 - \frac{8}{\pi^2}\sum_{j=0}^{\infty}\frac{1}{(2j+1)^2}\exp\left[-\frac{(2j+1)^2\pi^2 K^H t}{h^2}\right]$$

(8.3

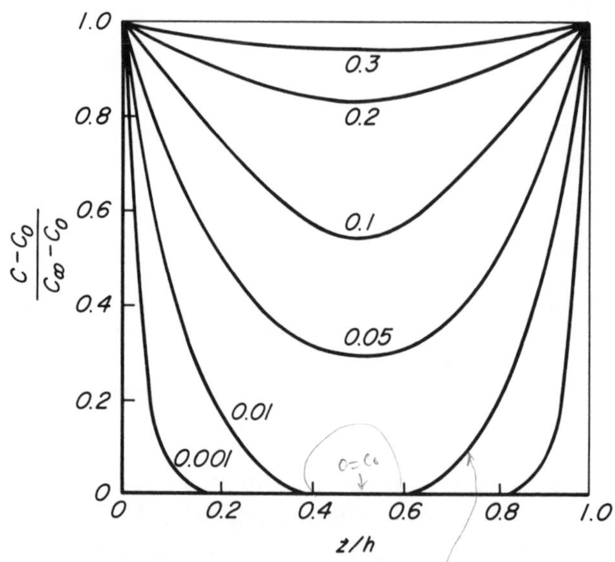

Figure 8.3 Moisture profile as a function of time. The numbers are the values of the nondimensional time $K^H t/h^2$.

Equation 8.33 is compared with experimental data in Figure 8.4. [2 Since the moisture diffusion is through the thickness, it does n depend on the type of laminate. Also, Equation 8.33 is applicab regardless of the type of diffusion.

For t sufficiently large Equation 8.33 can be approximated by th first term in the series,

$$\frac{\bar{c}-c_o}{c_\infty - c_o} = 1 - \frac{8}{\pi^2}\exp\left(-\frac{\pi^2 K^H t}{h^2}\right)$$

(8.3

On the other hand, for short times an approximation can be obtaine from an alternate solution [1] as

$$\frac{\bar{c}-c_o}{c_\infty - c_o} = 4\left(\frac{K^H t}{\pi h^2}\right)^{1/2}$$

(8.3

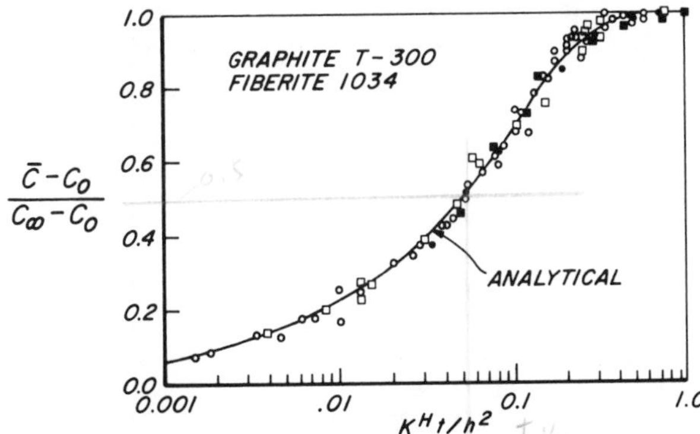

Figure 8.4 Experimental correlation of Equation 8.33 for graphite/epoxy laminates: unidirectional (○,●) and quasi-isotropic (□,■). [2] Open and filled symbols represent absorption and desorption, respectively.

us, the initial increase in moisture content is proportional to $(t/h^2)^{1/2}$. Equations 8.34 and 8.35 are frequently used to determine K^H from e measurements of moisture concentration. From Equation 8.34, the ne $t_{1/2}$ for which $(\bar{c}-c_o)/(c_\infty - c_o) = 1/2$, is given by

$$t_{1/2} = \frac{h^2}{\pi^2 K^H} \ln \frac{16}{\pi^2} \tag{8.36}$$

erefore, the diffusion coefficient is determined from the half-time of rption process as

$$K^H = \frac{0.04895\, h^2}{t_{1/2}} \tag{8.37}$$

The applicability of Equation 8.35 becomes apparent if the moisture ntent is plotted as a function of $\sqrt{t}$. A relationship between $\bar{c}$ and $\overline{t}$ is schematically shown in Figure 8.5. From the figure we choose, in e linear region, two moisture contents $\bar{c}_1$ and $\bar{c}_2$ corresponding to t_1 d t_2, respectively. Substituting these values into Equation 8.35 and lving the resulting equations for K^H, we can determine K^H as

$$K^H = \frac{\pi}{16} \left(\frac{\bar{c}_2 - \bar{c}_1}{c_\infty - c_o} \right)^2 \left(\frac{h}{\sqrt{t_2} - \sqrt{t_1}} \right)^2 \tag{8.38}$$

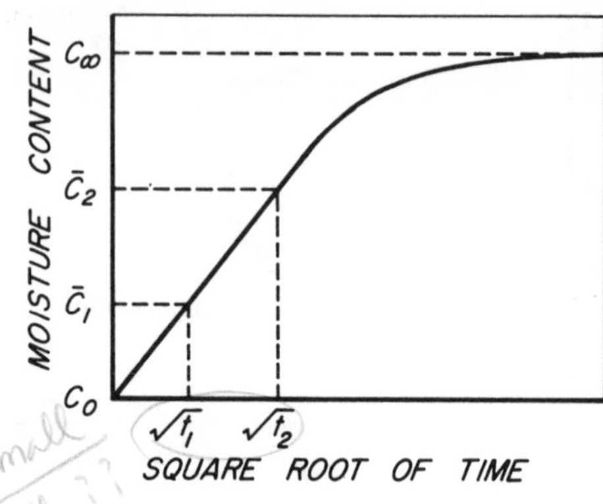

Figure 8.5 Determination of diffusion coefficient.

The equilibrium moisture concentration depends on the enviro- ment. In humid air it is related to the relative humidity ϕ in percent b a power law

$$c_\infty = a\left(\frac{\phi}{100}\right)^b \qquad (8.3$$

where a and b are material constants. A set of data bearing such rel tionship is shown in Figure 8.6.

The moisture diffusion coefficient strongly depends on temperatur The relationship can be described by an equation of the form

$$K^H = K_o^H \exp\left(-\frac{E_d}{RT}\right) \qquad (8.4$$

where K_o^H and E_d are the pre-exponential factor and activation energ respectively, and R is the gas constant (=1.987 cal/(mole · K)). For th graphite/epoxy composite of Figure 8.6 a relationship between K^H an T is shown in Figure 8.7. [3]

Typical hygrothermal properties of a graphite/epoxy composite a summarized in Table 8.2. From the table we can find, for example, th the equilibrium moisture content at 100% relative humidity is 1.8%. A room temperature (=23°C) the moisture diffusion coefficient is onl 2.62×10^{-8} mm^2/s whereas the thermal diffusivity in the transvers direction is 0.45 mm^2/s.

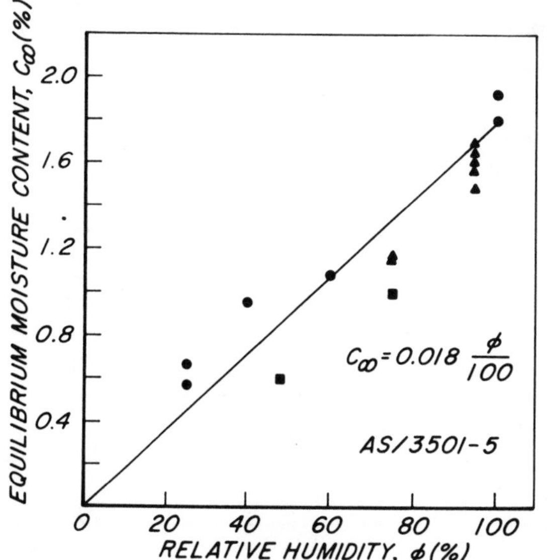

Figure 8.6 Equilibrium moisture content as a function of relative humidity for AS/3501-5. (● [3], ▲ [4], ■ [5]).

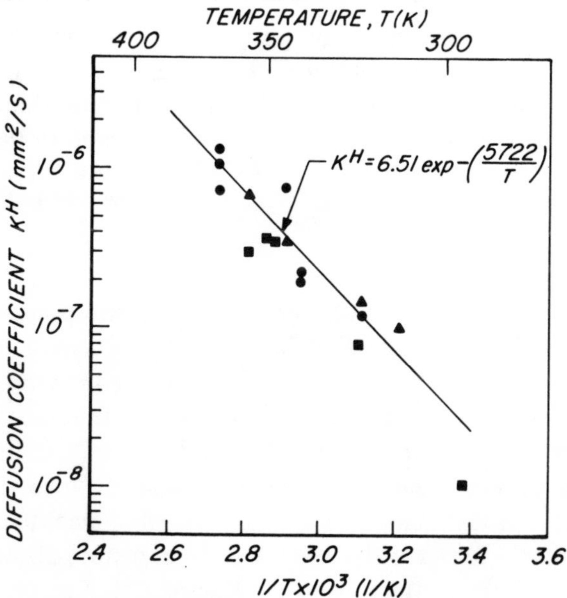

Figure 8.7 Transverse diffusion coefficient as a function of temperature for AS/3501-5. (● [3], ▲ [4], ■ [5]).

table 8.2
typical hygrothermal properties of unidirectional graphite/epoxy composite

ρ g/cm^3	C $J/(\mathrm{g}\cdot K)$	K_x^T $W/(\mathrm{m}\cdot K)$	$K_y^T = K_z^T$ $W/(\mathrm{m}\cdot K)$	α_x $(\mu\mathrm{m/m})/K$	$\alpha_y = \alpha_z$ $(\mu\mathrm{m/m})/K$
1.6	1.0	4.62	0.72	0.02	22.5
a	b	K_o^H mm$^2/s$	E_d/R K	β_x m/m	$\beta_y = \beta_z$ m/m
0.018	1	6.51	5722	0	0.44
T_o °C					
177					

2. stress-strain relations including hygrothermal strains

Just like any other material, composites deform when they abso moisture or when the temperature changes. In the linear theory t resulting nonmechanical strains are simply added to the mechanic strains induced by the stress to obtain the total strain. In the followi discussion of hygrothermoelastic constitutive relations, the temperatu and the moisture concentration are uniform throughout the materi volume.

Consider a unidirectional composite in a reference state where ter perature is T_o and $c = \sigma_i = 0$. Next the composite is brought into final state where temperature T is different from T_o and $c \neq 0$, $\sigma_i \neq$ Figure 8.8 shows pictorially the two different states the composite is i

To determine the resulting strain we assume that the material elastic; the response of the material does not depend on the history the input but only on the initial and final states. Since the order application of various changes is immaterial, we conceptually imagi that temperature is changed first, followed by moisture absorption. T application of stress is thus last, see Figure 8.8. Let us denote t strains induced by the temperature change and moisture absorption

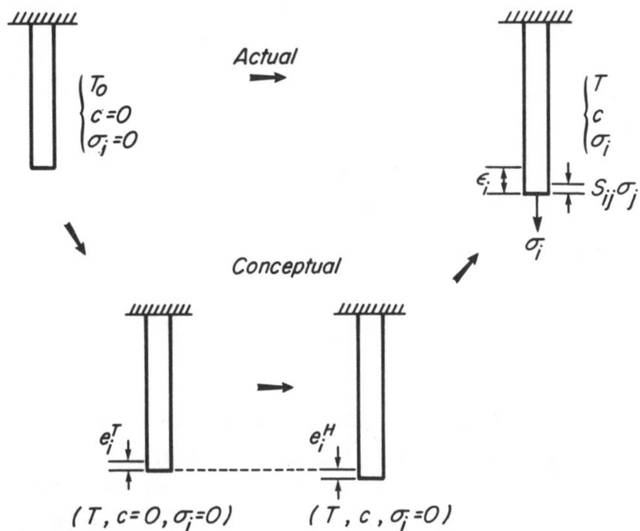

Figure 8.8 Decomposition of total strain into thermal, swelling and mechanical strains.

and e_i^H, respectively. The mechanical strain due to σ_i is given by $_j\sigma_j$. It should be noted that since the stress is applied at temperature T ıd moisture concentration c, $S_{ij}(T,c)$ is the compliance measured nder such condition. Thus, it will be different from $S_{ij}(T_o,c)$ which is ıeasured at (T_o,c). By the same line of reasoning we see that e_i^H is ıeasured at $(T,\sigma_i = 0)$ and e_i^T at $(c = 0, \sigma_i = 0)$. The final strain is the ım of the foregoing three types of strains:

$$\epsilon_i = e_i + S_{ij}\sigma_j \tag{8.41}$$

ere the nonmechanical strain e_i is the sum of e_i^T and e_i^H,

$$e_i = e_i^T + e_i^H \tag{8.42}$$

Because of the transverse isotropy characteristic of unidirectional omposites, not all components of the nonmechanical strain are inependent in the material symmetry axes. Specifically, we have

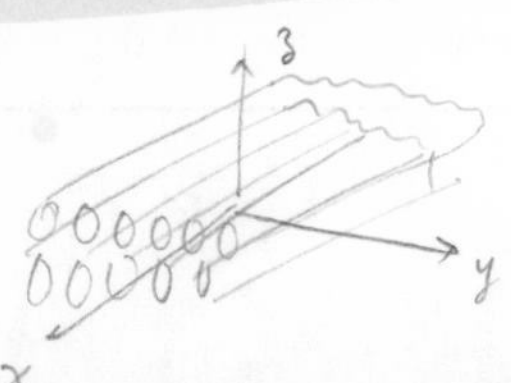

$$e_s^T = e_s^H = 0$$

$$e_y^T = e_z^T \qquad (8.4$$

$$e_y^H = e_z^H$$

The resulting volumetric strain is

$$\frac{\Delta V}{V} = e_x^T + 2e_y^T + e_x^H + 2e_y^H \qquad (8.4$$

Thus, unlike isotropic materials, composites do not allow determinati of nonmechanical linear strains from nonmechanical volumetric strai

In general, e_i^T and e_i^H are nonlinear functions of T and c, respe tively. In the linear theory we are interested in, however, the therm expansion coefficient α_i and the swelling coefficient β_i can be used calculate e_i^T and e_i^H, respectively. That is,

$$e_i^T = \alpha_i (T - T_o)$$

$$e_i^H = \beta_i c \qquad (8.4$$

Thus, α_i has the dimension K^{-1} whereas β_i has no dimension. Typic values of these coefficients for a graphite/epoxy are given in Table 8.

Thermal expansion coefficients of other composites are listed in Tab 8.3.

table 8.3

thermal expansion coefficients of typical unidirectional composites

Type	α_x $(\mu m/m)/K$	α_y $(\mu m/m)/K$
T300/5208	0.02	22.5
B(4)/5505	6.1	30.3
AS/3501	−0.3	28.1
Scotchply 1002	8.6	22.1
Kevlar 49/Epoxy	−4.0	79.0

fabrication stresses

etermination of fabrication stresses in a composite laminate requires, ɔove all, some understanding of the fabrication process involved. herefore, we shall start this section with a brief description of a 'pical fabrication process of polymer matrix composites.

Needless to say, fabrication of a composite laminate starts with fibers ıd a viscous matrix resin. The fibers are first impregnated with the sin and then wound with a backing sheet onto a mandrel in the form f a tape. The tape is then heated slightly to make the resin hard yet exible enough to handle; the resin is B-staged. Thus the tape has all ıe fibers in the same direction and is called the prepreg. The prepreg is ıen cut into sheets and these sheets are laid up with fibers in various irections to make a laminate. The laminate is put in a vacuum bag to ıueeze out the entrapped air and is slowly heated in an autoclave. As ıe temperature increases, the epoxy softens again and flows until a ıange in the internal structure starts to take place in the form of ıtanglement of polymer molecules, i.e., crosslinking occurs. As a sult, free motion of polymer molecules is prohibited and the epoxy egins to harden. At this point, usually around 270°F, a pressure in the ınge of 80–100 psi is applied to drive out volatiles. The temperature is ırther increased to 350°F and maintained there for 1–2 hours to finish ıe cure. The temperature is then lowered to the room temperature and ıe cured laminate is taken out of the autoclave. The typical cure rocedure just described is shown in Figure 8.9.

As the crosslinking takes place, the epoxy shrinks, i.e., chemical ırinkage occurs. The resulting deformation of a unidirectional com-osite in the transverse direction is much larger than in the longitudinal irection. Therefore, within the laminate the deformation of one ply is onstrained by the other plies with different fiber orientations, and ence residual stresses are built up in each ply. However, since most rosslinking takes place at the highest temperature, called the cure tem-erature, the epoxy can be still viscous enough to allow complete relax-tion of the residual stresses. Thus, the cure temperature can be taken s the stress-free temperature. In reality, however, the stress-free tem-erature will vary with the cure process employed because the property hange during cure is very much time-dependent. Yet, the cure temper-ture can serve as the stress-free temperature as long as almost all cure ıkes place at the cure temperature.

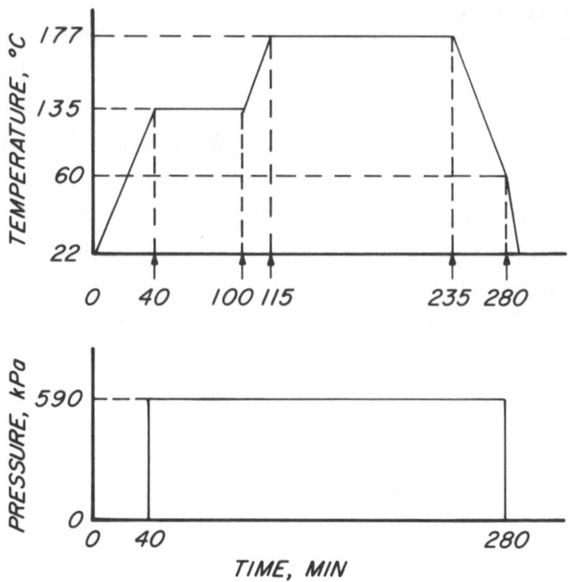

Figure 8.9 A typical cure cycle for a graphite/epoxy composite.

Now, consider, without loss of generality, a $[0/90]_s$ laminate bei cooled from the cure temperature to room temperature. Suppose fo moment that the 0-degree plies and the 90-degree plies can defor unconstrained by each other, Figure 8.10. For convenience, only o 0-degree ply and one 90-degree ply are shown in the figure. As tempe ture is lowered, the 0-degree ply deforms by e_x^T while the 90-degree p undergoes a thermal strain e_y^T in the same direction. Since e_x^T and are different from each other, there will be a geometrical mismat between the 0-degree and 90-degree plies. In the actual $[0/90]_s$ la inate, however, such mismatch is not allowed. Therefore, residu stresses σ_x^R and σ_y^R are internally exerted to the 0-degree and 90-degr plies, respectively, to bring about the geometrical compatibility. T final strain e_i^{oT} of the laminate is called the laminate curing strain ar depends on e_x^T and e_y^T as well as on the elastic moduli.

The procedure just described can be put in a more general form. this end, the constitutive relation for ply, Equation 8.41, is first co verted to

$$\sigma_i = Q_{ij}(\epsilon_j - e_j^T) \qquad (8.4$$

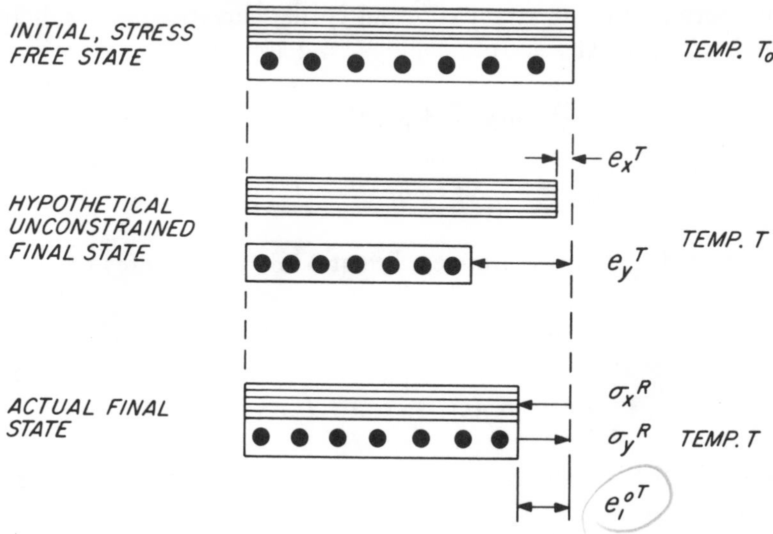

Figure 8.10 Build-up of residual stresses after fabrication.

ere the subscripts are associated with the laminate reference co-rdinates.

Following the same procedure as in Chapter 6, we obtain

$$
\begin{aligned}
N_i &= A_{ij}\epsilon_j^o + B_{ij}k_j - N_i^T \\
M_i &= B_{ij}\epsilon_j^o + D_{ij}k_j - M_i^T
\end{aligned}
\qquad (8.47)
$$

here the laminate moduli A_{ij}, B_{ij}, D_{ij} are defined in Chapter 6. The ew quantities N_i^T and M_i^T are defined by

$$
\begin{aligned}
N_i^T &= \int Q_{ij}e_j^T\,dz \\
M_i^T &= \int Q_{ij}e_j^T\,z\,dz
\end{aligned}
\qquad (8.48)
$$

ith the understanding that the integrations are from $-h/2$ to $h/2$. nce N_i^T and M_i^T have the same dimensions as N_i and M_i, respectively, ey are called the thermal stress resultant and thermal moment.

The curing in-plane strain e_i^{oT} and curing curvature k_i^T result from N_i^T and M_i^T in the absence of N_i and M_i and are given by

$$\begin{aligned} e_i^{oT} &= \alpha_{ij} N_j^T + \beta_{ij} M_j^T \\ k_i^T &= \beta_{ij} N_j^T + \delta_{ij} M_j^T \end{aligned} \tag{8.49}$$

With e_i^{oT} and k_i^T thus determined, Equation 8.47 can be rewritten as

$$\begin{aligned} N_i &= A_{ij}(\epsilon_j^o - e_j^{oT}) + B_{ij}(k_j - k_j^T) \\ M_i &= B_{ij}(\epsilon_j^o - e_j^{oT}) + D_{ij}(k_j - k_j^T) \end{aligned} \tag{8.50}$$

Through the thickness of the laminate, the curing strain ϵ_i^T is given by

$$\epsilon_i^T = e_i^{oT} + z\, k_i^T \tag{8.51}$$

The residual stress σ_i^R at z is then obtained from Equation 8.46 by substituting ϵ_i^T for ϵ_i.

$$\sigma_i^R = Q_{ij}(\epsilon_j^T - e_j^T) \tag{8.52}$$

The corresponding residual strain ϵ_i^R is thus

$$\epsilon_i^R = \epsilon_i^T - e_i^T \tag{8.53}$$

Since ϵ_i^T is the strain in the absence of N_i and M_i, the strain ϵ_i^M due to N_i and M_i is given by

$$\epsilon_i^M = \epsilon_i - \epsilon_i^T \tag{8.54}$$

Thus Equation 8.46 can be written as

$$\begin{aligned} \sigma_i &= Q_{ij}(\epsilon_j^M + \epsilon_j^R) \\ &= Q_{ij}\epsilon_j^M + \sigma_j^R \end{aligned} \tag{8.55}$$

Equation 8.55 indicates that the stress at a point within the laminate is the sum of the stress caused by N_i and M_i, and the residual stress.

4. residual stresses resulting from environmental change

In the preceding section the moisture concentration was zero. However, when a laminate absorbs moisture, the resulting swelling strain must be added to the thermal strain. The equations derived in the preceding section are still valid if we substitute e_i for e_i^T and if we use the stiffness at (T,c), $Q_{ij}(T,c)$, instead of $Q_{ij}(T,0)$. The superscript T is now replaced by N and the equations for the nonmechanical in-plane strain and curvature are

$$\begin{aligned} e_i^o &= \alpha_{ij} N_j^N + \beta_{ij} M_j^N \\ k_i^N &= \beta_{ij} N_j^N + \delta_{ij} M_j^N \end{aligned} \tag{8.56}$$

where

$$\begin{aligned} N_i^N &= \int Q_{ij}(T,c) e_j dz = \int Q_{ij}(T,c) e_j^T dz + \int Q_{ij}(T,c) e_j^H dz \\ M_i^N &= \int Q_{ij}(T,c) e_j z dz = \int Q_{ij}(T,c) e_j^T z dz + \int Q_{ij}(T,c) e_j^H z dz \end{aligned} \tag{8.57}$$

The evaluation of the integrals in Equation 8.57 is much simpler if c is uniform throughout the thickness. As pointed out earlier, the temperature reaches equilibrium almost instantaneously as compared to the moisture. In case of a uniform temperature and moisture distribution, both e_j^T and e_j^H are uniform in each ply and their variation from ply to ply is only due to the change of the fiber direction. Furthermore, a reduction of Equation 8.57 is possible if we use the equivalent stress that would produce the nonmechanical strain,

$$\sigma_i^N = Q_{ij} e_j \tag{8.58}$$

In the material symmetry axes of each ply, Equation 8.58 reduces to

$$\begin{aligned} \sigma_x^N &= Q_{xx} e_x + Q_{xy} e_y \\ \sigma_y^N &= Q_{xy} e_x + Q_{yy} e_y \\ \sigma_s^N &= 0 \end{aligned} \tag{8.59}$$

Note that Equation 8.59 does not change from ply to ply. For the ply with θ orientation, see Figure 8.11, the components of the equivalent stress are given by (see Table 2.2)

$$\begin{cases} \sigma_1^N = p^N + q^N \cos 2\theta \\ \sigma_2^N = p^N - q^N \cos 2\theta \\ \sigma_6^N = q^N \sin 2\theta \end{cases} \tag{8.60}$$

where

$$\begin{cases} p^N = \frac{1}{2}(\sigma_x^N + \sigma_y^N) = \frac{1}{2}(Q_{xx} + Q_{xy})\, e_x + \frac{1}{2}(Q_{xy} + Q_{yy})\, e_y \\ q^N = \frac{1}{2}(\sigma_x^N - \sigma_y^N) = \frac{1}{2}(Q_{xx} - Q_{xy})\, e_x + \frac{1}{2}(Q_{xy} - Q_{yy})\, e_y \end{cases} \tag{8.61}$$

Figure 8.12 shows ply-to-ply variations of the nonmechanical stresses in Equation 8.60.

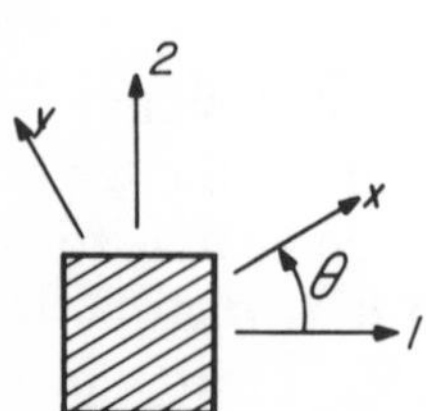

Figure 8.11 Coordinates for a typical ply orientation from the laminate axes 1-2.

Substituting Equation 8.60 into Equation 8.57, we have the nonmechanical stress resultants as

$$\begin{cases} N_1^N = p^N h + q^N V_{1A} \\ N_2^N = p^N h - q^N V_{1A} \\ N_6^N = q^N V_{3A} \end{cases} \tag{8.62}$$

Similarly, the nonmechanical moments can be written as

$$\begin{cases} M_1^N = q^N V_{1B} \\ M_2^N = -q^N V_{1B} \\ M_6^N = q^N V_{3B} \end{cases} \tag{8.63}$$

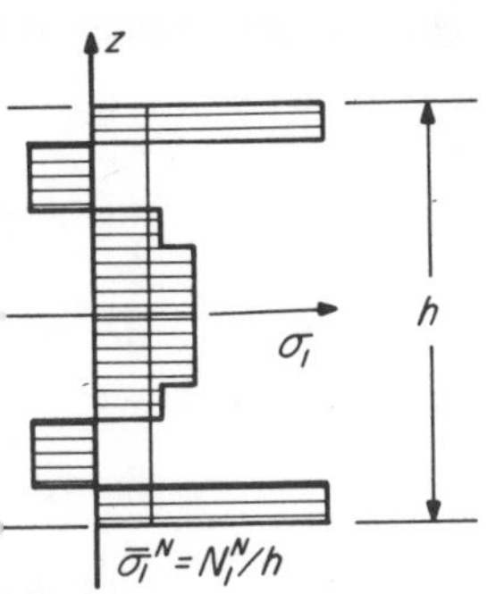

igure 8.12 Ply-to-ply variaions of nonmechanical stresses. he average stress is the nonnechanical stress resultant. The aminate and the expansional trains are assumed to be symnetric; otherwise, nonmechanical noments will be induced.

The V's in Equations 8.62 and 8.63 have been defined in Equations 6.79 through 6.82. Note that p^N does not contribute at all to M_i^N.

For symmetric laminates, we have

$$M_i^N = B_{ij} = 0 \tag{8.64}$$

If the expansion coefficients, Equation 8.45, can be used, the in-plane strain e_i^o is given by

$$e_i^o = \alpha_i^o (T-T_o) + \beta_i^o c \tag{8.65}$$

where

$$\begin{aligned} \alpha_i^o &= a_{ij} \int Q_{jk} \alpha_k \, dz \\ \beta_i^o &= a_{ij} \int Q_{jk} \beta_k \, dz \end{aligned} \tag{8.66}$$

It is interesting to note that, even when the expansion coefficients of ınidirectional composite are independent of T and c, such is not the ase with the expansion coefficients of laminate, α_i^o and β_i^o, because Q_{ij} lepends on T and c. When e_i^T and e_i^H are not proportional to $(T-T_o)$ and c, α_i^o and β_i^o in Equation 8.66 can be regarded as the instantaneous expansion coefficients because the change of Q_{ij} with T and c can be neglected when compared with that of e_i.

Equations 8.56 and 8.57 show that $e_i^o = k_i^N = 0$ if $e_i = 0$ for each ply. Since the residual stress is given by (see Equation 8.52)

$$\sigma_i^R = Q_{ij} (\epsilon_j^N - e_j) \tag{8.67}$$

where

$$\epsilon_j^N = e_j^o + z k_j^N \tag{8.68}$$

the condition of zero nonmechanical strain $e_i = 0$ in each constituent ply renders laminates free of residual stresses.

For graphite/epoxy laminates α_x and β_x are negligible and hence, th condition of zero residual stress is given by

$$e_y^T + e_y^H = 0 \tag{8.69}$$

Substituting Equations 8.39 and 8.45 into Equation 8.69, we derive relation between $(T - T_o)$ and ϕ so that there is no residual stress. The result is

$$T_o - T = a \left(\frac{\phi}{100}\right)^b \frac{\beta_y}{\alpha_y} \tag{8.70}$$

Equation 8.70 is shown graphically in Figure 8.13 for a graphite/epoxy laminate whose properties are given in Table 8.2. Note that the nonmechanical strain e_y is positive in the region to the right of the line and is negative to the left of the line.

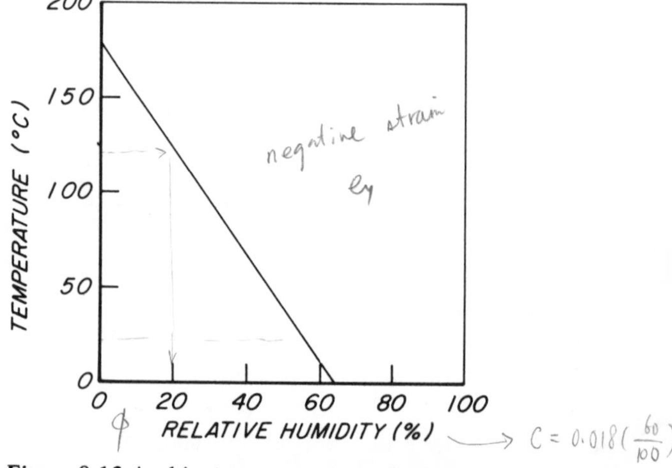

Figure 8.13 Ambient temperature and relative humidity required for a graphite/epoxy composite to be free of residual stresses.

The swelling coefficient β_i can also be obtained from a moisture absorption test of a unidirectional composite of thickness h subjected to a relative humidity ϕ on both sides. The resulting moisture distribution is given by Equation 8.30. The nonmechanical stress resultant ue to swelling is obtained from Equation 8.57 as

$$N_i^H = \int Q_{ij} e_j^H \, dz = \int Q_{ij} \beta_j \, c dz$$
$$= h Q_{ij} \beta_j \, \bar{c} \tag{8.71}$$

here the last equality follows from the assumption that Q_{ij} is inependent of c. For the unidirectional composite, we have

$$\alpha_{ij} = \frac{1}{h} Q_{ij}^{-1} \tag{8.72}$$

herefore, the laminate swelling strain becomes

$$e_i^{oH} = \beta_i \bar{c} \tag{8.73}$$

hus β_i can be determined by measuring e_i^{oH} and $\bar{c}$ during a moisture
osorption test of a unidirectional composite.

The nonmechanical curvature k_i^N can be translated into the non-
echanical out-of-plane displacement w^N by using the curvature-
splacement relations, Equation 5.9. To this end we take the laminate
Figure 8.2 to be a rectangular plate as shown in Figure 8.14. The
mensions a and b are much larger than the thickness h so that one-
mensional diffusion through the thickness is still applicable. The solu-
on to Equation 5.9 is

$$w^N = -\frac{1}{2}(k_1^N x_1^2 + k_2^N x_2^2 + k_6^N x_1 x_2) + b_1 x_1 + b_2 x_2 + b_3 \tag{8.74}$$

The integration constants b_1, b_2, b_3 are to be determined from the
oundary conditions. For example, referring to Figure 8.14, we assume
at the three conrners represented by (0,0), (a,0) and (0, b) rest on a
at surface; i.e.,

$$w^N = 0 \text{ at } (0,0), (a,0) \text{ and } (0,b) \tag{8.75}$$

To satisfy the boundary condition 8.75, the constants must be chosen as

$$b_1 = \frac{1}{2} k_1^N a$$

$$b_2 = \frac{1}{2} k_2^N b \tag{8.76}$$

$$b_3 = 0$$

Therefore, the final displacement is

$$w^N = \frac{1}{2} k_1^N (a - x_1) x_1 + \frac{1}{2} k_2^N (b - x_2) x_2 - \frac{1}{2} k_6^N x_1 x_2 \tag{8.77}$$

Thus, the effect of hygrothermal deformation can be easily seen from the out-of-plane displacement w^N.

Equation 8.74 is valid as long as w^N is not too large so that the assumptions of the linear theory are applicable. Thus this equation should be used only for those plates that do not have too large width-to-thickness ratio.

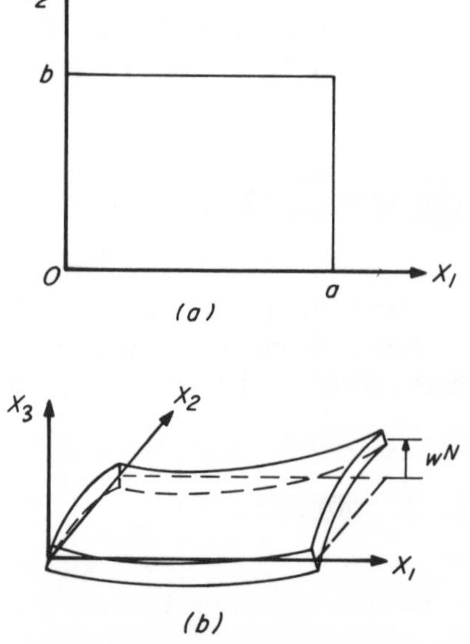

Figure 8.14 Out-of-plane deflection of an unsymmetric, rectangular laminate.

5. unsymmetric cross-ply laminates

Jnsymmetric cross-ply laminates are frequently used to show the effect of hygrothermal strains. Consider a rectangular $[0/90]_T$ laminate vhose dimensions are the same as those in Figure 8.14. Further, we ssume that the temperature T and the moisture concentration c are uniform throughout the laminate. The stiffnesses of this laminate have been derived in Sections 4.4 and 6.4. Thus, only nonmechanical stress esultants and moments need to be calculated in this section.

The nonmechanical stress resultants follow from Equation 8.62 as

$$\begin{aligned} N_1^N &= \frac{h}{2}(Q_{xx} + Q_{xy})\, e_x + \frac{h}{2}(Q_{yy} + Q_{xy})\, e_y \\ N_2^N &= N_1^N \\ N_6^N &= 0 \end{aligned} \tag{8.78}$$

vhere, since T and c are uniform, the nonmechanical strains are given y

$$\begin{aligned} e_x &= \alpha_x\,(T - T_o) + \beta_x c \\ e_y &= \alpha_y\,(T - T_o) + \beta_y c \end{aligned} \tag{8.79}$$

Similarly, the nonmechanical moments are

$$\begin{aligned} M_1^N &= -\frac{h^2}{8}\,(Q_{xx} - Q_{xy})\, e_x - \frac{h^2}{8}\,(Q_{xy} - Q_{yy})\, e_y \\ M_2^N &= -M_1^N \\ M_6^N &= 0 \end{aligned} \tag{8.80}$$

Therefore, there are only two unknowns e_1^o and k_1^N which must atisfy (see Equations 8.47 and 6.100)

$$\begin{aligned} N_1^N &= (A_{11} + A_{12})\, e_1^o + B_{11} k_1^N \\ M_1^N &= B_{11} e_1^o + (D_{11} - D_{12})\, k_1^N \end{aligned} \tag{8.81}$$

The other strains and curvatures are

$$e_2^o = e_1^o, \qquad e_6^o = 0$$
$$k_2^N = -k_1^N, \qquad k_6^N = 0 \tag{8.8}$$

The solutions to Equation 8.81 are

$$e_1^o = \frac{(D_{11} - D_{12}) N_1^N - B_{11} M_1^N}{(A_{11} + A_{12})(D_{11} - D_{12}) - B_{11}^2}$$
$$k_1^N = \frac{(A_{11} + A_{12}) M_1^N - B_{11} N_1^N}{(A_{11} + A_{12})(D_{11} - D_{12}) - B_{11}^2} \tag{8.8}$$

Equation 8.83 can also be expressed in terms of the ply stiffness usir the results of Sections 4.4 and 6.4. Thus, the final equations are

$$e_1^o = \frac{1}{Q}[(Q_{xx}^2 + 7Q_{xx}Q_{yy} - Q_{xx}Q_{xy} + Q_{xy}Q_{yy} - 8Q_{xy}^2) e_x$$
$$+ (Q_{yy}^2 + 7Q_{xx}Q_{yy} - Q_{yy}Q_{xy} + Q_{xx}Q_{xy} - 8Q_{xy}^2) e_y]$$
$$k_1^N = \frac{24}{hQ}(Q_{xx}Q_{yy} - Q_{xy}^2)(e_x - e_y) \tag{8.84}$$

where

$$Q = Q_{xx}^2 + 14Q_{xx}Q_{yy} + Q_{yy}^2 - 16Q_{xy}^2 \tag{8.85}$$

The residual stresses in the material symmetry axes of each ply ar the same in the 0-degree and 90-degree plies. They are given by Equa tions 8.67 and 8.68 as

$$\sigma_x^R = Q_{xx}(e_1^o - e_x + |z| k_1^N) + Q_{xy}(e_1^o - e_y - |z| k_1^N)$$
$$\sigma_y^R = Q_{xy}(e_1^o - e_x + |z| k_1^N) + Q_{yy}(e_1^o - e_y - |z| k_1^N) \tag{8.86}$$

Note that $|z|$ is the distance from the mid surface to a point of interest

As expected, if $e_x = e_y$, k_1^N vanishes and $e_1^o = e_x$. Consequently, there are no residual stresses in the plies.

For a rectangular plate shown in Figure 8.12, the out-of-plane displacement is given by Equation 8.77. However, since $k_6^N = 0$, we have

$$w^N = \frac{1}{2} k_1^N \left[(a - x_1)\, x_1 - (b - x_2) x_2 \right] \tag{8.87}$$

Equation 8.87 describes an anticlastic surface. The displacement w^N attains different maximum magnitudes depending on a and b. That is,

$$\begin{aligned} |w^N|_{\max} &= \frac{1}{2} |k_1^N| \left(\frac{a}{2}\right)^2 \text{ at } \left(\frac{a}{2}, 0\right) \quad \text{if } a > b, \\ &= \frac{1}{2} |k_1^N| \left(\frac{b}{2}\right)^2 \text{ at } \left(0, \frac{b}{2}\right) \quad \text{if } a < b. \end{aligned} \tag{8.88}$$

5. antisymmetric angle-ply laminates

Just like unsymmetric cross-ply laminates, antisymmetric angle-ply laminates are also susceptible to nonmechanical warping. The simplest example is that of a rectangular $[-\theta/\theta]_T$ laminate. Again, its dimensions are as shown in Figure 8.14, and T and c are uniform throughout the laminate. For the stiffnesses of this laminate we refer to Sections 4.5 and 6.5.

The nonmechanical stress resultants follow from Equation (8.62) as

$$\begin{aligned} N_1^N &= p^N h + q^N h \cos 2\theta \\ N_2^N &= p^N h - q^N h \cos 2\theta \\ N_6^N &= 0 \end{aligned} \tag{8.89}$$

where p^N and q^N are given by Equation 8.61, and because of the antisymmetry, M_1^N and M_2^N vanish and the only nonzero moment is

$$M_6^N = \frac{h^2}{4} q^N \sin 2\theta \tag{8.90}$$

Therefore, Equation 8.47 with the superscript T replaced by N becom

$$
\begin{aligned}
N_1^N - B_{16}k_6^N &= A_{11}e_1^o + A_{12}e_2^o \\
N_2^N - B_{26}k_6^N &= A_{12}e_1^o + A_{22}e_2^o \qquad (8.9 \\
k_6^N &= M_6^N/D_{66}
\end{aligned}
$$

The in-plane stiffnesses listed in Table 4.8 are repeated here:

$$
\begin{aligned}
A_{11} &= h(U_1 + U_2 \cos 2\theta + U_3 \cos 4\theta) \\
A_{22} &= h(U_1 - U_2 \cos 2\theta + U_3 \cos 4\theta) \qquad (8.9 \\
A_{12} &= h(U_4 - U_3 \cos 4\theta)
\end{aligned}
$$

The remaining stiffnesses in Equation 8.91 are from Section 6.5:

$$
\begin{aligned}
B_{16} &= \frac{h^2}{4}\left(\frac{U_2}{2} \sin 2\theta + U_3 \sin 4\theta\right) \\
B_{26} &= \frac{h^2}{4}\left(\frac{U_2}{2} \sin 2\theta - U_3 \sin 4\theta\right) \qquad (8.9 \\
D_{66} &= \frac{h^3}{12}(U_5 - U_3 \cos 4\theta)
\end{aligned}
$$

With all the stiffnesses known; the nonmechanical in-plane strains are

$$
\begin{aligned}
e_1^o &= \frac{A_{22}(N_1^N - B_{16}k_6^N) - A_{12}(N_2^N - B_{26}k_6^N)}{A_{11}A_{22} - A_{12}^2} \\
& \qquad (8.94 \\
e_2^o &= \frac{A_{11}(N_2^N - B_{26}k_6^N) - A_{12}(N_1^N - B_{16}k_6^N)}{A_{11}A_{22} - A_{12}^2}
\end{aligned}
$$

The remaining strains and curvatures are easily shown to vanish:

$$
e_6^o = k_1^N = k_2^N = 0 \qquad (8.95
$$

In the material symmetry axes of the θ-degree ply, the final non-
echanical strains are

$$
\begin{aligned}
\epsilon_x^N &= m^2 e_1^o + n^2 e_2^o + m\, n\, k_6^N\, z \\
\epsilon_y^N &= n^2 e_1^o + m^2 e_2^o - m\, n\, k_6^N\, z \\
e_s^N &= -2m\, n\, e_1^o + 2m\, n\, e_2^o + (m^2 - n^2)\, k_6^N\, z
\end{aligned}
\tag{8.96}
$$

here $m = \cos\theta$ and $n = \sin\theta$.

he residual stresses are thus obtained from

$$
\begin{aligned}
\sigma_x^R &= Q_{xx}(\epsilon_x^N - e_x) + Q_{xy}(\epsilon_y^N - e_y) \\
\sigma_y^R &= Q_{xy}(\epsilon_x^N - e_x) + Q_{yy}(\epsilon_y^N - e_y) \\
\sigma_s^R &= Q_{ss}\, \epsilon_s^N
\end{aligned}
\tag{8.97}
$$

Finally, the nonmechanical displacement w^N depends only on k_6^N:

$$
w^N = -\frac{1}{2} k_6^N\, x_1 x_2 \tag{8.98}
$$

hus, the largest displacement of the plate in Figure 8.14 occurs at the
orner (a,b):

$$
|w^N|_{max} = \frac{1}{2} |k_6^N| ab \tag{8.99}
$$

effect of residual stress on failure

s we saw in Chapter 7, the strength ratios for a unidirectional com-
osite can be obtained by solving a quadratic equation in stress space or
strain space. All material constants in F's and G's are known for a
iven material including the assumed value for the interaction term.
hen, for a given state of stress or strain, the strength ratios R and R'
re the two roots of the appropriate quadratic equation. For the
trength ratios of a laminate, we must first establish the on-axis ply

strain. Then the strength ratio can be calculated ply by ply.

If residual strains are to be included, we must redefine the strengt ratio based on the mechanical strain,

$$\epsilon_{i(a)}^{M} = R\,\epsilon_i^{M} \qquad (8.10$$

This ratio signifies the amount of applied strain that can be increase before failure occurs. The strain which produces stresses in the plies the sum of the mechanical strain and the residual strain, as shown Equation 8.55 and Figure 8.15. Therefore, the failure criterion 7.2 becomes

$$G_{ij}(\epsilon_{i(a)}^{M} + \epsilon_i^{R})(\epsilon_{j(a)}^{M} + \epsilon_j^{R}) + G_i(\epsilon_{i(a)}^{M} + \epsilon_i^{R}) = 1 \qquad (8.10$$

Substituting Equation 8.100 into Equation 8.101, we obtain the equ tion for the mechanical strength ratio as follows:

$$G_{ij}(R\epsilon_i^{M} + \epsilon_i^{R})(R\epsilon_j^{M} + \epsilon_j^{R}) + G_i(R\epsilon_i^{M} + \epsilon_i^{R}) - 1 = 0 \qquad (8.102$$

or, in short,

$$aR^2 + bR + c = 0 \qquad (8.10$$

The positive roots are

$$R = -\frac{b}{2a} + \left[\left(\frac{b}{2a}\right)^2 - \frac{c}{a}\right]^{1/2}$$

$$R' = -\frac{b}{2a} + \left[\left(\frac{b}{2a}\right)^2 - \frac{c}{a}\right]^{1/2} \qquad (8.104$$

A comparison of Equation 8.102 with Equation 7.50 shows that, i the presence of residual strains, the failure surface is simply dislocate by the amount of the residual strains in the opposite direction. Suc translation of failure surface is schematically shown in Figure 8.16 for constituent ply in a laminate.

Since all plies of the laminate are made of the same material, th on-axis free nonmechanical strain e_i will remain the same for a give temperature and moisture concentration. For a symmetric laminate, th

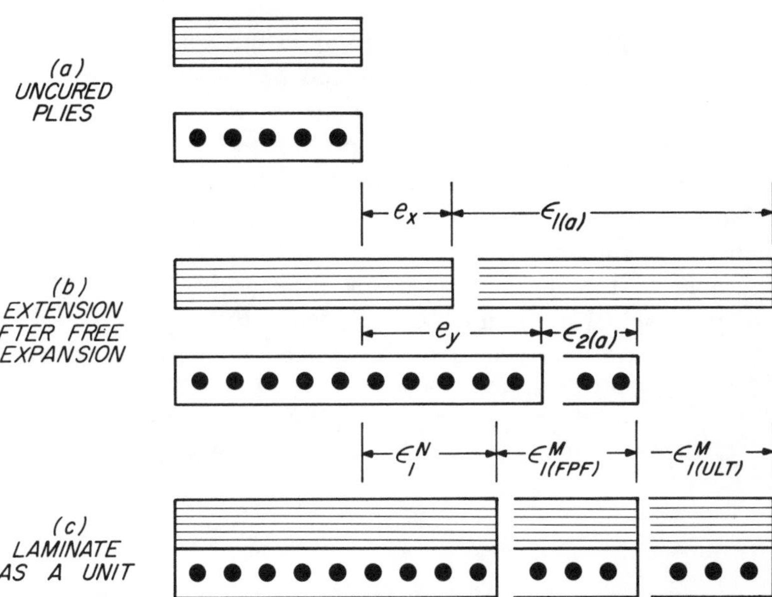

Figure 8.15 Relation between total strain of a laminate and ply strain. Total strain in (*c*) is based on as-cured plies in (*a*). The strengths of unidirectional composites are measured from cured and expanded plies in (*b*). So strength analysis must be based on $\epsilon_{1\,(a)}$ and $\epsilon_{2\,(a)}$ shown in (*b*).

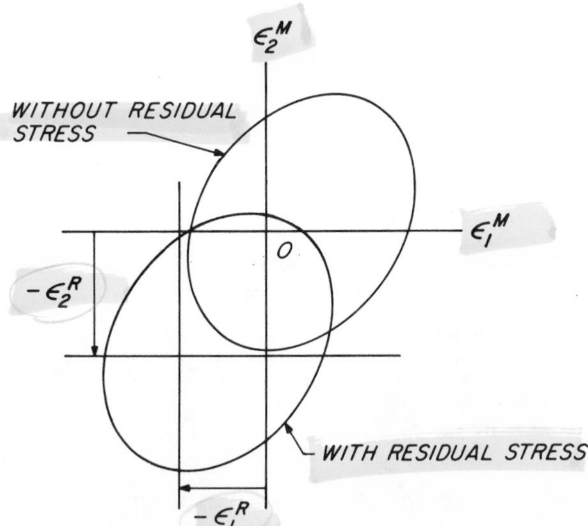

Figure 8.16 Translation of failure surface caused by residual stress in a constituent ply of a laminate. The residual stress moves the failure surface by $-\epsilon_i^R$. Note that the shape of the failure surface in the strain space is not changed by the residual stress.

nonmechanical ply strains are equal to e_i^o in the laminate axes, but tl
on-axis nonmechanical ply strains depend on the ply orientations. S
for multidirectional laminates, there are as many different on-ax
residual strains as there are ply orientations. Figure 7.9 shows tl
strength ratio for each ply of a laminate under a given loading cond
tion. Note again that the strength ratios are based on mechanical strai
as such, they will provide a measure of the margin of safety on hc
much increase the mechanical strain can sustain before failure. O
course, the residual strains must be included in Figure 7.9.

Up to now we have addressed the in-plane strength of a symmetr
laminate. For the flexural strength of a symmetric laminate, we face
linear variation of mechanical strain across the thickness of the lam
inate. Specifically, the mechanical strain due to the applied moment

$$\epsilon_i^M = zk_i \qquad (8.10$$

Therefore, the equation for the mechanical strength ratios becomes

$$G_{ij}(Rzk_i + \epsilon_i^R)(Rzk_i + \epsilon_i^R) + G_i(Rzk_i + \epsilon_i^R) - 1 = 0 \qquad (8.10$$

For a given k_i, the positive root of this equation is the product of t
strength ratio R and the coordinate z. A higher z would mean a low
strength ratio within each ply or ply group. Thus the outer surface
each ply, having higher z, would be the location to calculate t
strength ratio.

8. effects of temperature and moisture on properties of unidirectional composites

Polymers presently used in composites are susceptible to temperatu
and moisture. Consequently, the transverse and shear properties of un
directional composites, which are very much affected by matrix prope
ties, also degrade upon exposure to elevated temperature and upo
absorption of moisture. However, the change of longitudinal properti
in the same range of temperature and moisture variations is negligib
because of the excellent retention of mechanical properties by t
fibers.

Figure 8.17 shows typical changes in moduli of AS/3501 caused by mperature and moisture concentration. Similar changes in tensile and ear strengths of the same material are shown in Figure 8.18. the neral features of these figures can be summed up as follows.

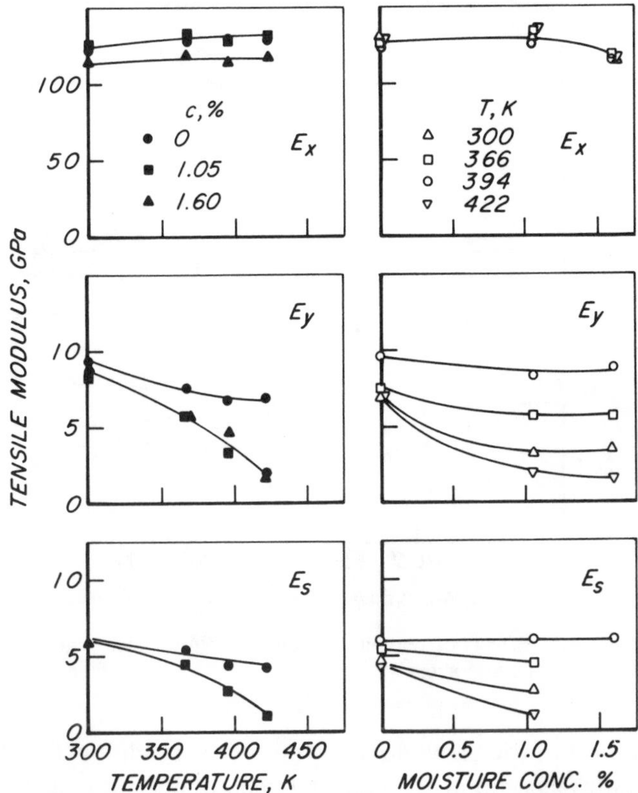

Figure 8.17 Effects of temperature and moisture concentration on tensile and shear moduli of AS/3501. (Data from [3]).

The longitudinal properties are not degraded by temperature and oisture. This behavior is a manifestation of excellent resistance of the aphite fiber to hygrothermal exposure.

The room temperature properties do not change much upon moisre absorption. However, this statement needs a qualification. Note at the transverse strength Y suffers a significant loss at higher

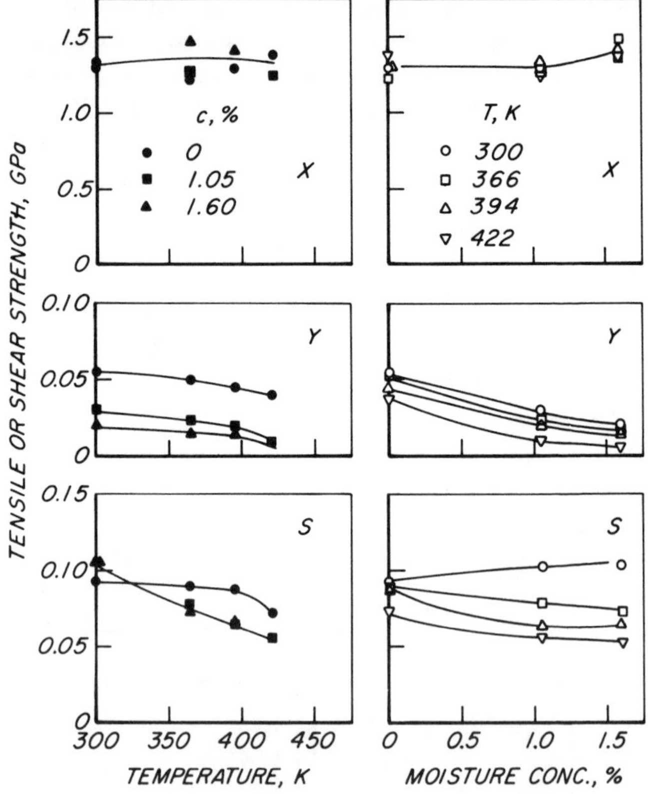

Figure 8.18 Effects of temperature and moisture concentration on tensile and shear strengths of AS/3501. (Data from [3]).

moisture concentrations while the transverse modulus E_T does n Since the interfacial bond strength between fibers and matrix affects but not E_T, we can conclude that the reduction in Y may be a result the weakening of the interfacial bond by absorbed moisture.

The most reduction in properties occurs when temperature a moisture are combined. Consequently, poor structural performance unidirectional composites is expected in hot and humid environmen

Whereas hot and humid environments are detrimental to the matr controlled properties of unidirectional composites, the same is n necessarily true for laminated composites because in such environmer laminates can be free of residual stress or may even benefit fr residual stress. For example, consider a $[0/90]_s$ laminate subjected t nsile load in the 0-degree direction. After fabrication the residual ress σ_y in the 90-degree ply is tensile. However, as temperature and oisture concentration increase, σ_y will decrease and may become mpressive. Thus, even though the transverse strength Y decreases, the)-degree ply may fail at higher applied load. The net effect can be edicted using pertinent property data in conjunction with Equation 102.

sample problems

A 1-mm thick $[0/\pm45/90]_s$ graphite/epoxy laminate is exposed on th sides to air at a temperature of 75°C and 90 percent relative midity. The initial moisture content is 0.5 percent. Estimate the time quired to reach one percent moisture content. Use the approximate uation for time sufficiently large.

In Equation 8.34 the following variables are known:

$$\begin{aligned} \bar{c} &= 0.01, \quad c_o = 0.005 \\ h &= 1\text{ mm} \end{aligned} \tag{8.107}$$

e equilibrium moisture concentration at 90 percent relative humidity obtained from Equation 8.39. Since $a = 0.018$ for the graphite/epoxy listed in Table 8.2, we have

$$c_\infty = 0.018 \times \frac{90}{100} = 0.0162 \tag{8.108}$$

xt, the diffusion coefficient at 75°C follows from Equation 8.40. ain, the appropriate constants are taken from Table 8.2. The result is

$$K^H = 6.51 \times \exp\left(-\frac{5722}{348}\right) = 4.706 \times 10^{-7}\ \text{mm}^2/s \tag{8.109}$$

erefore, the time required is

$$\begin{aligned} t &= \frac{h^2}{\pi^2 K^H}\left\{-\ln\left[\frac{\pi^2}{8}\left(1-\frac{\bar{c}-c_o}{c_\infty - c_o}\right)\right]\right\} \\ &= 22.8\ h \end{aligned} \tag{8.110}$$

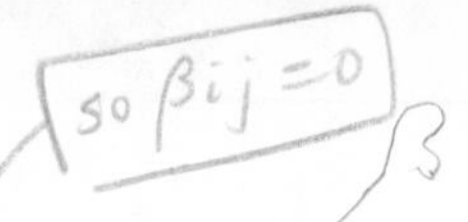

b. Determine the in-plane thermal expansion coefficients and t residual stresses for a $[0/\pm45/90]_s$ T300/5208 laminate at room te perature (= 22°C). The stress-free temperature is the cure temperatu and use the properties in Table 8.2.

The in-plane stiffnesses of this laminate are given in Section 4.6. T ply nonmechanical strains are

$$e_x = 0.02\ \Delta T\ \mu\text{m/m}, \quad e_y = 22.5\ \Delta T\ \mu\text{m/m} \qquad (8.11$$

where

$$\Delta T = 295 - 450 = -155\ \text{K} \qquad (8.11$$

The corresponding equivalent stresses follow from Equation 8.59

$$\begin{aligned} \sigma_x^T &= 68.89\ \Delta T\ \text{kPa}, \quad \sigma_y^T = 232.8\ \Delta T\ \text{kPa} \\ \sigma_s^T &= 0 \end{aligned} \qquad (8.11$$

These stresses in turn yield

$$\begin{aligned} p^T &= 150.8\ \Delta T\ \text{kPa}, \quad q^T = -81.94\ \Delta T\ \text{kPa} \\ r^T &= 0 \end{aligned} \qquad (8.11$$

The integrals of the trigonometric functions all vanish. Therefore, t nonmechanical stress resultants are simply given by

$$N_1^T = N_2^T = p^T\,h, \; N_6^T = 0 \qquad (8.11$$

Consequently, the thermal in-plane strains are

$$e_1^{oT} = e_2^{oT} = \frac{1-\nu}{E}\,p^T \qquad (8.11$$

Using the values of E and ν in Equation 4.63, we finally obtain

$$\begin{aligned} e_1^{oT} &= e_2^{oT} = 1.52\ \Delta T\ \mu\text{m/m} \\ e_6^{oT} &= 0 \end{aligned} \qquad (8.11$$

he corresponding in-plane thermal expansion coefficients are

$$\alpha_1^o = \alpha_2^o = 1.52\ (\mu\text{m/m})/\text{K}$$
$$\alpha_6^o = 0 \tag{8.118}$$

The residual stresses in each ply are obtained from Equation 8.52. ince we have

$$e_x^{oT} = e_y^{oT} = e_1^{oT},\ e_s^{oT} = 0 \tag{8.119}$$

or each ply, the residual stresses do not change from ply to ply. The nal results are

$$\sigma_x^R = 213\ \Delta T\ \text{kPa} = -33.0\ \text{MPa}$$
$$\sigma_y^R = -213\ \Delta T\ \text{kPa} = 33.0\ \text{MPa} \tag{8.120}$$
$$\sigma_s^R = 0$$

c. Determine the nonmechanical in-plane strains and curvatures of a $0_8/90_8]_T$ T300/5208 laminate at room temperature and 50% relative umidity. Use the properties in Table 8.2. The thickness of the minate is 2 mm.

The equilibrium moisture content at 50% relative humidity is 0.009. ince the temperature difference is −155 K, the nonmechanical strains re

$$e_x = 0.02 \times (-155) + 0 \times 9 = -3.1\ \mu\text{m/m}$$
$$e_y = 22.5 \times (-155) + 440 \times 9 = 472.5\ \mu\text{m/m} \tag{8.121}$$

The nonmechanical stress resultants are determined by Equation .78:

$$N_1^N = N_2^N = 5.682\ \text{kN}/m$$
$$N_6^N = 0 \tag{8.122}$$

On the other hand, Equation 8.80 gives the nonmechanical moments

$$M_1^N = -M_2^N = 2.036\ \text{N}$$

$$M_6^N = 0$$

(8.12

Finally, using the moduli in Section 6.4, we obtain the nonmechani strains and curvatures as follows:

$$e_1^o = e_2^o = 107\ \mu\text{m/m},\ e_6^o = 0$$

$$k_1^N = -k_2^N = -0.180\ \text{m}^{-1},\ k_6^N = 0$$

(8.12

Note that these answers can be obtained directly from Equation 8.8

d. A 100 mm × 100 mm plate is made of $[-45_8/45_8]_T$ T300/520 Determine the maximum out-of-plane deflection at room temperatu and 50% relative humidity. Use the results of the preceding problem.

The nonmechanical stresses corresponding to the nonmechani strains calculated in the preceding problem are

$$\sigma_x^N = 0.805\ \text{MPa},\ \sigma_y^N = 4.88\ \text{MPa}$$

$$\sigma_s^N = 0$$

(8.12

These stresses in turn yield

$$p^N = 2.84\ \text{MPa},\ q^N = -2.04\text{MPa}$$

$$r^N = 0$$

(8.12

Using $\theta = 45°$ in Equations 8.89 and 8.90, we calculate the nonvanis ing components of N_i^N and M_i^N as

$$N_1^N = N_2^N = 5.68\ \text{kN}/m$$

$$M_6^N = -2.04\,\text{N}$$

(8.12

The bending modulus D_{66} is given in Section 6.5 as

$$D_{66} = 31 \text{ Nm} \tag{8.128}$$

herefore, the nonmechanical curvature k_6^N becomes

$$k_6^N = \frac{-2.04}{31} = -0.066\text{m}^{-1} \tag{8.129}$$

he maximum deflection of the plate is obtained by substituting k_6^N to Equation 8.99 and noting that $a = b = 100$ mm:

$$|w^N|_{\max} = 0.33 \text{ mm} \tag{8.130}$$

e. A $[0/\pm 45/90]_s$ AS/3501 laminate is subjected to the stress resul-nts $N_1 = 30$ kN/m, $N_2 = 20$ kN/m at room temperature immediately ter fabrication. What is the strength ratio for each ply? If there were residual stresses, what would be the strength ratios? Use the thermal roperties in Table 8.2.

From Problem *b* the residual strains in the material symmetry co-rdinates are seen to be the same in all plies. Specifically, they are

$$\begin{aligned} \epsilon_x^R &= e_x^{oT} - e_x = -294.5\ \mu\text{m/m} \\ \epsilon_y^R &= e_y^{oT} - e_y = 4107.5\ \mu\text{m/m} \end{aligned} \tag{8.131}$$

he in-plane moduli are obtained by substituting U_1, U_4 and U_5 of able 3.6 into Equation 4.62:

$$\begin{aligned} E &= 54.83 \text{ GPa}, \\ \nu &= 0.284 \end{aligned} \tag{8.132}$$

he mechanical laminate strains are

$$\begin{aligned} \epsilon_1^{oM} &= \frac{1}{E}\left(\frac{N_1}{h} - \nu\frac{N_2}{h}\right) = 443.6\ \mu\text{m/m} \\ \epsilon_2^{oM} &= \frac{1}{E}\left(\frac{N_2}{h} - \nu\frac{N_1}{h}\right) = 209.4\ \mu\text{m/m} \end{aligned} \tag{8.133}$$

ote that we have used $h = 1$ mm.

Using G_{ij} in Table 7.3 we can calculate *a, b* and *c* in Equation 8.1
for each ply:

$$a = G_{xx}\epsilon_x^{M^2} + 2G_{xy}\epsilon_x^M\epsilon_y^M + G_{yy}\epsilon_y^{M^2} + G_{ss}\epsilon_s^{M^2}$$

$$b = G_x\epsilon_x^M + G_y\epsilon_y^M + 2\,[G_{xx}\epsilon_x^M\epsilon_x^R + G_{xy}\,(\epsilon_x^M\epsilon_y^R + \epsilon_x^R\epsilon_y^M) + G_{yy}\epsilon_y^M\epsilon_y^R + G_{ss}\epsilon_s^M\epsilon_s^R\,] \quad (8.13$$

$$c = -1 + G_{xx}\epsilon_x^{R^2} + 2G_{xy}\epsilon_x^R\epsilon_y^R + G_{yy}\epsilon_y^{R^2} + G_{ss}\epsilon_s^{R^2} + G_x\epsilon_x^R + G_y\epsilon_y^R$$

The mechanical strains are:

0-degree ply

$$\epsilon_x^M = \epsilon_1^{oM},\ \epsilon_y^M = \epsilon_2^{oM},\ \epsilon_s^M = 0 \quad (8.13$$

90-degree ply

$$\epsilon_x^M = \epsilon_2^{oM},\ \epsilon_y^M = \epsilon_1^{oM},\ \epsilon_s^M = 0 \quad (8.13$$

45-degree ply

$$\epsilon_x^M = \epsilon_y^M = \frac{1}{2}(\epsilon_1^{oM} + \epsilon_2^{oM}),\ \epsilon_s^M = -(\epsilon_1^{oM} - \epsilon_2^{oM}) \quad (8.13$$

The final results are given below in a tabular form:

Ply	$a(10^{-3})$	$b(10^{-3})$	$c(10^{-3})$	R	R'
0-degree	1.45	57.60	−344.1	5.2	45.0
90-degree	1.46	89.88	−344.1	3.6	65.2
45-degree	1.53	69.70	−344.1	4.5	50.1

In the absence of residual stresses, the results would be as follows:

Ply	$a(10^{-3})$	$b(10^{-3})$	$c(10^{-3})$	R	R'
0-degree	1.45	44.80	−1000	15.0	45.9
90-degree	1.46	66.22	−1000	11.9	57.3
45-degree	1.53	55.50	−1000	13.2	49.5

The strength ratios to be used under the given state of stress are given y R. Note that the strength ratios in tension, R, are substantially duced by the residual stresses for all plies. However, in compression e strength ratios, R', can increase in the presence of residual stresses.

). conclusions

s polymers undergo both dimensional and property changes under the ange of environment, so do composite materials utilizing polymers as atrix phase. For most structural composites, fibers are fairly in-nsitive to environmental changes. Thus, the environmental suscepti-lity of composites is mainly through the matrix phase.

While the thermal diffusion through composites can be described by e Fourier equation, the Fick's equation can be used to handle the oisture diffusion. Under most circumstances, these two equations can : used separately because the thermal diffusion takes place at a rate in e order of 10^6 times faster than the moisture diffusion.

Dimensional changes resulting from environmental changes are scribed by a modified set of linear equations. That is, the total strain inus the nonmechanical strain is linearly related to the stress. The onmechanical strain is measured from a stress-free reference state and e elastic moduli to be used are taken at the final environmental ondition. The theory is based on the assumption of elastic behavior; owever, a nonlinear dependence of the nonmechanical strains on the mperature change and moisture concentration is allowed.

The anisotropy of unidirectional composites also manifests itself in e hygrothermal behavior. Because of the directional dependence of grothermal expansion, residual stresses are induced in composite minates. These residual stresses can be calculated using the laminated plate theory developed in the preceding sections. Since the transver residual stresses in plies after fabrication are tensile while the residu stresses induced by moisture absorption are compressive, a combinatic of temperature and moisture concentration can be chosen to render laminate free of residual stresses.

The residual stresses in a laminate change the ply failure stresses; tl ply strength ratios depend on the residual stresses. Depending on tl direction of loading, the residual stresses can be beneficial or delete ous. Also, the transverse residual stresses are usually much lower ar even compressive in a hostile environment such as high temperature ar high humidity. Therefore, although material properties are degraded the hostile environment, such environment is rather beneficial from tl viewpoint of residual stresses. The true effect of residual stresses can l assessed only by analyzing the overall performance of the composi under a given service condition.

In this chapter we have presented an analysis method for the hygr thermal behavior of composite laminates. By necessity a few simplif ing assumptions had to be introduced. Any improvement over tl present theory quickly brings complexity and the necessity for mo information about the material behavior. In the absense of such add tional information, the present theory can still be used to analyze tl average behavior of composite laminates.

1. homework problems

In a moisture absorption test a dry 10 mm X 10 mm glass/epoxy laminate was immersed in water. The laminate was 2-mm thick, and the equilibrium moisture content was 2%. The additional moisture contents measured at three different times were as follows:

$$\bar{c} = 0.2\% \text{ at } t = 16\ h$$

$$= 0.4\% \text{ at } t = 64\ h$$

$$= 0.6\% \text{ at } t = 144\ h$$

What is the diffusion coefficient of the glass/epoxy laminate?

In one-dimensional diffusion through the thickness it takes 1 month for a 1-mm thick graphite/epoxy laminate to reach 90% of the equilibrium moisture content at 22°C. How long does it take for a 10-mm thick graphite/epoxy laminate to reach the same moisture content at 75°C?

For a graphite/epoxy laminate with the properties of Table 8.2, determine the change in volume when the temperature is increased by 100 K and the equilibrium moisture concentration by 1 percent.

. Show that for a $[-\theta/\theta]_s$ laminate the inplane nonmechanical strains are given by

$$e_1^o = \frac{h[J_1(A_{22} - A_{12}) + J_2(A_{12} + A_{22})\cos 2\theta]}{A_{11}A_{22} - A_{12}^2}$$

$$e_2^o = \frac{h[J_1(A_{11} - A_{12}) - J_2(A_{11} + A_{12})\cos 2\theta]}{A_{11}A_{22} - A_{12}^2}$$

where

$$J_1 = U_1 p_e + U_2 q_e + U_4 p_e$$

$$J_2 = U_2 p_e + q_e (U_1 - U_4 + 2U_3)$$

$$p_e = \frac{1}{2}(e_x + e_y)$$

$$q_e = \frac{1}{2}(e_x - e_y)$$

In the above equations, h is the thickness and the U's are functio of Q_{ij} defined in Section 3.1.

e. A unidirectional composite has the following properties:

$$E_x = 910 \text{ MPa}, \qquad E_y = 7.24 \text{ MPa}$$

$$\nu_x = 0.36, \qquad E_s = 1.81 \text{ MPa}$$

When the composite is immersed in benzene until an equilibriu state is reached, the swelling strains are

$$e_x = 0.02, \qquad e_y = 0.75, \qquad e_s = 0$$

Now a $[-\theta/\theta]_s$ laminate is made of the composite just describe What is the angle θ which gives a minimum swelling strain e_1^o whe the laminate is subjected to the same environment? Is the minimu swelling strain positive or negative?

f. Determine the residual stresses in a $[0/90]_s$ AS/3501 laminate whe the laminate is subjected to an 80% relative humidity at 30°C. Als determine the strength ratios when a stress resultant N_1 of 10 kN/ is applied. Assume the laminate is in an equilibrium state.

g. Show that the residual stresses in $[0/\pm 60]_s$ laminate are the same those in $[0/\pm 45/90]_s$ and $[0/90]_s$ laminates.

omenclature

= Specific heat, in J/g·K
= Specific moisture concentration, in g/g
= Moisture content or average specific moisture concentration
= Nonmechanical strain of a ply
= Nonmechanical in-plane strain of a laminate
d = Activation energy, in J/mole
= Moisture concentration, in g/m^3
H = Moisture diffusion coefficient in the transverse direction ($= K_y^H = K_z^H$)
H_o = Pre-exponential factor for K^H, in m^2/s
T_{ij} = Thermal conductivity, in w/(m·K)
T_i = Heat flux, in w/m^2
H_i = Moisture flux, in g/(m^2·s)
= Gas constant [= 8.319 J/(mole·K)]
$,R'$ = Strength ratios
= Temperature, in K
= Time
$_o$ = Stress-free temperature
$_i$ = Thermal expansion coefficient, in (μm/m)/K
= Swelling coefficient
ub $,i$ = Partial differentiation with respect to x_i
ub o = Initial equilibrium
ub ∞ = Final equilibrium
up H = Hygro- (moisture-related)
up M = Mechanical
up N = Nonmechanical
up R = Residual

eferences

. J. Crank, *Mathematics of Diffusion*, Oxford University Press, London, 1956.
. C.-H. Shen and G. S. Springer, "Moisture Absorption and Desorption of Composite Materials," *J. Composite Materials*, Vol. 10, 1976, pp. 2–20.
. C.-H. Shen and G. S. Springer, "Moisture Absorption of Graphite-Epoxy Composites Immersed in Liquids and in Humid Air," *J. Composite Materials*, Vol. 13, 1979, pp. 131–147.
. C. E. Browning, G. E. Husman, and J. M. Whitney, "Moisture Effects in Epoxy Matrix Composites," *Composite Materials: Testing and Design* (Fourth Conference), ASTM STP 617, American Society for Testing and Materials, 1977, pp. 481–496.

5. R. DeIsai and J. B. Whiteside, "Effect of Moisture on Epoxy Resins a Composites," *Advanced Composite Materials–Environmental Effects*, AST STP 658, American Society for Testing and Materials, 1978, pp. 2–20.

chapter 9
micromechanics

lastic moduli and hygrothermal expansion coefficients of unidirec-onal composites can be predicted from the properties and volume actions of the constituents. Easy-to-use formulas are presented for ich predictions. The strength prediction is difficult, and is limited to ecific composite materials under specific failure modes. A general icromechanics theory comparable to the elastic moduli is not avail-ble at this time.

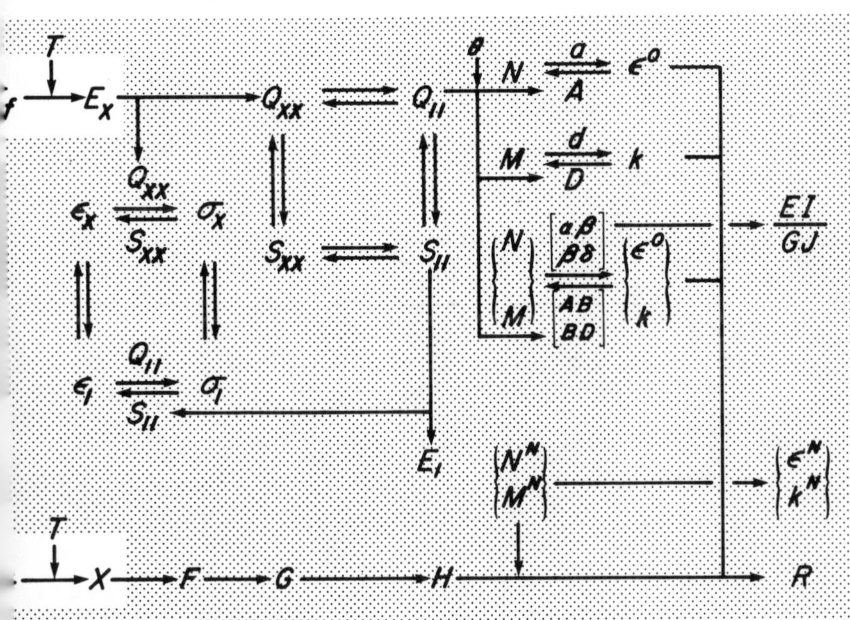

1. general remarks

In the preceding chapters unidirectional composites have been treated as anisotropic, in particular, transversely isotropic, homogeneous material. Upon magnification, however, these composites reveal a heterogeneous structure – fibers embedded in matrix. A typical cross section of a Kevlar/epoxy composite is shown in Figure 9.1. In the figure the Kevlar fibers are approximately 12 μm in diameter. Typical properties of some fibers are listed in Table 9.1.

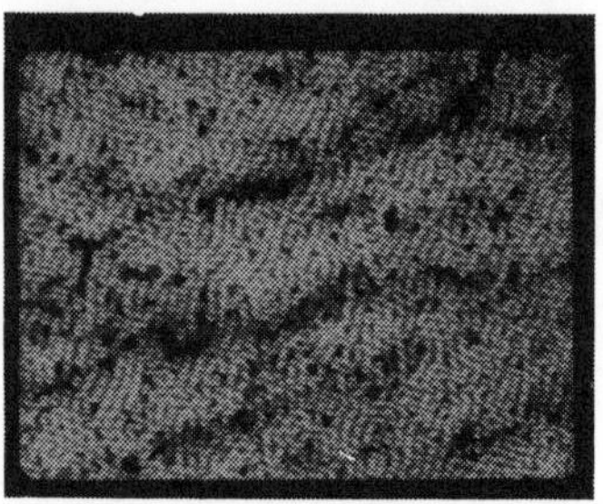

Figure 9.1 Cross section of a Kevlar/epoxy composite. Fibers are 12 μm in diameter.

In structural composites fibers are stiff and strong, and serve as the load-bearing constituent. On the other hand, matrix is soft and weak, and its direct load bearing is negligible. However, the role of matrix is very important for the structural integrity of composites: matrix protects fibers from hostile environments and localizes the effect of broken fibers.

Micromechanics is a study of mechanical properties of unidirectional composites in terms of those of constituent materials. In particular, the properties to be discussed are elastic moduli, hygrothermal expansion coefficients and strengths.

table 9.1
typical fiber properties

Fiber	Diameter μm	Density g/cm^3	Modulus GPa	Strength GPa
Graphite (T300, AS)	7	1.75	230	2.80
Boron (4-mil)	100	2.6	410	3.45
Glass (E)	16	2.6	72	3.45
Kevlar (49)	12	1.44	120	3.62

In discussing composite properties it is important to define a volume ement which is small enough to show the microscopic structural tails, yet large enough to represent the overall behavior of the com- osite. Such a volume element is called the representative volume ement. A simple representative volume element can consist of a fiber nbedded in a matrix block.

Once a representative volume element is chosen, proper boundary nditions are prescribed. Ideally, these boundary conditions must present the in situ states of stress and strain within the composite. hat is, the prescribed boundary conditions must be the same as those the representative volume element were actually in the composite.

Finally, a prediction of composite properties follows from the solu- on of the foregoing boundary value problem. Although the procedure volved is conceptually simple, the actual solution is rather difficult. onsequently, many assumptions and approximations have been intro- uced, and therefore, various solutions are available. In this chapter, owever, we limit our discussion to the simplest model without loss of nerality in the procedures involved.

density of composite

onsider a composite of mass M and volume V, schematically shown in igure 9.2. Here V is the volume of a representative volume element. nce the composite is made of fibers and matrix, the mass M is the sum the total mass M_f of fibers and the mass M_m of matrix:

$$M = M_f + M_m \tag{9.1}$$

quation 9.1 is valid regardless of voids which may be present. How- er, the composite volume V includes the volume V_v of voids so that

$$V = V_f + V_m + V_v \tag{9.2}$$

ividing Equations 9.1 and 9.2 by M and V, respectively, leads to the llowing relations for the mass fractions and volume fractions:

$$m_f + m_m = 1 \tag{9.3}$$

$$v_f + v_m + v_v = 1 \tag{9.4}$$

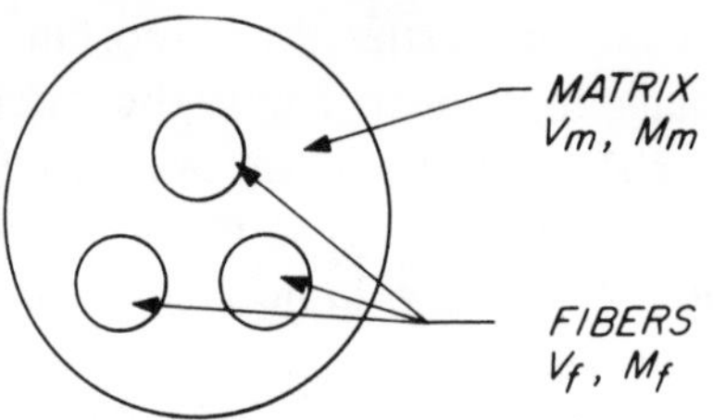

Figure 9.2 A representative volume element. The total volume and mass of each constituent are denoted by V and M, respectively. The subscripts m and f stand for matrix and fiber, respectively.

In this section the subscripts f, m, v are exclusively used to deno fiber, matrix and void, respectively. Thus these subscripts do not follo the rules of index notation.

The composite density ρ follows from Equations 9.1 and 9.2 as

$$\rho = \frac{M}{V} = \frac{(\rho_f V_f + \rho_m V_m)}{V} = \rho_f v_f + \rho_m v_m \qquad (9.$$

In terms of mass fractions, ρ becomes

$$\rho = \frac{1}{m_f/\rho_f + m_m/\rho_m + v_v/\rho} \qquad (9.$$

Equation (9.6) is frequently used to determine the void fraction

$$v_v = 1 - \rho\left(\frac{m_f}{\rho_f} + \frac{m_m}{\rho_m}\right) \qquad (9.$$

The mass fraction of fibers can be measured by removing the matri In the case of glass/epoxy composites the matrix can be burnt o without affecting the glass fibers. As for boron/epoxy and graphit epoxy composites, acids are usually used to dissolve the matrix. On m_f is thus determined, the mass fraction m_m of matrix simply follo from Equation 9.3.

composite stresses and strains

Equation 9.5 the composite density ρ is seen to be equal to the ensities of the constituents averaged over the composite volume. The omposite stresses and composite strains are defined similarly.

Suppose the stress field in the representative element is σ_i. The com-osite stress $\overline{\sigma}_i$ is defined by

$$\overline{\sigma}_i = \frac{1}{V}\int_V \sigma_i \, dV = \frac{1}{V}\left[\int_{V_f} \sigma_i \, dV + \int_{V_m} \sigma_i \, dV + \int_{V_v} \sigma_i \, dV\right] \tag{9.8}$$

e now introduce the volume-average stresses $\overline{\sigma}_{fi}$ and $\overline{\sigma}_{mi}$ in the fibers nd matrix, respectively,

$$\overline{\sigma}_{fi} = \frac{1}{V_f}\int_{V_f} \sigma_i \, dV, \; \overline{\sigma}_{mi} = \frac{1}{V_m}\int_{V_m} \sigma_i \, dV \tag{9.9}$$

ince no stress is transmitted in the voids, i.e., $\sigma_i = 0$ within V_v, quation 9.8 can be written as

$$\overline{\sigma}_i = v_f \, \overline{\sigma}_{fi} + v_m \, \overline{\sigma}_{mi} \tag{9.10}$$

quation 9.10 thus gives the composite stress $\overline{\sigma}_i$ in terms of the average onstituent stresses $\overline{\sigma}_{fi}$ and $\overline{\sigma}_{mi}$

Similarly to the composite stress, the composite strain is defined as e volume-average strain, and is obtained as

$$\overline{\epsilon}_i = v_f \, \overline{\epsilon}_{fi} + v_m \, \overline{\epsilon}_{mi} + v_v \, \overline{\epsilon}_{vi} \tag{9.11}$$

nlike the stress, the strain in voids does not vanish. The void strain is efined in terms of the boundary displacements of the voids. However, nce the void fraction is usually negligible, i.e. less than 1%, in com-osites of acceptable quality, we neglect the last term on the right-hand de. Thus, in the following discussions we shall use

$$\overline{\epsilon}_i = v_f \, \overline{\epsilon}_{fi} + v_m \, \overline{\epsilon}_{mi} \tag{9.12}$$

ith the understanding that $(v_f + v_m)$ is unity.

Note that Equations 9.10 and 9.11 simply follow from the defi tions of the composite stress and strain that the composite variables a the volume averages. Thus, these equations are valid regardless of t material behavior.

Determination of the composite stress and strain requires t knowledge of the stress and strain fields within composite. However, shall show that they can be determined directly from the bounda tractions and boundary displacements.

Consider a representative volume element in the form of a rectang lar prism, as shown in Figure 9.3. The fibers in the element are pe fectly bonded to the matrix. Suppose a stress $\tilde{\sigma}_x$ is applied on t boundaries at $x = 0$ and L_1.

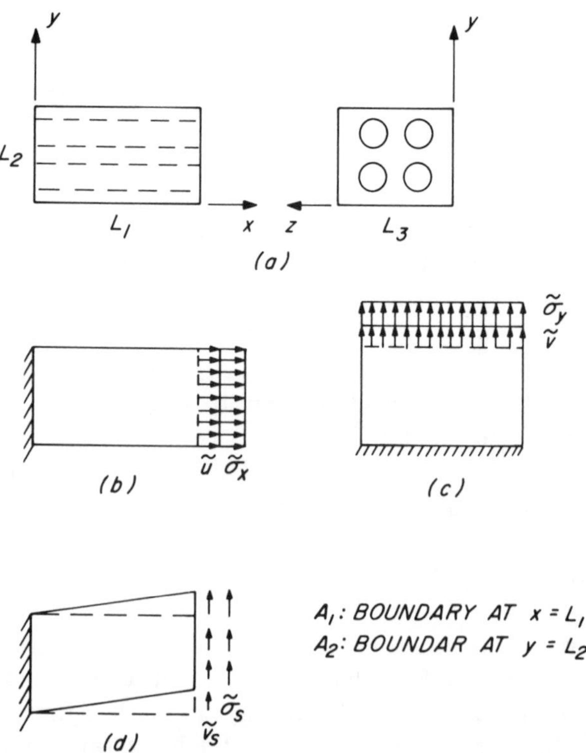

Figure 9.3 Force and displacement boundary conditions applied to a representative volume element. Initial shapes are indicated by broken lines in (*b*), (*c*) and (*d*).

By definition the composite stress $\bar{\sigma}_x$ is given by

$$\bar{\sigma}_x = \frac{1}{L_1 L_2 L_3} \int_{L_1} \int_{A(x)} \sigma_x \, dA \, dx \tag{9.13}$$

here $A(x)$ is the cross section at x normal to the x-axis. Of course, the 'ea of $A(x)$ is constant and equal to $L_2 L_3$. The equilibrium condition quires that the resultant of σ_x be equal to the resultant of the applied ress $\tilde{\sigma}_x$,

$$\int_{A(x)} \sigma_x \, dA = \int_{A_1} \tilde{\sigma}_x \, dA \tag{9.14}$$

ıbstituting Equation 9.14 into Equation 9.13, we can show that $\bar{\sigma}_x$ is qual to the average applied stress,

$$\bar{\sigma}_x = \frac{1}{L_2 L_3} \int_{A_1} \tilde{\sigma}_x \, dA \tag{9.15}$$

milarly, the other stress components $\bar{\sigma}_2$ and $\bar{\sigma}_6$ are equal to the spective average applied stresses,

$$\begin{aligned} \bar{\sigma}_y &= \frac{1}{L_1 L_3} \int_{A_2} \tilde{\sigma}_y \, dA \\ \bar{\sigma}_s &= \frac{1}{L_2 L_3} \int_{A_1} \tilde{\sigma}_s \, dA \end{aligned} \tag{9.16}$$

In case when the applied stresses are uniform, i.e.,

$$\begin{aligned} \tilde{\sigma}_x &= \sigma_x^o \quad \text{on } A_1 \\ \tilde{\sigma}_y &= \sigma_y^o \quad \text{on } A_2 \\ \tilde{\sigma}_s &= \sigma_s^o \quad \text{on } A_1 \end{aligned} \tag{9.17}$$

where σ_i^o are constants, the composite stresses become equal to th applied stresses. That is,

$$\bar{\sigma}_i = \sigma_i^o \qquad (9.1$$

Referring back to the representative volume element of Figure 9. we now show that the composite strains are related to the averag relative boundary displacements. First, consider the composite strain $\bar{\epsilon}$ which, for the representative volume element, is given by

$$\bar{\epsilon}_x = \frac{1}{L_1 L_2 L_3} \int_{A(x)} \int_{L_1} \epsilon_x \, dx \, dA \qquad (9.1$$

At each point in the cross section $A(x)$, we evaluate the integral of ϵ over the length L_1 to obtain

$$\int_{L_1} \epsilon_x \, dx = \int_{L_1} (\partial u/\partial x) \, dx = \widetilde{u} \qquad (9.2$$

Here, $\widetilde{u}$ is the value of u at $x = L_1$ and we have assumed $\widetilde{u} = 0$ at $x =$ Of course, $\widetilde{u}$ depends on y and z and represents the displacement of th boundary at $x = L_1$. Substitution of Equation 9.20 into Equation 9.1 thus results in an equation relating the composite strain to the averag boundary displacement,

$$\bar{\epsilon}_x = \frac{1}{L_1} \frac{1}{L_2 L_3} \int_{A_1} \widetilde{u} \, dA \qquad (9.2$$

The remaining strains can be easily found to be

$$\bar{\epsilon}_y = \frac{1}{L_2} \frac{1}{L_1 L_3} \int_{A_2} \widetilde{v} \, dA$$

$$\bar{\epsilon}_s = \frac{1}{L_1} \frac{1}{L_2 L_3} \int_{A_1} \widetilde{v}_s \, dA \qquad (9.2$$

ote that $\widetilde{v}$ in the equation for $\overline{\epsilon}_y$ is the displacement of A_2 whereas $\widetilde{v}_s$ the equation for $\overline{\epsilon}_s$ is the displacement of A_1 in the y direction.

If the boundary displacements are uniform, i.e.,

$$\begin{aligned} \widetilde{u} &= \epsilon_x^o L_1 \quad \text{on } A_1 \\ \widetilde{v} &= \epsilon_y^o L_2 \quad \text{on } A_2 \\ \widetilde{v}_s &= \epsilon_s^o L_1 \quad \text{on } A_1 \end{aligned} \tag{9.23}$$

here ϵ_i^o are constants, the composite strains reduce to

$$\overline{\epsilon}_i = \epsilon_i^o \tag{9.24}$$

Equations 9.15, 9.16, 9.21, and 9.22 allow us to determine the com-osite stresses and composite strains directly from the boundary condi-ons without knowing the stress and strain distributions within the presentative volume element. These results are very helpful in tests here the applied loads and relative displacements are directly mea-ured. Equations 9.18 and 9.24 are frequently used in the derivation of omposite moduli.

The conservation of energy requires that the strain energy stored ithin the representative volume element be equal to the work done y the applied stresses,

$$\int_V \sigma_i \epsilon_i dV = \int_{A_1} \widetilde{\sigma}_x \widetilde{u}\, dA + \int_{A_2} \widetilde{\sigma}_y \widetilde{v}\, dA + \int_{A_1} \widetilde{\sigma}_s \widetilde{v}_s\, dA \tag{9.25}$$

If either the applied stresses or the boundary displacements are uni-orm (see Equations 9.17 and 9.23), then the strain energy density ithin the representative volume element is simply given by

$$W = \frac{1}{2V} \int_V \sigma_i \epsilon_i \, dV = \frac{1}{2} \overline{\sigma}_i \overline{\epsilon}_i \tag{9.26}$$

hus the strain energy density of composite is the same in form as that f a homogeneous material.

4. elastic moduli of composite

In a state of plane stress which is applicable to thin laminates, the required composite stress-strain relations are

$$\bar{\epsilon}_i = S_{ij}\,\bar{\sigma}_j \; (i, j = 1, 2, 6) \tag{9.27}$$

$$S_{ij} = \begin{bmatrix} 1/E_x & -\nu_x/E_x & 0 \\ -\nu_x/E_x & 1/E_y & 0 \\ 0 & 0 & 1/E_s \end{bmatrix}$$

Thus our goal is to determine the four components of the composite compliance matrix S_{ij} in terms of the structural details of the composite and the compliance matrices of the constituents.

There is a total of 18 variables – 3 components each of stress and strain for the fiber, matrix, and composite, respectively. Since we are seeking 3 stress-strain relations of Equation 9.27, we need 15 equations relating those 18 variables. Six of these required equations are provided from the definitions of the composite stress and strain, Equations 9.11 and 9.12. Six additional equations are the constitutive relations of the constituents,

$$\bar{\epsilon}_{fi} = S_{fij}\,\bar{\sigma}_{fj}, \; \bar{\epsilon}_{mi} = S_{mij}\,\bar{\sigma}_{mj} \tag{9.28}$$

where

$$S_{fij} = \begin{bmatrix} 1/E_f & -\nu_f/E_f & 0 \\ -\nu_f/E_f & 1/E_f & 0 \\ 0 & 0 & 1/G_f \end{bmatrix}$$

$$S_{mij} = \begin{bmatrix} 1/E_m & -\nu_m/E_m & 0 \\ -\nu_m/E_m & 1/E_m & 0 \\ 0 & 0 & 1/G_m \end{bmatrix} \tag{9.29}$$

›t we need three more equations which must be chosen so that (1) resses are in equilibrium, (2) strains are related to stresses through nstitutive equations, (3) strains are related to displacements, and (4) e given boundary conditions are satisfied. However, finding the stress d strain fields which satisfy all four conditions described above in a alistic representative volume element is rather complicated. Therere, here we shall choose a simple representative volume element and nple boundary conditions so that the above conditions are easily tisfied.

Consider a composite laminate with fibers in the x direction, Figure 4. The representative volume element of this composite is chosen to a fiber embedded in a matrix plate. The fiber is assumed to have a ctangular cross section with the same thickness as the matrix plate. ıerefore, the width ratio $w_f/(w_f + w_m)$ is chosen to be the same as e fiber volume fraction of the composite itself.

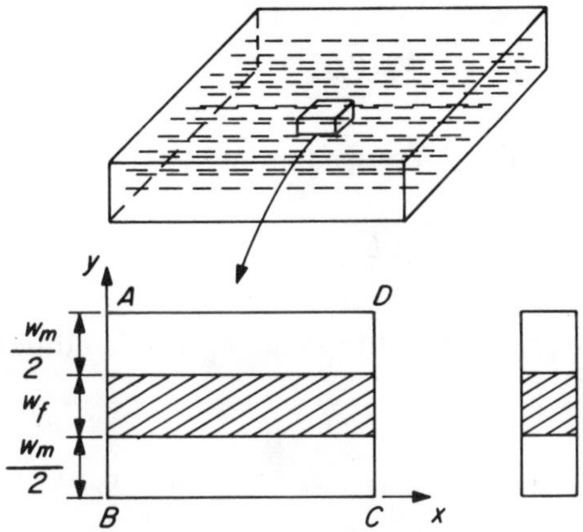

Figure 9.4 A simple representative volume element. A perfect bond is assumed between the fiber and the matrix.

Suppose boundary AB is fixed and boundary CD is given a unifor[m] displacement $\overline{BC}\,\epsilon_x^o$. Boundaries AD and BC are free. Thus the impose[d] boundary conditions are

$$\begin{aligned} \tilde{u} &= \overline{BC}\,\epsilon_x^o \quad \text{on } CD \\ &= 0 \quad \text{on } AB \\ \tilde{\sigma}_y &= 0 \quad \text{on } AD \text{ and } BC \\ \tilde{\sigma}_s &= 0 \quad \text{on all boundaries} \end{aligned} \tag{9.30}$$

Using Equations 9.15–18 and 9.21–24, we easily find that

$$\begin{aligned} \bar{\epsilon}_x &= \bar{\epsilon}_{fx} = \bar{\epsilon}_{mx} = \epsilon_x^o \\ \bar{\sigma}_y &= \bar{\sigma}_{fy} = \bar{\sigma}_{my} = 0 \\ \bar{\sigma}_s &= \bar{\sigma}_{fs} = \bar{\sigma}_{ms} = 0 \end{aligned} \tag{9.3[1]}$$

Equation 9.31 can now be used to determine the composite stress-stra[in] relations. From Equations 9.10 and 9.23 we see that

$$\bar{\sigma}_x = v_f\bar{\sigma}_{fx} + v_m\bar{\sigma}_{mx} = (v_fE_f + v_mE_m)\bar{\epsilon}_x \tag{9.3[2]}$$

Therefore, the longitudinal Young's modulus E_x becomes

$$E_x = v_fE_f + v_mE_m \tag{9.3[3]}$$

Next, the transverse composite strain $\bar{\epsilon}_y$ is related to the longitudin[al] composite strain $\bar{\epsilon}_x$ by

$$\bar{\epsilon}_y = v_f\bar{\epsilon}_{fy} + v_m\bar{\epsilon}_{my} = -(v_f\nu_f + v_m\nu_m)\bar{\epsilon}_x \tag{9.3[4]}$$

The longitudinal Poisson's ratio is thus

$$\nu_x = v_f\nu_f + v_m\nu_m \tag{9.3[5]}$$

To determine the transverse Young's modulus E_y, we apply the following boundary conditions

$$\begin{aligned} \tilde{\sigma}_x &= 0 \quad \text{on } AB \text{ and } CD \\ \tilde{\sigma}_y &= \sigma_y^o \quad \text{on } AD \text{ and } BC \\ \tilde{\sigma}_s &= 0 \quad \text{on all boundaries} \end{aligned} \tag{9.36}$$

Again, using Equations 9.15-18 and 9.21-24, and neglecting the shear stresses on the fiber/matrix interface we find that

$$\begin{aligned} \bar{\sigma}_x &= \bar{\sigma}_{fx} = \bar{\sigma}_{mx} = 0 \\ \bar{\sigma}_y &= \bar{\sigma}_{fy} = \bar{\sigma}_{my} = \sigma_y^o \\ \bar{\sigma}_s &= \bar{\sigma}_{fs} = \bar{\sigma}_{ms} = 0 \end{aligned} \tag{9.37}$$

Therefore, the composite strain $\bar{\epsilon}_y$ is given by

$$\bar{\epsilon}_y = v_f \bar{\epsilon}_{fy} + v_m \bar{\epsilon}_{my} = \left(\frac{v_f}{E_f} + \frac{v_m}{E_m} \right) \bar{\sigma}_y \tag{9.38}$$

The resulting transverse Young's modulus is obtained from

$$\frac{1}{E_y} = \frac{v_f}{E_f} + \frac{v_m}{E_m} \tag{9.39}$$

Finally, the boundary conditions for the determination of the shear modulus are

$$\begin{aligned} \tilde{\sigma}_x &= 0 \quad \text{on } AB \text{ and } CD \\ \tilde{\sigma}_y &= 0 \quad \text{on } AD \text{ and } BC \\ \tilde{\sigma}_s &= \sigma_s^o \quad \text{on all boundaries.} \end{aligned} \tag{9.40}$$

The procedures to be followed next are similar to those for E_y. Therefore, the shear modulus is given without derivation:

$$\frac{1}{E_s} = \frac{v_f}{G_f} + \frac{v_m}{G_m} \tag{9.41}$$

Equations 9.33, 9.35, 9.39 and 9.41 are called the rule-of-mixture equations for composite moduli.

The boundary conditions to be imposed on the representative volume element must simulate the in situ state of stress as closely as possible. When σ_y^o was applied on boundaries AD and BC we assumed no stress on boundaries AB and CD. However, such boundary condition is not realistic because $\bar{\epsilon}_{fx}$ is not the same as $\bar{\epsilon}_{mx}$ unless $\nu_f/E_f = \nu_m/E_m$. The resulting difference in displacements cannot be sustained in actual composites.

To remedy the foregoing contradiction, we modify the boundary conditions 9.36 as follows:

$$\begin{aligned} \tilde{u} &= C \cdot \overline{BC} \text{ on } CD \\ &= 0 \qquad\quad \text{on } AB \\ \tilde{\sigma}_y &= \sigma_y^o \qquad\quad \text{on } AD \text{ and } BC \\ \tilde{\sigma}_s &= 0 \qquad\quad \text{on all boundaries} \end{aligned} \tag{9.42}$$

The constant C is to be determined from the condition $\bar{\sigma}_x = 0$.

The application of Equations 9.15–18 and 9.21–24 allow the following equations:

$$\begin{aligned} \bar{\epsilon}_x &= \bar{\epsilon}_{fx} = \bar{\epsilon}_{mx} = C \\ \bar{\sigma}_y &= \bar{\sigma}_{fy} = \bar{\sigma}_{my} = \sigma_y^o \\ \bar{\sigma}_s &= \bar{\sigma}_{fs} = \bar{\sigma}_{ms} = 0 \end{aligned} \tag{9.43}$$

Combining Equation 9.28 with Equation 9.43 and recalling $\bar{\sigma}_x = 0$, we first obtain $\bar{\sigma}_{fx}$ and $\bar{\sigma}_{mx}$ as

$$
\begin{aligned}
\bar{\sigma}_{fx} &= \frac{\nu_f E_m - \nu_m E_f}{v_f E_f + v_m E_m} v_m \bar{\sigma}_y \\
\bar{\sigma}_{mx} &= \frac{\nu_m E_f - \nu_f E_m}{v_f E_f + v_m E_m} v_f \bar{\sigma}_y
\end{aligned}
\tag{9.44}
$$

Thus the transverse Young's modulus E_y is obtained from

$$
\frac{1}{E_y} = \frac{v_f}{E_f} + \frac{v_m}{E_m} - v_f v_m \frac{\nu_f^2 E_m/E_f + \nu_m^2 E_f/E_m - 2\nu_f \nu_m}{v_f E_f + v_m E_m}
\tag{9.45}
$$

The determination of C leads to the transverse Poisson's ratio ν_y because

$$
C = -\frac{\nu_y}{E_y} \sigma_y^o
\tag{9.46}
$$

It can be shown that ν_y determined from Equation 9.46 satisfies the symmetry condition

$$
\frac{\nu_y}{E_y} = \frac{\nu_x}{E_x}
\tag{9.47}
$$

The first two terms on the right-hand side of Equation 9.45 are the ame as Equation 9.39 based on a uniaxial state of average stress. The hird term represents the effect of lateral constraint imposed by the train compatibility and leads to a higher transverse modulus.

Equations 9.33 and 9.35 provide accurate predictions for the longi-udinal Young's modulus and Poisson's ratio. However, Equations 9.39 nd 9.41 give lower values than experimentally observed for the trans-erse Young's modulus and shear modulus, respectively, as can be seen n Figures 9.5 and 9.6 for a glass/epoxy composite. Equation 9.45

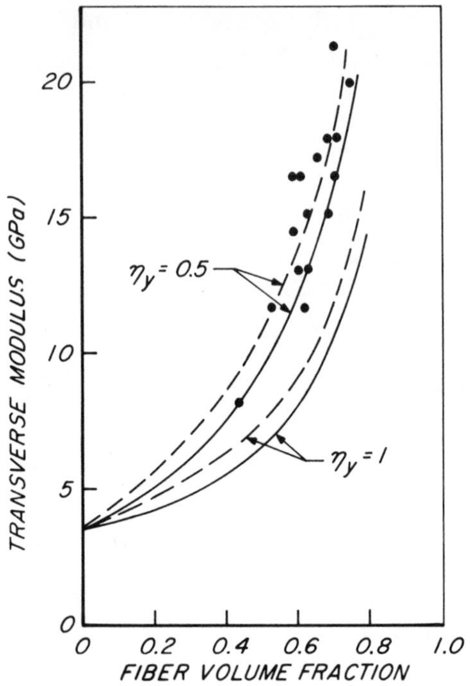

Figure 9.5 Transverse modulus versus fiber volume fraction for a glass/epoxy composite. The solid lines represent Equation 9.49, and the broken lines Equation 9.45. Equation 9.45 can be modified to include η_y, and Equation 9.39 is represented by the solid line with $\eta_y = 1$. The elastic moduli used are: $E_f = 73.1$ GPa, $E_m = 3.45$ GPa, $\nu_f = 0.22$ and $\nu_m = 0.35$. (Data from [1]).

yields a higher modulus than Equation 9.39. However, both predictio are considerably lower than the data. A simplistic method of correcti for such discrepancy is discussed in the next section.

5. modified rule-of-mixtures equations for transverse and shear moduli

In the preceding section the representative volume element consisted two plates of the same thickness, each representing a fiber and matri respectively. However, in actual composites, fibers are complete

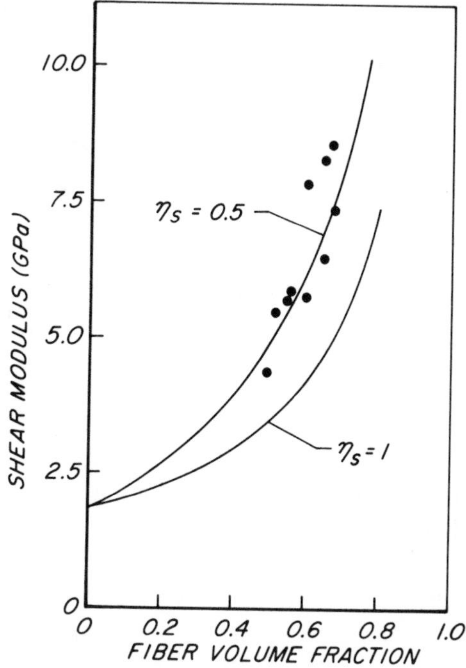

Figure 9.6 Shear modulus versus fiber volume fraction for a glass/epoxy composite. The solid lines represent Equation 9.49, and $\eta_s = 1$ corresponds to Equation 9.41. The shear moduli used are: $G_f = 30.2$ GPa and $G_m = 1.8$ GPa. (Data from [2]).

surrounded by the matrix. Thus a more realistic representative volume element will be a concentric cylinder as shown in Figure 9.7. Also, the boundary conditions should be changed to simulate the in situ state of stress the new representative volume element would be in. The exact determination of stresses is rather complicated and is beyond the scope of this book. Therefore, in the following we shall describe a semi-empirical approach to provide better estimates of moduli than the simple rule-of-mixtures equations can.

Noting that matrix is softer than fiber, we assume that

$$\begin{aligned} \bar{\sigma}_{my} &= \eta_y \bar{\sigma}_{fy}, \quad 0 < \eta_y \leq 1 \\ \bar{\sigma}_{ms} &= \eta_s \bar{\sigma}_{fs}, \quad 0 < \eta_s \leq 1 \end{aligned} \tag{9.48}$$

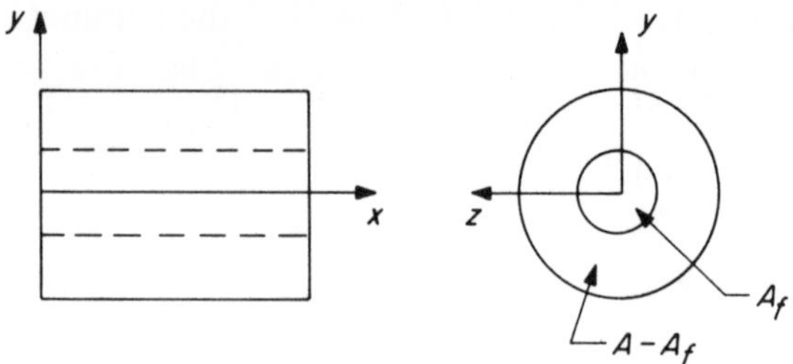

Figure 9.7 A concentric cylinder model. Boundary conditions to be applied are different from those for the representative volume element of Figure 9.4.

The above equations imply that the average matrix stress has the sam sign as, but lower in magnitude than, the average fiber stress, in tran verse or shear loading. The stress partitioning parameters η_y and η measure the relative magnitudes of the average matrix stresses as com pared to the average fiber stresses.

The equations for the moduli can now be derived following the sam procedure as in the preceding section but using Equation 9.48. Th results are

$$\frac{1}{E_y} = \frac{1}{v_f + \eta_y v_m} \left(v_f \frac{1}{E_f} + \eta_y v_m \frac{1}{E_m} \right)$$

$$\frac{1}{E_s} = \frac{1}{v_f + \eta_s v_m} \left(v_f \frac{1}{G_f} + \eta_s v_m \frac{1}{G_m} \right) \tag{9.4}$$

The above equations are the same as Equations 9.39 and 9.41 if $\eta_y v_m$ and $\eta_s v_m$, respectively, are replaced by v_m.

The moduli predicted by Equation 9.49 with $\eta_y = \eta_s = 0.5$ ar compared with the experimental data in Figures 9.5 and 9.6. Thes equations are seen to provide much better estimates of moduli than d the simple rule-of-mixtures equations of the preceding section.

According to Equation 9.49 the moduli increase as the stress part tioning parameters decrease. Decreasing parameters indicate an increas ing load sharing by fibers. Since fiber is stiffer than matrix, the resultin composite moduli therefore increase.

So far we have not concerned ourselves with the dependence of th partitioning parameters on constituent properties. The experiment orrelations in Figures 9.5 and 9.6 show that the parameters are weakly ependent, if at all, on the fiber volume fraction. The results of some dvanced methods [3, 4] based on the concentric cylinder model of 'igure 9.7 can indeed be used to show that η_s is independent of ν_f. 'urthermore,

$$\eta_s = \frac{1}{2}\left(1 + \frac{G_m}{G_f}\right) \tag{9.50}$$

Because of the axisymmetry of the concentric cylinder model, it is asier to determine moduli other than E_y. Specifically, the moduli to e determined are the transverse plane strain bulk modulus k_y and the ransverse shear modulus G_y. When the only nonvanishing strain comonents are $\bar{\epsilon}_y = \bar{\epsilon}_z = \bar{\epsilon}$, k_y yields

$$\bar{\sigma}_y = \bar{\sigma}_z = 2k_y\bar{\epsilon} \tag{9.51}$$

'he transverse shear modulus relates the shear strain to the shear stress 1 the yz plane:

$$\bar{\sigma}_{yz} = G_y\bar{\epsilon}_{yz} \tag{9.52}$$

)nce k_y and G_y are known, the transverse Young's modulus E_y is iven by

$$E_y = \frac{4k_y G_y}{k_y + mG_y} \tag{9.53}$$

/here

$$m = 1 + \frac{4k_y \nu_x{}^2}{E_x} \tag{9.54}$$

The equations for k_y and G_y are similar to those for E_y and E_s. 'hat is,

$$\frac{1}{k_y} = \frac{1}{\nu_f + \eta_k \nu_m}\left(\nu_f \frac{1}{k_{fy}} + \eta_k \nu_m \frac{1}{k_m}\right) \tag{9.55}$$

(continues)

$$\frac{1}{G_y} = \frac{1}{v_f + \eta_G v_m}\left(v_f \frac{1}{G_{fy}} + \eta_G v_m \frac{1}{G_m}\right) \qquad \text{(9.55) (concluded)}$$

where

$$\eta_k = \frac{1}{2(1-\nu_m)}\left(1 + \frac{G_m}{k_{fy}}\right)$$

$$\eta_G = \frac{1}{4(1-\nu_m)}\left(3-4\nu_m + \frac{G_m}{G_{fy}}\right) \qquad \text{(9.56)}$$

In the foregoing equations the subscript y has been used to denote th transverse properties of fiber in case the fiber is transversely isotropi In this regard we note that graphite and aramid fibers are not isotrop but rather transversely isotropic. The equations derived so far a summarized in Table 9.2 where P stands for property.

table 9.2
formulas for composite moduli

$$P = \frac{1}{v_f + \eta v_m}\left(v_f P_f + \eta v_m P_m\right)$$

Engineering constant	P	P_f	P_m	η
E_x	E_x	E_f	E_m	1
ν_x	ν_x	ν_f	ν_m	1
E_s	$\frac{1}{E_s}$	$\frac{1}{G_f}$	$\frac{1}{G_m}$	η_s*
k_y	$\frac{1}{k_y}$	$\frac{1}{k_{fy}}$	$\frac{1}{k_m}$	η_k**
G_y	$\frac{1}{G_y}$	$\frac{1}{G_{fy}}$	$\frac{1}{G_m}$	η_G**

*See Equation 9.50
**See Equation 9.56

If the fiber is isotropic, the subscript y can be dropped. Furthermore, quation 9.53 for the fiber becomes

$$k_f = \frac{G_f}{1-2\nu_f} \tag{9.57}$$

ierefore, η_k reduces to

$$\eta_k = \frac{1}{2(1-\nu_m)}\left[1+(1-2\nu_f)\frac{G_m}{G_f}\right] \tag{9.58}$$

Equations 9.49 and 9.53 together with Equations 9.50 and 9.56 are own graphically in Figures 9.8 and 9.9, respectively, for three different composite systems: glass/epoxy, boron/epoxy and graphite/oxy. The properties of the epoxy are

$$E_m = 3.45 \text{ GPa}, \ \nu_m = 0.35 \tag{9.59}$$

addition to the Young's moduli listed in Table 9.1, Poisson's ratio the fibers is

$$\nu_f = 0.2 \tag{9.60}$$

For the glass and boron fibers, E_f and ν_f are sufficient. However, for e graphite fiber we further have

$$\begin{aligned} E_{fy} &= 16.6 \text{ GPa}, \ G_f = 8.27 \text{ GPa} \\ G_{fy} &= 5.89 \text{ GPa} \end{aligned} \tag{9.61}$$

The stress partitioning parameters are shown as functions of G_f/G_m Figure 9.10. The matrix Poisson's ratio used is 0.35 and fibers are sumed to be isotropic with $\nu_f = 0.2$ for η_k. The decreasing stress rtitioning parameters indicate less load sharing by the matrix as $/G_m$ increases.

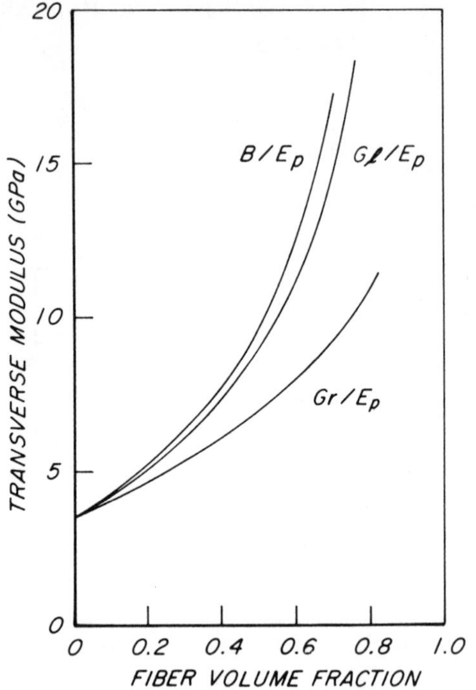

Figure 9.8 Transverse moduli predicted by Equation 9.53. The properties used are in Table 9.1 and Equations 9.59 through 9.61.

For most structural composites the modulus ratio G_m/G_f is mu smaller than unity. Therefore, the corresponding stress partitioni parameters can be approximated by

$$\eta_s = \frac{1}{2}$$

$$\eta_k = \frac{1}{2(1-\nu_m)} \qquad (9.6$$

$$\eta_G = \frac{3-4\nu_m}{4(1-\nu_m)}$$

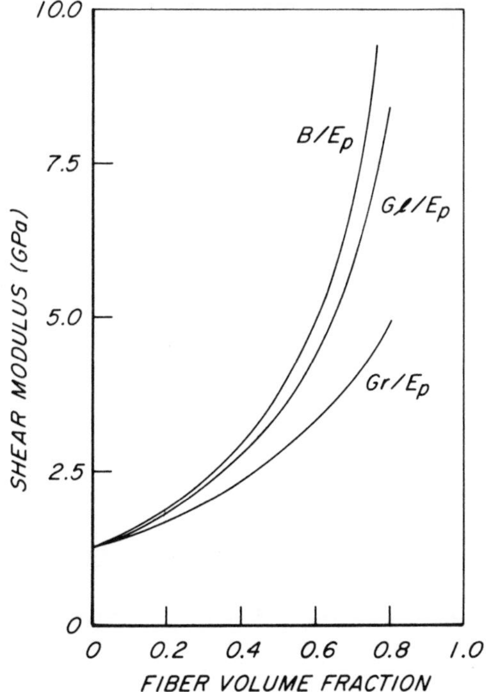

Figure 9.9 Shear moduli predicted by Equations 9.49 and 9.50. The properties used are in Table 9.1 and Equations 9.59 through 9.61.

ıus, it is seen that these parameters depend only on the Poisson's ratio the matrix. For most epoxies ν_m is close to 0.35. Therefore, the final lues of the remaining η's are

$$\begin{aligned} \eta_k &= 0.77 \\ \eta_G &= 0.62 \end{aligned} \tag{9.63}$$

om Figure 9.10 we see that the error in using the limiting values of s is less than 5 percent as long as the modulus ratio G_f/G_m is larger an 20, which is the case for glass/epoxy.

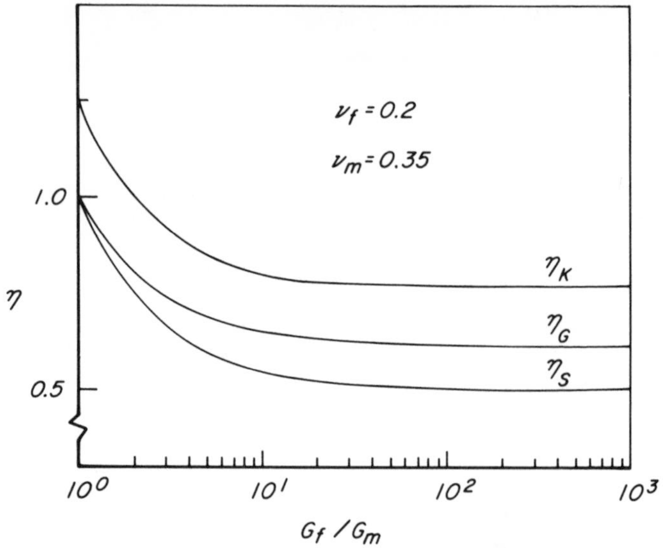

Figure 9.10 Stress partitioning parameters predicted by Equations 9.50, 9.56 and 9.58. Fibers are assumed to be isotropic.

6. hygrothermal properties

Hygrothermal properties of composite as a homogeneous, anisotrop material were discussed in Chapter 8. In this section we shall investiga how constituent properties affect the macroscopic hygrothermal b havior of composite.

Consider a composite which is completely dry. In the dry state, t total mass M of a composite body is the sum of those of the fibers a of the matrix,

$$M = M_m + M_f \quad (9.6$$

Now the same composite absorbs moisture and reaches an equil rium state. The new mass M' after the moisture absorption is given

$$M' = M + M_{mw} + M_{fw} + M_{vw} \quad (9.6$$

Here M_{mw} and M_{fw} are the masses of moisture absorbed in the fibe and matrix, respectively. The last term on the right-hand side

Equation 9.65 is the amount of moisture entrapped in voids. The moisture concentration in the composite is then

$$c = \frac{M'-M}{M} = c_m m_m + c_f m_f + M_{vw}/M \tag{9.66}$$

where m's are again mass fractions, and c's are moisture concentrations in the constituents. Specifically, we have

$$c_f = \frac{M_{fw}}{M_f}, \quad c_m = \frac{M_{mw}}{M_m} \tag{9.67}$$

Equation 9.66 can also be rewritten in terms of mass densities or specific gravities:

$$\begin{aligned} c &= (c_m v_m \rho_m + c_f v_f \rho_f + v_v \rho_w)/\rho \\ \text{and} \quad &= (c_m v_m s_m + c_f v_f s_f + v_v)/s \end{aligned} \tag{9.68}$$

Note that ρ_w is the density of moisture.

Unlike temperature the moisture concentration varies from fiber to matrix. Inorganic fibers such as graphite and boron do not absorb moisture and hence, $c_f = 0$. However, the moisture concentration in most epoxies can be as high as 8%.

As discussed in the preceding chapter, a composite undergoes a nonmechanical strain upon temperature change and absorption of moisture. It has been pointed out that such a nonmechanical strain is measured from the stress-free state and consists of the thermal strain and the swelling strain. Here we shall see how the constituent properties affect the nonmechanical strains of a unidirectional composite.

In the presence of nonmechanical strains e_{fi} and e_{mi}, the constitutive relations of the constituents, Equation 9.28, are replaced by

$$\bar{\epsilon}_{fi} = S_{fij}\bar{\sigma}_{fj} + e_{fi}, \quad \bar{\epsilon}_{mi} = S_{mij}\bar{\sigma}_{mj} + e_{mi} \tag{9.69}$$

Note that, since the constituents are isotropic, we have

$$\begin{aligned} e_{fx} &= e_{fy} = e_f, \quad e_{fs} = 0 \\ e_{mx} &= e_{my} = e_m, \quad e_{ms} = 0 \end{aligned} \tag{9.70}$$

In Equation 9.69 the volume averages are taken only over the tot strains and the stresses, but not over the nonmechanical strains becaus the latter are uniform in each constituent.

Now we use the rule-of-mixtures assumptions and recall that

$$\begin{aligned} \bar{\epsilon}_x &= \bar{\epsilon}_{fx} = \bar{\epsilon}_{mx} \\ \bar{\sigma}_y &= \bar{\sigma}_{fy} = \bar{\sigma}_{my} = 0 \end{aligned} \tag{9.71}$$

The nonmechanical strains e_x and e_y of the composite are then obtained by solving Equations 9.10, 9.12 and 9.69 in conjunction wit Equation 9.71 for $\bar{\epsilon}_x$ and $\bar{\epsilon}_y$ in terms of e_f and e_m. The results are

$$\begin{aligned} e_x &= \frac{v_f E_f e_f + v_m E_m e_m}{v_f E_f + v_m E_m} \\ e_y &= v_f e_f + v_m e_m + v_f \nu_f e_f + v_m \nu_m e_m - (v_f \nu_f + v_m \nu_m) e_x \end{aligned} \tag{9.72}$$

In the absence of any applied stress, e_x and e_y are the composit strains resulting from e_f and e_m. From Equation 9.71 we see that ther are no residual stresses in the transverse direction in the constituent Since $\bar{\epsilon}_{fx} = \bar{\epsilon}_{mx} = e_x$, the residual stresses in the longitudinal directio are obtained by substituting Equation 9.72 into Equation 9.69 an solving the resulting equations for the stresses. Thus the residual stres in the fiber is given by

$$\bar{\sigma}_{fx}^R = v_m E_m E_f \frac{e_m - e_f}{v_f E_f + v_m E_m} \tag{9.73}$$

whereas in the matrix we have

$$\bar{\sigma}_{mx}^R = v_f E_f E_m \frac{e_f - e_m}{v_f E_f + v_m E_m} \tag{9.74}$$

If the thermal strains are proportional to temperature change, we ca use the thermal expansion coefficients

$$\begin{aligned} e_m^T &= \alpha_m (T - T_o) \\ e_f^T &= \alpha_f (T - T_o) \end{aligned} \tag{9.75}$$

where T is the final temperature of interest and T_o is the stress-free temperature, usually the cure temperature. However, the dependence of moduli on temperature does not allow the use of thermal expansion coefficients for composite. Therefore, we define average thermal expansion coefficients by

$$\alpha_i = \frac{e_i}{T - T_o} \tag{9.76}$$

Substitution of Equations 9.72 and 9.75 into Equation 9.76 yields

$$\alpha_x = \frac{v_f E_f(T)\alpha_f + v_m E_m(T)\alpha_m}{v_f E_f(T) + v_m E_m(T)} \tag{9.77}$$

$$\alpha_y = v_f \alpha_f + v_m \alpha_m + v_f \nu_f \alpha_f + v_m \nu_m \alpha_m - (v_f \nu_f + v_m \nu_m)\,\alpha_x$$

Equation 9.77 can also be used as instantaneous thermal expansion coefficients when the change of moduli with temperature can be neglected in comparison with that of strains. Equation 9.77 is shown graphically in Figure 9.11 for a glass/epoxy composite.

The determination of swelling strain requires not only the average moisture concentration in composite but also the moisture concentrations in constituent phases. Assuming zero void fraction, we solve Equation 9.68 for c_m

$$c_m = \frac{s}{v_m s_m + v_f s_f c_{fm}}\, c \tag{9.78}$$

where c_{fm} is the ratio; s is specific gravity.

$$c_{fm} = c_f / c_m \tag{9.79}$$

Substituting into Equation 9.72 the swelling coefficients of constituents defined by

$$e_f^H = \beta_f c_f, \quad e_m^H = \beta_m c_m \tag{9.80}$$

and using Equation 9.78, we obtain the average swelling coefficients of composite as

$$\beta_x = \frac{v_f E_f c_{fm} \beta_f + v_m E_m \beta_m}{(v_f E_f + v_m E_m)(v_m s_m + v_f s_f c_{fm})} s \tag{9.81}$$

$$\beta_y = \frac{v_f (1 + \nu_f) c_{fm} \beta_f + v_m (1 + \nu_m) \beta_m}{v_m s_m + v_f s_f c_{fm}} s - (v_f \nu_f + v_m \nu_m) \beta_x$$

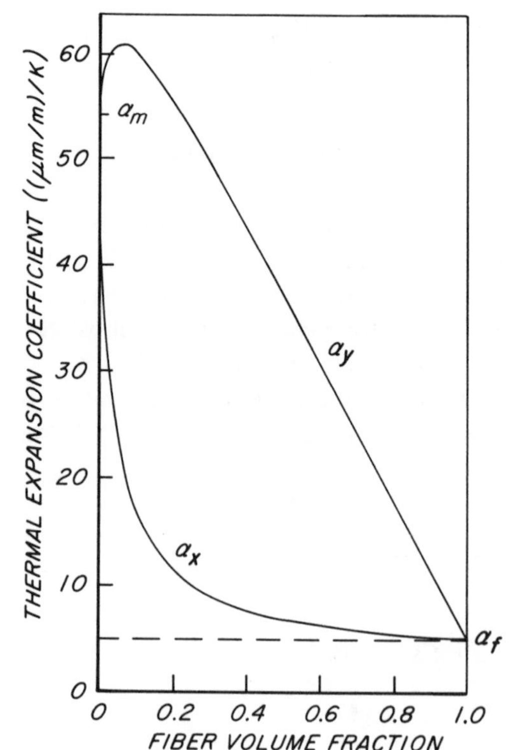

Figure 9.11 Thermal expansion coefficients of glass/epoxy composite. The properties used are: $\alpha_f = 5.0$ $(\mu m/m)/K$, $\alpha_m = 54$ $(\mu m/m)/K$, $E_f = 72$ GPa, $\nu_f = 0.2$, $E_m = 2.76$ GPa, $\nu_m = 0.35$.

A considerable simplification of the foregoing equations is possible or graphite/epoxy composites for which

$$c_f = \alpha_f \approx 0, \quad E_m/E_f \ll 1 \tag{9.82}$$

The final results are given below without derivation:

$$\begin{aligned} \alpha_x &= \beta_x = 0 \\ \alpha_y &= \nu_m(1+\nu_m)\alpha_m \\ \beta_y &= \frac{1+\nu_m}{s_m}\beta_m s \end{aligned} \tag{9.83}$$

7. strengths

As discussed in Chapter 7, unidirectional composites possess excellent strength and stiffness in the longitudinal direction because load is carried mostly by fibers. In the other loading conditions the load sharing is about equal between fibers and matrix; therefore, composite strengths are comparable to those of the matrix used.

Another parameter which plays a very important role in the strength of composites is the interface between fiber and matrix. The assumption of perfect interfacial bond under which elastic properties were discussed in the preceding sections was appropriate because the stresses involved were rather small. However, since failure of a material is initiated at the weakest point, a weak interface will certainly lead to a premature failure when a substantial load sharing is expected by the interface.

Load sharing by constituent phases depends on the type of loading. Therefore, we shall discuss strengths of unidirectional composite under five different loadings: longitudinal tension and compression, transverse tension and compression, and shear.

Consider a unidirectional composite subjected to a uniaxial tension in the fiber direction. Since $\bar{\epsilon}_x = \bar{\epsilon}_{fx} = \bar{\epsilon}_{mx}$ in the present case, the stresses in the constituent phases will be as schematically shown in Figure 9.12. This figure has been constructed based on the following observations. First, fiber is linear elastic up to fracture. Second, matrix

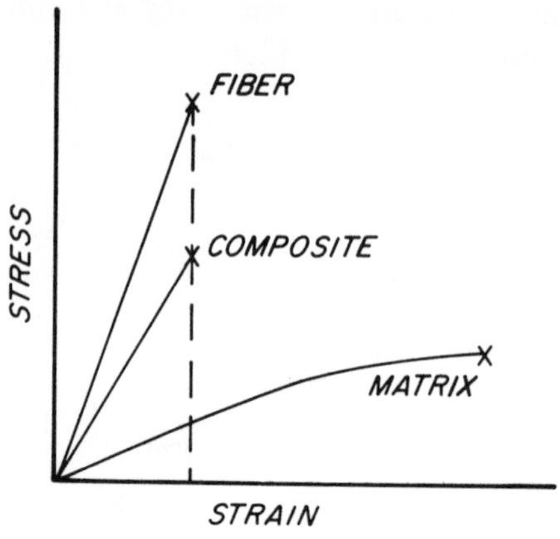

Figure 9.12 Typical stress-strain relations of fiber, matrix and composite. The composite failure strain is close to the fiber failure strain. The matrix is nonlinear above the fiber failure strain.

is linear initially; however, it behaves nonlinearly as strain increases. The strain at which nonlinearity starts to appear is greater than the fracture strain of fiber.

Since not all fibers are expected to be of equal strength and equally stressed, some fibers will fail before others. When these fibers break there are three modes of further damage growth depending on the properties of the matrix and interface.

If the matrix is brittle and the interface strong, the cracks created by the fiber breaks will propagate through the matrix across the neighboring fibers leading to the composite failure. If the interface is weak, then interfacial failure can be initiated at the fiber breaks and the fiber-matrix debonding will grow along the broken fibers. A longitudinal damage growth is also possible in the form of matrix yielding between fibers if the matrix is ductile with low yield stress. As far as the composite strength is concerned, the latter two modes of damage growth have a similar effect. Therefore, we shall simply divide the failure mode under a longitudinal tension into the transverse crack propagation mode and the longitudinal damage growth mode.

The transverse crack propagation mode is in fact what is observed in brittle, homogeneous materials. In this failure mode, the strength of stronger fibers cannot be fully utilized, and hence the composite strength is not an optimum.

In the other extreme case of complete longitudinal damage growth node, broken fibers are simply separated from intact ones as far as load haring is concerned, and the composite will behave like a dry bundle of ibers. We shall now show that the resulting strength is less than the verage fiber strength.

Suppose a bundle of many fibers of equal length L is subjected to a train ϵ. In the bundle the same strain ϵ is applied to every fiber. Suppose the fraction of unbroken fibers at the strain ϵ is given by

$$R(\epsilon) = \exp\,[-L(\epsilon/\epsilon_o)^{\alpha}] \tag{9.84}$$

where α and ϵ_o are constants. The above function is shown schematically in Figure 9.13. The nominal stress of the bundle, which is the load divided by the original cross-sectional area of the fibers, is equal to

$$\sigma = E_f \epsilon R(\epsilon) \tag{9.85}$$

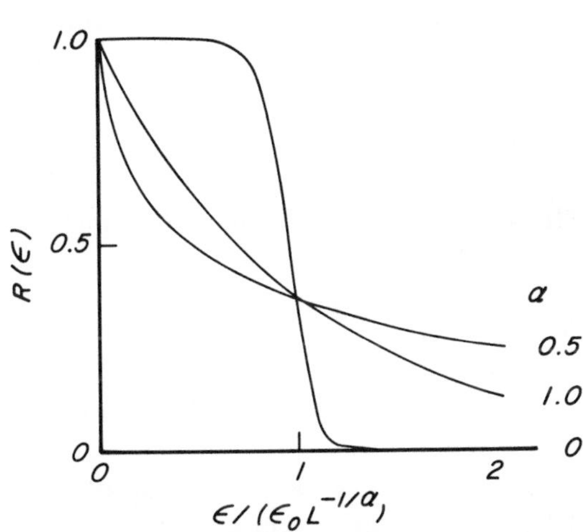

Figure 9.13 Fraction of unbroken fibers at different strain levels. The function $(1-R(\epsilon))$ is known as a Weibull distribution for the failure strain.

The bundle is said to have failed when it cannot support any further increase in load. That is, the bundle failure strain Y_b satisfies

$$\left.\frac{d\sigma}{d\epsilon}\right|_{\epsilon = Y_b} = 0 \tag{9.86}$$

Using Equations 9.84 and 9.85 in Equation 9.86 leads to

$$Y_b = \epsilon_o (L\alpha)^{-1/\alpha} \tag{9.87}$$

The strength in stress of this bundle of length L is thus

$$X_b(L) = E_f \epsilon_o (L\alpha e)^{-1/\alpha} \tag{9.88}$$

where e is the base of the natural logarithm.

Equation 9.84 can also be regarded as the probability of a single fiber surviving the strain ϵ. Thus $(1-R)$ is the cumulative distribution of failure strain and is known as a Weibull distribution (see Appendix 9.1). The average fiber failure strain Y_f is thus given by

$$Y_f = \int_o^\infty \epsilon \left(-\frac{dR}{d\epsilon}\right) d\epsilon = \epsilon_o L^{-1/\alpha} \Gamma(1 + 1/\alpha) \tag{9.89}$$

where $\Gamma(\cdot)$ is the gamma function. The corresponding average strength of the fiber of length L is

$$X_f(L) = E_f Y_f = E_f \epsilon_o L^{-1/\alpha} \Gamma(1 + 1/\alpha) \tag{9.90}$$

We can now study the ratio of the bundle strength to the average fiber strength,

$$\frac{X_b(L)}{X_f(L)} = \frac{E_f Y_b R(Y_b)}{E_f Y_f} = \frac{1}{(\alpha e)^{1/\alpha} \Gamma(1 + 1/\alpha)} \tag{9.91}$$

Figure 9.14 shows the ratio X_b/X_f as a function of $1/\alpha$ together with some experimental data. The factor $1/\alpha$ is a measure of scatter, almost equal to the coefficient of variation (see Appendix 9.2). Thus the bundle strength can be substantially lower than the fiber strength if the fiber strength exhibits a large scatter.

Thus far we have shown that the longitudinal strength will be lower than maximum if the failure mode is dominantly one of the transverse crack propagation or of the longitudinal damage growth. An optimum strength is realized somewhere between these two extremes; that is,

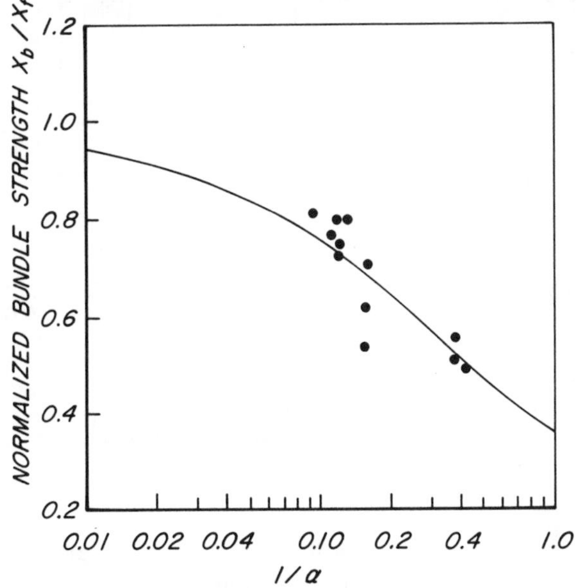

Figure 9.14 The bundle strength is lower than the average fiber strength. The reduction increases with $1/\alpha$; i.e., the larger the scatter in fiber strength, the larger the reduction. The data are for E and S glass fibers, taken from [5, 6].

oth transverse and longitudinal damage growths are localized to the iber breaks to lead to an optimum composite strength. The three ypical failure modes are shown pictorially in Figure 9.15.

In the optimum condition the effect of broken fiber is limited to a mall region of length δ as shown in Figure 9.16. Thus, unless two fiber reaks are within this region, one broken fiber does not know the xistence of the other broken fiber. We then assume that the composite ails when all the fibers within this region fail. Thus the composite trength X can be taken as the strength of a fiber bundle of length δ aultiplied by v_f plus the matrix contribution $v_m \bar{\sigma}^*_{mx}$ [7]

$$X = v_f X_b(\delta) + v_m \bar{\sigma}^*_{mx} = \frac{1}{(\alpha e)^{1/\alpha}\,\Gamma(1 + 1/\alpha)} \left(\frac{L}{\delta}\right)^{1/\alpha} v_f X_f + v_m \bar{\sigma}^*_{mx} \tag{9.92}$$

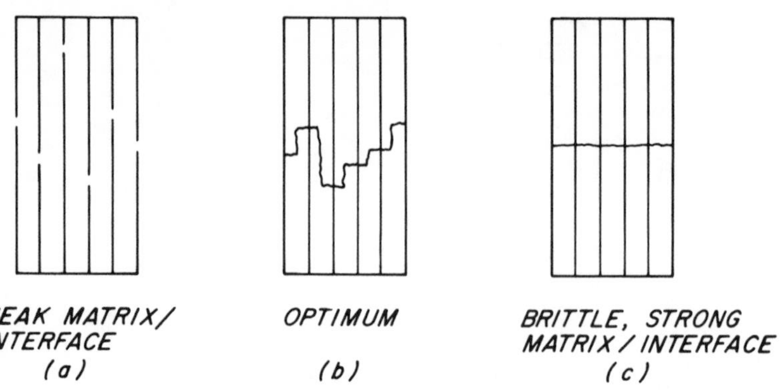

Figure 9.15 Three typical failure modes of unidirectional composite; (*a*) Longitudinal damage growth typical of dry fiber bundle; (*b*) mixed failure mode desired; (*c*) transverse crack propagation mode typical of brittle, homogeneous materials.

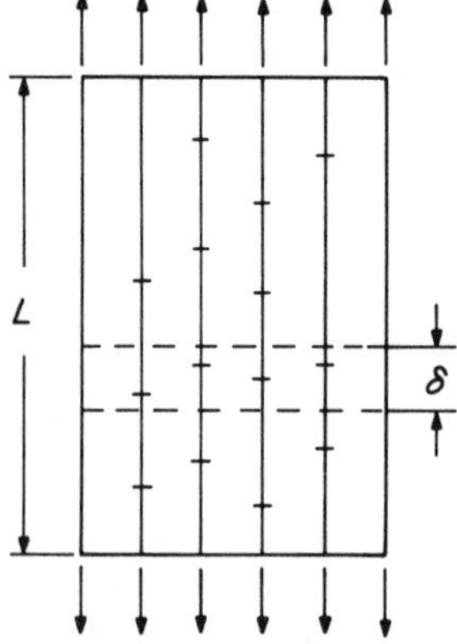

Figure 9.16 Zone of interaction between fiber failures. The composite falls when sufficient number of fibers fail within the zone of length δ.

Note that X_f is the average strength of fiber of length L, not δ, and that $\bar{\sigma}^*_{mx}$ is the average matrix stress at the time of the composite failure.

The ratio $(X - v_m \bar{\sigma}^*_{mx})/(v_f X_f)$ determined by Equation 9.92 is shown as function of L/δ for three different values of $1/\alpha$ in Figure 9.17. If the fiber strength has a small scatter, i.e., small $1/\alpha$, the composite strength is close to $(v_f X_f + v_m \bar{\sigma}^*_m)$ regardless of L/δ. Otherwise, $(X - v_m \bar{\sigma}^*_{mx})/(v_f X_f)$ increases with L/δ.

Strength usually exhibits a larger scatter than modulus because the former depends much more on defects within a material than the latter. Since the probability of finding defects increases with the volume of the material tested, the strength distribution will depend on the volume. In the case of fibers of the same cross-sectional area, the volume is proportional to the length, and hence we can have Equation 9.84. An example of length-dependence of strength is shown in Figure 9.18 for glass fibers.

To determine δ, which is required to use Equation 9.92, consider a

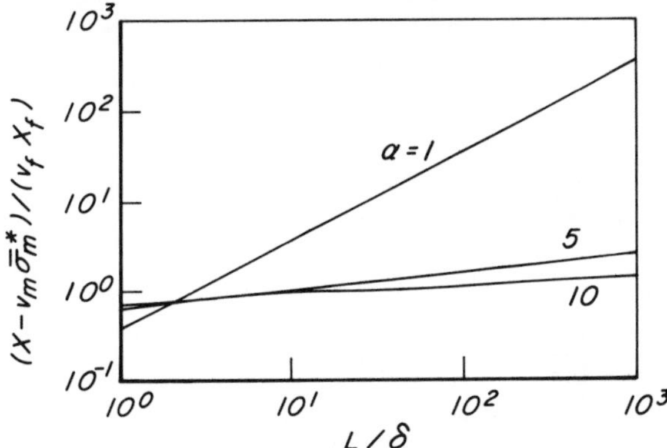

Figure 9.17 The strengthening efficiency of fiber depends on α and δ. α is the shape parameter of the fiber strength distribution and δ is the length of the fiber failure interaction zone. X_f is the average strength of fiber of length L.

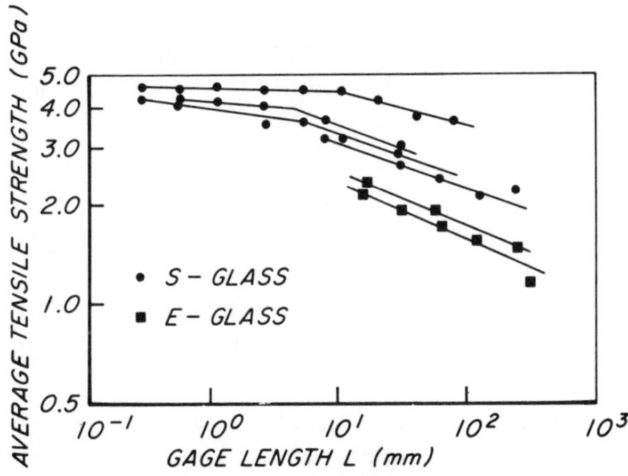

Figure 9.18 Length effects on average tensile strength of glass fibers. The average tensile strength decreases as the gage length increases. (Data from [8]).

broken fiber in a matrix, Figure 9.19. As the fiber is pulled away from the matrix, the interfacial shear stress at the tip of the fiber increases. Assuming a rigid-perfectly plastic behavior at the interface, we see that the shear stress is equal to the yield stress τ_y if there is a plastic slip at

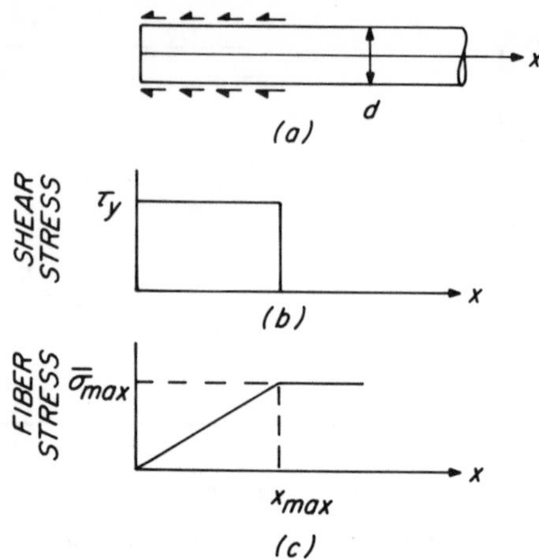

Figure 9.19 Simplified stress distributions around and inside a broken fiber. A rigid-perfectly plastic behavior is assumed at the interface.

the interface. Denoting by σ the stress in the fiber a distance x from the fiber end, Figure 9.19, we use the equilibrium condition for the fiber to obtain

$$\frac{x}{d} = \frac{\sigma}{4\tau_y} \tag{9.93}$$

Thus the maximum distance over which interfacial yielding occurs is limited by the maximum value of σ which the fiber can sustain without failure in the interval $(0, x)$.

The average of σ_{max} is obtained from Equations 9.140 and 9.141 in Appendix 9.1 using x_{max} in place of L. Since $\sigma_o = E_f \epsilon_o$, Equation 9.90 can be used to replace σ_o by X_f. The final result is

$$\bar{\sigma}_{max} = X_f \left[\frac{L(\alpha+1)}{x_{max}} \right]^{1/\alpha} \tag{9.94}$$

The distance x_{max} corresponding to $\bar{\sigma}_{max}$ then follows from substitution of Equation (9.94) into Equation (9.93):

$$\frac{x_{max}}{d} = \left(\frac{X_f}{4\tau_y} \right)^{\alpha/(\alpha+1)} [(\alpha+1)L/d]^{1/(\alpha+1)} \tag{9.95}$$

Noting that $\delta = 2x_{\max}$, we finally obtain

$$\frac{\delta}{d} = 2\left(\frac{X_f}{4\tau_y}\right)^{\alpha/(\alpha+1)} [(\alpha+1)L/d]^{1/(\alpha+1)} \tag{9.96}$$

If the fiber strength has very little scatter, i.e., α is very large, then δ/d approaches $X_f/(2\tau_y)$,

$$\frac{\delta}{d} = \frac{X_f}{2\tau_y} \tag{9.97}$$

The composite strength given by Equation 9.92 reduces to

$$\begin{aligned} X &= v_f X_f + v_m \bar{\sigma}^*_{mx} \\ &= (v_f + v_m E_m/E_f) X_f \end{aligned} \tag{9.98}$$

where the second equality follows from the linear behavior of the matrix up to the composite failure.

Equation 9.98 is called the rule-of-mixtures for the longitudinal strength. Although it has been derived under the assumption of deterministic fiber strength, it has proven to provide a reasonable estimate for actual composites. It is possible that the combined effect of all the parameters in Equations 9.92 and 9.96 makes Equation 9.98 a reasonable approximation.

Since Equation 9.98 is based on the assumption of fiber failure triggering the composite failure, the composite strength can be less than the matrix strength X_m if fiber volume fraction is too small. We can determine this minimum fiber volume fraction by substituting X_m for X and solving the resulting equation for v_f; i.e.,

$$v_f = \frac{E_f}{E_f - E_m}\left(\frac{X_m}{X_f} - \frac{E_m}{E_f}\right) \tag{9.99}$$

Since $X_m/X_f > E_m/E_f$, see Figure 9.12, there is always a minimum fiber volume fraction required to strengthen matrix by fiber reinforcements.

In longitudinal tension the role of the matrix was rather secondary. In longitudinal compression, however, the matrix provides lateral support for fibers to carry compressive load without buckling. Without such support fibers can hardly resist any compression because of their small diameter.

Consider a composite subjected to compression in the fiber direction, Figure 9.20. Fibers may have initial curvature as shown in the figure [9]. Suppose the initial deflection of fibers can be described by

$$v_o = f_o \sin \frac{\pi x}{l} \tag{9.100}$$

where l is the half wavelength. When a compressive stress σ is applied, the final deflection of fibers become

$$v = f \sin \frac{\pi x}{l} \tag{9.101}$$

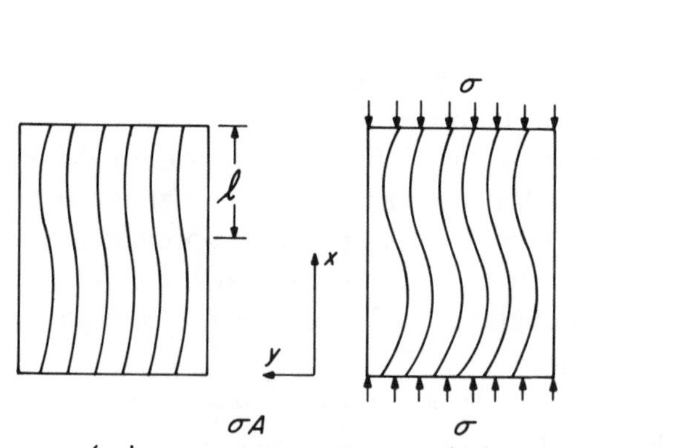

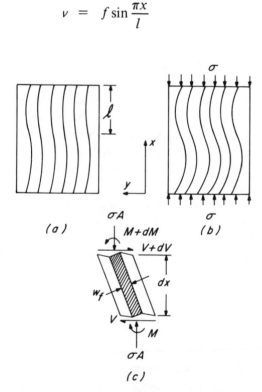

Figure 9.20 Local buckling at a point in composite; (*a*) initial state; (*b*) under a compressive stress σ; (*c*) free body diagram for a representative volume element of cross-sectional area A.

Let us take an infinitesimal representative volume element of length dx and cross-sectional area A at distance x. A free body diagram for this element is shown in Figure 9.20(c). The balance of moments requires that

$$\frac{dM}{dx} - V + \sigma A \frac{dv}{dx} = 0 \tag{9.102}$$

The bending moment M is borne by the fiber while the shear V is the result of the overall shear deformation of the representative volume element. Thus

$$\begin{aligned} M &= E_f I_f \frac{d^2}{dx^2}(v - v_o) \\ V &= A E_s I \frac{d}{dx}(v - v_o) \end{aligned} \tag{9.103}$$

Here $E_f I_f$ is the bending stiffness of the fiber and E_s is the effective shear modulus of the composite.

Substituting Equation 9.103 into Equation 9.102 and noting that $I_f/A = v_f w_f^2/12$, we obtain an equation for σ

$$\sigma = \left[E_s + \frac{\pi^2}{12} v_f E_f \left(\frac{w_f}{l}\right)^2\right]\left(1 - \frac{f_o}{f}\right) \tag{9.104}$$

In actual composites, w_f/l is much smaller than unity. The second term inside the brackets can therefore be neglected with the final result

$$\sigma = E_s \left(1 - \frac{f_o}{f}\right) \tag{9.105}$$

Equation 9.105 describes a relation between the compressive stress and the amplitude of fiber deflection f. As σ increases, f will increase reaching a critical value f_c at which the composite fails. Thus the compressive strength X' is given by

$$X' = E_s \left(1 - \frac{f_o}{f_c}\right) \tag{9.106}$$

There are two possible sources of composite failure: local shear failure and bending failure of fibers. The maximum local shear stress can be calculated from Equation 9.103. The f_c is then the value of f when this maximum local shear stress reaches the shear strength S of the composite:

$$f_c = f_o + \frac{l}{\pi} \frac{S}{E_s} \tag{9.107}$$

On the other hand, fibers can fail in bending when the maximum bending stress reaches the fiber strength X_f. In this case we obtain

$$f_c = f_o + \frac{2l}{w_f} \frac{l}{\pi^2} \frac{X_f}{E_f} \tag{9.108}$$

Since $2l/w_f$ is much larger than unity, f_c will be larger when calculated from Equation 9.108 than from Equation 9.107. Therefore, the compressive failure of composite is caused by a local shear failure and the resulting compressive strength is

$$X' = E_s \frac{1}{1 + (\pi f_o/l)/(S/E_s)} \tag{9.109}$$

When fibers are perfectly straight, f_o/l vanishes, and hence the compressive strength becomes equal to the shear modulus. In actuality, however, X' is always less than E_s. For boron/epoxy composites, the ratio X'/E_s is slightly less than 0.5 whereas it is less than 0.25 for graphite/epoxy composites. Other than the initial deflection of fibers, there is another reason for such discrepancy: the nonlinearity in shear of unidirectional composites. In such cases we can show that E_s in Equation 9.109 must be replaced by the secant shear modulus at failure. Note that the secant modulus is the ratio of the stress to the corresponding strain. Since the secant modulus at failure is lower than E_s, with difference increasing with ductility, the resulting compressive strength will be lower.

In transverse tension or compression the load sharing by matrix is of the same order of magnitude as that of fibers. In the elastic range the average matrix stress was seen to be about half the average fiber stress.

nce matrix is much weaker than fiber, matrix will fail first, causing e composite to fail.

We recall that the composite stress is related to the average stresses in e constituents by

$$\bar{\sigma}_y = \nu_f \bar{\sigma}_{fy} + \nu_m \bar{\sigma}_{my} \tag{9.110}$$

$$= [1 + \nu_f (1/\eta_y - 1)] \, \bar{\sigma}_{my} \tag{9.111}$$

he stress σ_{my} in matrix is not uniformly distributed; it reaches a aximum at the fiber-matrix interface. Therefore, failure will be initiated at the fiber-matrix interface when the stress there reaches the atrix tensile strength X_m or the interface strength X_{int}, whichever is naller. Introducing a stress concentration factor K_{my} defined by

$$K_{my} = \frac{(\sigma_{my})_{max}}{\bar{\sigma}_{my}} \tag{9.112}$$

e can express the transverse tensile strength of composite as

$$Y = \frac{1 + \nu_f (1/\eta_y - 1)}{K_{my}} X_m \tag{9.113}$$

f course, X_m should be replaced by X_{int} if $X_{int} < X_m$.

An exact determination of η_y and K_{my} is difficult because the ehavior of matrix is nonlinear near failure, and is therefore beyond the ope of this book. However, we can make the following predictions ased on Equation 9.113.

If the matrix is linear elastic up to failure, then the factor $[1 + (1/\eta_y - 1)]/K_{my}$ is known to be less than unity and decreases with creasing fiber volume fraction. Therefore, the transverse tensile rength will always be less than the matrix tensile strength, the difference increasing with ν_f.

On the other hand, if the matrix is ductile, stress concentrations can e relaxed so that K_{my} becomes closer to unity. As a result the transerse tensile strength can be greater than the matrix tensile strength. owever, if there is a premature interface failure before the matrix goes into the nonlinear range to allow a stress relaxation, the composit behavior will be the same as if the matrix is brittle.

An experimental evidence for the observations just described shown for glass/epoxy composites in Figure 9.21. Here the strengt ratio Y/X_m is seen to be as high as 2.3. The curve represents a constar composite strength of 30 MPa. Thus, the composite strength itself doe not depend much on the matrix strength. At lower X_m the matrix usually more ductile; consequently, a much better utilization of th matrix strength is realized. As X_m increases, the matrix becomes mor and more brittle and the stress concentration increases. As a result, th composite strength is lower than the matrix strength.

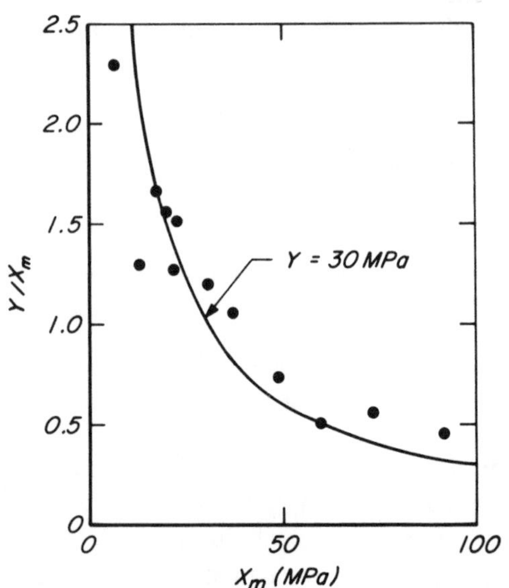

Figure 9.21 The composite-to-matrix strength ratio in transverse tension decreases as the matrix strength increases. Note that the composite strength can be higher than the matrix strength. The data are for E-glass and S-glass/epoxy composites.

The fracture surface in transverse tension is normal to the loadin However, in transverse compression it is approximately 45° to the loac ing. Thus, the transverse compression failure is in fact a shear failure o the 45° plane. The transverse compressive strength is four to seve mes as high as the transverse tensile strength.

Finally, the mechanisms of shear failure are similar to those of trans-erse tension failure. The shear strength S can be studied using an quation similar to Equation 9.113:

$$S = \frac{1 + v_f(1/\eta_s - 1)}{K_{ms}} S_m \tag{9.114}$$

here S_m and K_{ms} are respectively the matrix shear strength and the atrix stress concentration factor in shear. Again, if the interface rength S_{int} is lower than S_m, S_{int} should be used in place of S_m.

Typical failure modes corresponding to the strengths discussed so far, ccept for the longitudinal tension, are schematically shown in Figure 22. The longitudinal compression failure is accompanied by a brush-ke failure surface. In transverse tension or shear the failure surface is arallel to the fibers and normal to the specimen surfaces. The failure rface in transverse compression is still parallel to the fibers but makes angle of about 45° with the loading direction. Strength values of ose composites discussed in the preceding chapter are listed in able 7.1.

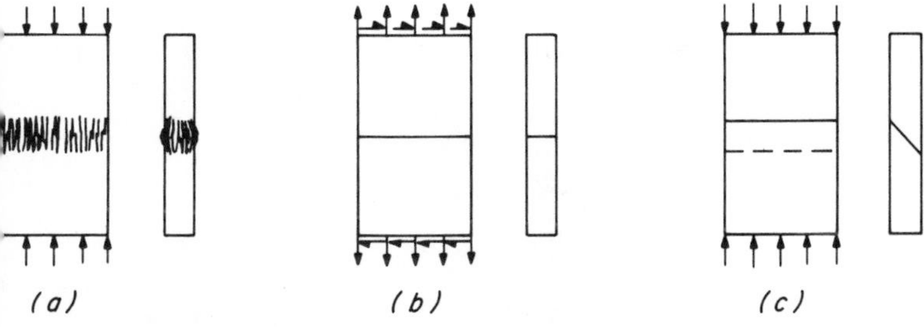

Figure 9.22 Schematic views of typical failure modes: (*a*) longitudinal compression; (*b*) transverse tension and shear; and (*c*) transverse compression.

sample problems

The weight of the matrix in a void-free glass/epoxy composite is easured to be 36% of the weight of the composite. What is the fiber lume fraction? The specific gravities of the epoxy and glass are 1.2 d 2.6, respectively.

Since $\nu_v = 0$ and $m_f = \rho_f \nu_f / \rho$, a comparison of Equations 9.3 and 9 yields

$$\nu_f = \frac{\rho_m (1 - m_m)}{\rho_f m_m + \rho_m (1 - m_m)} \tag{9.115}$$

Substituting the known values of the variables into Equation 9.115, w obtain the fiber volume fraction as

$$\nu_f = \frac{1.2 \times 0.64}{2.6 \times 0.36 + 1.2 \times 0.64} = 0.45 \tag{9.116}$$

b. Using the representative volume element of Figure 9.4, calcula the stresses $\bar{\sigma}_{fx}$ and $\bar{\sigma}_{mx}$ in a 60 vol % glass/epoxy composite resultir from the application of a transverse tension of 10 MPa. Assume that $\bar{\epsilon}_{f} = \bar{\epsilon}_{mx}$, and use the properties in Table 9.1 and Equations 9.59 an 9.60.

The equations to be used are in Equation 9.44. The variables are:

$$\begin{aligned} E_f &= 72 \text{ GPa}, \ \nu_f = 0.2 \\ E_m &= 3.45 \text{ GPa}, \ \nu_m = 0.35 \\ v_f &= 0.6, \ v_m = 0.4 \end{aligned} \tag{9.117}$$

Thus the stresses $\bar{\sigma}_{fx}$ and $\bar{\sigma}_{mx}$ are

$$\begin{aligned} \bar{\sigma}_{fx} &= -2.2 \text{ MPa} \\ \bar{\sigma}_{mx} &= 3.3 \text{ MPa} \end{aligned} \tag{9.118}$$

Note that, as expected, we have

$$v_f \bar{\sigma}_{fx} + v_m \bar{\sigma}_{mx} = 0 \tag{9.119}$$

c. Using the equations in Table 9.2 and the properties in Table 9. and Equations 9.59 and 9.60, determine the moduli E_x, ν_x, E_y and E for a 60 vol % boron/epoxy composite.

The composite moduli are determined simply by substituting the onstituent properties into the equations in Table 9.2. First, the shear nd plane strain bulk moduli are

$$G_f = \frac{E_f}{2(1+\nu_f)} = 170.8 \text{ GPa}, \quad k_f = 284.7 \text{ GPa}$$

$$G_m = \frac{E_m}{2(1+\nu_m)} = 1.28 \text{ GPa}, \quad k_m = 4.27 \text{ GPa} \tag{9.120}$$

Next, we calculate the stress partitioning parameters:

$$\eta_s = 0.504, \quad \eta_k = 0.776$$

$$\eta_G = 0.618 \tag{9.121}$$

Finally, the composite moduli are

$$E_x = 0.6 \times 410 + 0.4 \times 3.45 = 247.4 \text{ GPa}$$

$$\nu_x = 0.6 \times 0.2 + 0.4 \times 0.35 = 0.26$$

$$E_s = \frac{0.6 + 0.504 \times 0.4}{0.6/170.8 + 0.504 \times 0.4/1.28} = 4.98 \text{ GPa}$$

$$k_y = \frac{0.6 + 0.776 \times 0.4}{0.6/284.7 + 0.776 \times 0.4/4.27} = 12.17 \text{ GPa} \tag{9.122}$$

$$G_y = \frac{0.6 + 0.618 \times 0.4}{0.6/170.8 + 0.618 \times 0.4/1.28} = 4.31 \text{ GPa}$$

$$m = 1.013$$

$$E_y = 12.69 \text{ GPa}$$

d. For the same composite as in Problem *c*, determine the composit moduli using the rule-of-mixtures equations in Section 4. Compare th results with those in Problem *c*.

The moduli E_x and ν_x do not change. E_y and E_s are determine from Equations 9.39 and 9.41, respectively:

$$E_y = 8.52 \text{ GPa}, \quad E_s = 3.16 \text{ GPa} \tag{9.12}$$

These moduli are lower than those calculated in Problem *c*.

e. For graphite fiber the shape parameter for the strength distribu tion is 7.68. What is the expected ratio of the bundle strength to th average fiber strength?

The strength ratio is given by Equation 9.91. Using $\alpha = 7.68$ an noting that $\Gamma(1 + 1/7.68) = 0.94$, we obtain

$$\frac{X_b}{X_f} = \frac{1}{(7.68\, e)^{1/7.68}\,(0.94)} = 0.72 \tag{9.124}$$

f. The average strength of 25-mm long graphite fiber is 2.80 GPa wit a coefficient of variation of 15%. The fiber is used with an epox having a tensile strength of 50 MPa and a modulus of 3.45 GPa to mak a 50 vol % graphite/epoxy composite. Assuming the interfacial yiel stress τ_y to be half the matrix tensile strength, (*a*) determine the lengt δ of the fiber failure interaction zone; (*b*) determine the composit strength using δ; and (*c*) compare the strength obtained in (*b*) with tha from the rule-of-mixtures. The composite is also 25-mm long.

The shape parameter α is calculated from Equation 9.145 and th fiber diameter is found in Table 9.1. Thus, the required variables are:

$$\begin{aligned}
\alpha &= 7.68, & d &= 7\ \mu\text{m} \\
X_f &= 2.80\text{ GPa}, & \tau_y &= 25\text{ MPa} \\
L &= 25\text{ mm}, & v_f &= 0.5 \\
v_m &= 0.5 \\
\bar{\sigma}_m^* &= 2800 \times \frac{3.45}{230} = 42\text{ MPa}
\end{aligned} \tag{9.125}$$

) The length δ is determined from Equation 9.96:

$$\delta = 0.88 \text{ mm} \tag{9.126}$$

) The composite strength follows from Equation 9.92

$$X = 1550 + 21 = 1571 \text{ MPa} \tag{9.127}$$

:) The rule of mixtures for the strength is Equation 9.98. Therefore, e have

$$X = 1400 + 21 = 1421 \text{ MPa} \tag{9.128}$$

1 the present case there is about 10% difference between Equations .92 and 9.98.

. conclusions

he properties of a composite depend on the geometrical arrangement nd the properties of its constituents. The exact analysis of such strucıre-property relationship is rather complex because of many variables ıvolved. Therefore, we have introduced a few simplifying assumptions egarding the structural details and the state of stress within composite.

We have seen that the concept of a representative volume element nd the selection of appropriate boundary conditions are very imortant in the discussion of micromechanics. The composite stress and train are defined as the volume averages of the stress and strain fields, espectively, within the representative volume element. By finding relaions between the composite stresses and the composite strains in terms f the constituent properties, we can derive expressions for the comosite moduli. In addition, we have shown that the results of advanced nethods can be put in a form similar to the rule-of-mixtures equations.

Estimating the hygrothermal expansion coefficients is not much diferent from what we do for the elastic moduli. A major difference in nalyzing the thermal expansion and swelling is that in the former the emperature does not change from fiber to matrix whereas in the latter he moisture concentration can vary drastically. Thus any analytical nodeling should take into account such nonuniformity of the moisture oncentration.

Prediction of composite strengths is rather difficult because there a many unknown variables and also because failure critically depends o defects. However, we can qualitatively explain the effects of co stituents including fiber-matrix interface on composite strength Certainly, failure modes can change depending on the material com binations. Thus, an analytical model developed for one material com bination cannot be expected to work for a different one. Ideally, truly analytical model will be applicable to any material combinatio However, such an analytical model is not available at present. Ther fore, we have chosen to provide models each of which is applicable onl to a known failure mode. Yet, they can explain many of the effects the constituents.

appendix 9.1

Consider a fiber of length L over which stress field is given by

$$\sigma = \sigma_a f(x) \qquad (9.12$$

where σ_a is the stress at $x = L$, so that $f(L) = 1$.

Now we divide the fiber into N segments so that in each segment th stress can be assumed to be uniform. Suppose the probability of the it segment surviving the stress $\sigma_{(i)}$ is given by $[R_o(\sigma_{(i)})]^{\Delta x(i)}$. Here $\Delta x_($ is the length of the ith segment.

The probability of survival of the entire fiber is given by

$$R = \prod_{i=1}^{N} [R_o(\sigma_{(i)})]^{\Delta x(i)} \qquad (9.130$$

Taking natural logarithms of both sides, we rewrite Equation 9.130 as

$$\ln R = \sum_{i=1}^{N} \Delta x_{(i)} \ln R_o(\sigma_{(i)}) \qquad (9.131$$

The right-hand side can be converted to an integral by taking N infinitely large:

$$\ln R = \int_o^L \ln R_o(\sigma)dx \tag{9.132}$$

Therefore, R becomes

$$R = \exp\left[\int_o^L \ln R_o(\sigma)dx\right] \tag{9.133}$$

Suppose R_o is a Weibull distribution,

$$R_o(\sigma) = \exp\left[-(\sigma/\sigma_o)^\alpha\right] \tag{9.134}$$

where α and σ_o are called the shape parameter and scale parameter, respectively. Substitution of Equation 9.134 into 9.133 yields

$$R = \exp\left[-\int_o^L (\sigma/\sigma_o)^\alpha\, dx\right] \tag{9.135}$$

Using Equation 9.129, we can rewrite Equation 9.135 as

$$R(\sigma_a) = \exp\left[-\left(\frac{\sigma_a}{\sigma_{ao}}\right)^\alpha\right] \tag{9.136}$$

where

$$\sigma_{ao} = \sigma_o\left[\int_o^L f^\alpha(x)dx\right]^{-1/\alpha} \tag{9.137}$$

Equation 9.136 can be used to determine the probability of survival for any given stress field σ. For example, suppose σ is uniform over the fiber length. Then Equation 9.136 reduces to

$$R(\sigma) = \exp\left[-L(\sigma/\sigma_o)^\alpha\right] \tag{9.138}$$

This equation can be easily converted to Equation 9.84 by noting tha $\sigma = E\epsilon$.

Next, suppose σ varies linearly with x so that

$$f(x) = \frac{x}{L} \tag{9.139}$$

Substituting Equation 9.139 into 9.137, we obtain

$$\sigma_{ao} = \sigma_o \left(\frac{\alpha+1}{L}\right)^{1/\alpha} \tag{9.140}$$

Thus, the average of σ_a that can be applied to the point of fiber failur is given by (see Appendix 9.2)

$$\bar{\sigma}_a = \sigma_{ao}\Gamma(1 + 1/\alpha) \tag{9.141}$$

appendix 9.2

Let us consider the probability of survival $R(\sigma)$ at stress σ,

$$R(\sigma) = \exp\left[-\left(\frac{\sigma}{\sigma_o}\right)^\alpha\right] \tag{9.142}$$

The function $R(\sigma)$ is also the probability that the strength is greate than σ. Therefore, the probability density function $f(\sigma)$ for strength i

$$f(\sigma) = -\frac{dR}{d\sigma} = \frac{\alpha}{\sigma_o^\alpha}\sigma^{\alpha-1}\exp\left[-\left(\frac{\sigma}{\sigma_o}\right)^\alpha\right] \tag{9.143}$$

The mean $\bar{\sigma}$ and the standard deviation s for the strength distributior are respectively given by

$$\bar{\sigma} = \int_o^\infty \sigma f(\sigma) d\sigma = \sigma_o \Gamma\left(1 + \frac{1}{\alpha}\right)$$

(9.144)

$$s = \int_o^\infty (\sigma - \bar{\sigma}) f(\sigma) d\sigma = \sigma_o\left[\Gamma\left(1 + \frac{2}{\alpha}\right) - \Gamma^2\left(1 + \frac{1}{\alpha}\right)\right]^{1/2}$$

where $\Gamma(\cdot)$ is the gamma function. The coefficient of variation (C.V.) is thus

$$\text{C.V.} = \frac{s}{\bar{\sigma}} = \left[\frac{\Gamma\left(1 + \dfrac{2}{\alpha}\right)}{\Gamma^2\left(1 + \dfrac{1}{\alpha}\right)} - 1 \right]^{1/2} \approx \frac{1.2}{\alpha} \tag{9.145}$$

10. homework problems

a. A glass/epoxy specimen weighing 1 g was burnt and the weight of the remaining fibers was found to be 0.5 g. Knowing that the densities of glass and epoxy are 2.6 and 1.25, respectively, determine the density of the composite in the absence of voids. If the actual density of the composite was measured to be 1.55, what is the void fraction?

b. For the composite bar shown, determine the average strain $\bar{\epsilon}_x$.

c. For the composite bar shown, determine the average stress $\bar{\sigma}_x$.

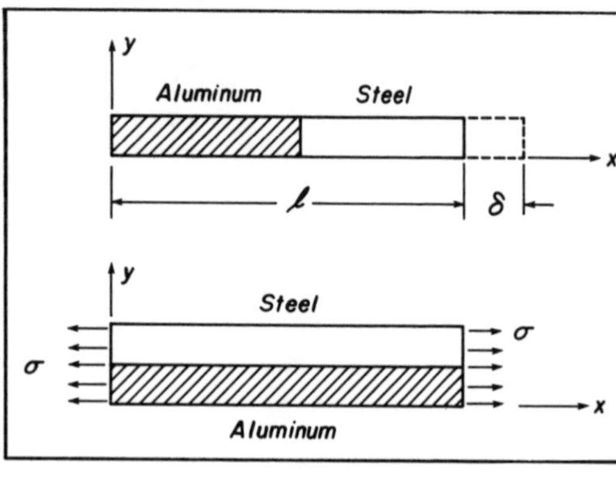

d. Derive Equation 9.26.

e. Prove Equation 9.47.

f. For a 60 vol % Gr/Ep composite determine the moduli E_x, E_y, E_s, and ν_x using the equations in Table 9.2. Compare the results with those from the rule-of mixtures equations of Section 4. Use the properties in Table 9.1 and Section 5.

g. Derive Equation 9.49.

h. The maximum water absorption in a typical epoxy is 6%. What is the maximum amount of water in a graphite/epoxy ($v_f = 0.70$)? Specific gravities of the epoxy and composite are 1.25 and 1.6, respectively

i. Derive Equation 9.72.

j. Determine the curing stress $\bar{\sigma}^R_{mx}$ in the matrix of a 60 vol % glass/epoxy. Use the properties in Figure 9.11.

k. Determine the swelling coefficients for a 70 vol % graphite/epoxy The swelling coefficient of epoxy is 0.35, and graphite does not absorb moisture.

l. In Figure 8.19 the longitudinal tensile strength does not suffer any reduction at elevated temperatures. Would the same trend be observed of the longitudinal compressive strength? If not, explain why.

m. The shear strength of a composite with surface-treated fibers increases with fiber volume fraction whereas the opposite is true when fibers without surface treatment are used. Provide the reasons for such difference. Is such difference an indication that a better fiber-matrix bond is achieved through fiber surface treatment?

The density of a bulk epoxy at room temperature is ρ_{mo}. The density ρ_m of the same epoxy in a composite is different from ρ_{mo} because of the presence of residual stresses. Determine the ratio ρ_m/ρ_{mo} in terms of the curing strains e_f^T and e_m^T, the constituent elastic moduli, and the fiber volume fraction v_f. Assume that $\bar{\epsilon}_{fx} = \bar{\epsilon}_{mx}$ and $\bar{\sigma}_{fy} = \bar{\sigma}_{my}$.

nomenclature

A, A_1, A_2	= Cross-sectional areas of representative volume element
c	= Specific moisture concentration, in g/g
d	= Fiber diameter
f, f_o, f_c	= Amplitudes of fiber deflection
G	= Shear modulus
K_{ms}, K_{my}	= Matrix stress concentration factors in shear and transverse tension, respectively ($= (\sigma_m)_{max}/\bar{\sigma}_m$)
k	= Plane strain bulk modulus
L, L_1, L_2, L_3	= Dimensions of representative volume element
L	= Fiber length
l	= Half wavelength of fiber deflection
M	= Mass
m	= Mass fraction
$R(\epsilon)$	= Probability of survival at strain ϵ or fraction of fibers surviving ϵ
s	= Specific gravity
u, v, v_s	= Boundary displacements
V	= Volume
v_f, v_m, v_v	= Volume fractions
v, v_o	= Final and initial fiber deflections, respectively
w	= Width of a constituent phase in representative volume element
S	= Shear strength
X, X'	= Longitudinal tensile and compressive strengths, respectively
Y, Y'	= Transverse tensile and compressive strengths, respectively
α	= Shape parameter for Weibull distribution of fiber failure strain
$\Gamma(\cdot)$	= Gamma function
δ	= Length of fiber failure interaction zone
ϵ_o	= Scale parameter for Weibull distribution of fiber failure strain
ϵ_i^o	= Constant strain components
$\bar{\epsilon}_i$	= Volume-average strain components

$_G$ = Stress partitioning parameter in transverse shear
$_k$ = Stress partitioning parameter in plane strain hydrostatic tension and compression
$_y$ = Stress partitioning parameter in transverse tension and compression
$_s$ = Stress partitioning parameter in shear
= Mass density
$_o$ = $E_f \epsilon_o$
^o_i = Constant stress components
$_i$ = Boundary stress components
$_i$ = Volume-average stress components
ub b = Bundle
ub f = Fiber
ub m = Matrix
ub v = Void
ub y = Transverse
ub int = Interface

eferences

. S. W. Tsai, "Structural Behavior of Composite Materials," NASA CR-71, 1964.
. J. V. Noyes and B. H. Jones, "Analytical Design Procedures for the Strength and Elastic Properties of Multilayer Fiber Composites," *Proc. AIAA/ASME 9th Structures, Dynamics and Materials Conference*, Paper 68-336,1968.
. Z. Hashin, *Theory of Fiber Reinforced Materials*, NASA CR-1974, 1972.
. J. J. Hermans, "The Elastic Properties of Fiber Reinforced Materials When the Fibers are Aligned," *Proc. Koninkl. Nederl. Akad. Wetenschappen*, B70, 1967, p. 1.
. S. W. Tsai, D. F. Adams and D. R. Doner, "Effect of Constituent Material Properties on the Strength of Fiber-Reinforced Composite Materials," AFML-TR-66-190, Air Force Materials Laboratory, 1966.
. A. E. Armenakas, S. K. Garg, C. A. Sciammarella, V. Sualbonas, "Strength of Glass-Fiber Bundles and Composites Subjected to Slow Rates of Straining," AFML-TR-69-314, Air Force Materials Laboratory, 1970.
. B. W. Rosen, "Mechanics of Composite Strengthening," *Fiber Composite Materials*, Chapter 3, American Society for Metals, 1965.
. L. R. McCreight, et al., *Ceramic and Graphite Fibers and Whiskers, Refractory Materials*, Vol. 1, Academic Press, 1965.
. A. S. D. Wang, "A Non-Linear Microbuckling Model Predicting the Compressive Strength of Unidirectional Composites," ASME Paper No. 78-WA/Aero-1, 1978.

appendix A
transformation equations

. general transformation

The transformation equation of a tensor depends on its rank. In fact, the definition of a tensor is one that follows a particular transformation equation. Then, by definition, the quantity is a tensor of a given rank. Stress and strain are second rank tensors. Modulus and compliance are fourth rank tensors. Their transformation equations are:

$$T'_{ij} = a_{ik} a_{jf} T_{kf} \tag{A.1}$$

$$T'_{ijkf} = a_{im} a_{jn} a_{ko} a_{fp} T_{mnop} \tag{A.2}$$

where T' are the transformed components of T. We use uncontracted notation here. The number of indices now correspond to the tensorial rank. The a's are direction cosines of the new, transformed coordinate system relative to the old, original coordinate system. The usual range and summation conventions apply. When we have n dimensions,

$$i,j, \ldots = 1,2, \ldots n \tag{A.3}$$

The beauty and simplicity of tensors lie in their generality that only one transformation equation will suffice; i.e.,

$$T'_{ijk\ldots} = a_{im} a_{jn} a_{ko} \ldots T_{mno\ldots} \tag{A.4}$$

There is no conceptional difficulty if we want to define the transformation equation of a sixth rank tensor in four dimensional space. (It may take a little time writing it down.)

2. specialized transformation

We have not followed the general tensorial approach in this book Instead we use specialized equations of various quantities. We want t list the reasons:

- We decided to use contracted notation in order to reduce th number of indices. The general definition of transformation mus be altered. In the contracted notation, we no longer can treat od rank tensors such as vectors.
- We decided to use engineering shear strains. The transformatio equation for strain must be altered accordingly. The strain trans formation in Table 2.5 is not the same as the stress transformatio in Table 2.1.
- We also decided to differentiate between the behavioral quantitie from material properties. Stress, strain and their integrals (stres resultant, etc.) are the behavioral quantities, and are all secon rank tensors. Stiffness and compliance and their integrals (e.g. in-plane modulus) are material properties, and are fourth ran tensors. Strength parameters F_{ij} and G_{ij} are also fourth ran tensors. We have properties which are second rank tensors such a thermal and moisture expansion coefficients and the strengt parameters F_i and G_i.

The separation above is useful for composite materials. For th behavioral quantities, we are usually interested in the on-axis stress o strain using the transformation from a fixed reference, off-axis co ordinate system, say, the 1-2 axes. We normally go from the off-axis t the on-axis of a ply. For the material properties, we are interested i the off-axis properties from the on-axis orientations. It is the opposit of the stress, strain, etc. We let the material rotate while we stay fixe at the same reference coordinate system. The rotation is θ for uni directional composites; γ for laminated composites; see Figure 4.12 fo the rigid body rotation. The angle θ is also used for the transformatio of stress and strain. The latter θ has the opposite meaning because it i intended to go from the off-axis to the on-axis. The transformatio equations are listed separately in this Appendix. The opposite meanin of the angle of transformation is included in the equations. No sig change is necessary.

3. transformation of σ_i, N_i, M_i

Listed in this section are the transformation of stress and their integrals from the off-axis, 1-2 coordinate system to the on-axis, 1′-2′ coordinate system.* Three formulations for the stress transformation will be given; viz., the power, and the multiple angle and the invariant functions. The transformation of the stress integrals can be done by direct substitution of N_i or M_i for σ_i.

table A.1
"stress" transformation in power functions

	σ_1	σ_2	σ_6
σ_1'	m^2	n^2	$2mn$
σ_2'	n^2	m^2	$-2mn$
σ_6'	$-mn$	mn	m^2-n^2

$m = \cos\theta$, $n = \sin\theta$

table A.2
"stress" transformation in multiple angle functions

	p	q	r
σ'	1	$\cos 2\theta$	$\sin 2\theta$
σ_2'	1	$-\cos 2\theta$	$-\sin 2\theta$
σ_6'		$-\sin 2\theta$	$-\cos 2\theta$

where

$$p = \frac{1}{2}(\sigma_1 + \sigma_2)$$

$$q = \frac{1}{2}(\sigma_1 - \sigma_2) \tag{A.5}$$

$$r = \sigma_6$$

*The on-axis coordinates were the x-y for the unidirectional composite; and the x_i-y_i coordinates for the i-th ply in a laminate.

table A.3
"stress" transformation in invariant functions

	I	R
σ_1'	1	$\cos 2(\theta-\theta_o)$
σ_2'	1	$-\cos 2(\theta-\theta_o)$
σ_6'		$-\sin 2(\theta-\theta_o)$

where
$$
\begin{aligned}
I &= p = \frac{1}{2}(\sigma_1 + \sigma_2) \\
R &= \sqrt{q^2 + r^2} \\
\theta_o &= \frac{1}{2}\tan^{-1}\frac{r}{q} = \frac{1}{2}\sin^{-1}\frac{r}{R} = \frac{1}{2}\cos^{-1}\frac{q}{R}
\end{aligned}
\tag{A.6}
$$

4. transformation of ϵ_i, ϵ_i^o, k_i

table A.4
"strain" transformation in power functions

	ϵ_1	ϵ_2	ϵ_6
ϵ'	m^2	n^2	mn
ϵ_2'	n^2	m^2	$-mn$
ϵ_6'	$-2mn$	$2mn$	m^2-n^2

table A.5
"strain" transformation in multiple angle functions

	p	q	r
ϵ_1'	1	$\cos 2\theta$	$\sin 2\theta$
ϵ_2'	1	$-\cos 2\theta$	$-\sin 2\theta$
ϵ_6'		$-2\sin 2\theta$	$2\cos 2\theta$

where
$$p = \frac{1}{2}(\epsilon_1 + \epsilon_2)$$
$$q = \frac{1}{2}(\epsilon_1 - \epsilon_2) \qquad \text{(A.7)}$$
$$r = \frac{1}{2}\epsilon_6$$

table A.6
"strain" transformation in invariant functions

	I	R
ϵ_1'	1	$\cos 2(\theta - \theta_o)$
ϵ_2'	1	$-\cos 2(\theta - \theta_o)$
ϵ_6'		$-2\sin 2(\theta - \theta_o)$

where
$$I = p = \frac{1}{2}(\epsilon_1 + \epsilon_2)$$
$$R = \sqrt{q^2 + r^2} \qquad \text{(A.8)}$$
$$\theta_o = \frac{1}{2}\tan^{-1}\frac{r}{q} = \frac{1}{2}\sin^{-1}\frac{r}{R} = \frac{1}{2}\cos^{-1}\frac{q}{R}$$

5. transformation of Q_{ij}, A_{ij}, B_{ij}, D_{ij}, G_{ij}, H_{ij}

These are material properties of the fourth rank tensor. The transformation goes from the 1′-2′ to the 1-2 coordinate system. This transformation can also be viewed as subjecting the composite material through a rigid body rotation while the observer remains fixed at the 1-2 coordinate system. We do not distinguish between the on-axis and off-axis here because the starting point ($\theta = \gamma = 0$) may not be on the material symmetry axes. The material can be anisotropic. Angle θ or γ is measured from the 1-2 to the 1′-2′ axes, having positive value if the rotation is counterclockwise.

table A.7
"stiffness" transformation in power functions

	Q'_{11}	Q'_{22}	Q'_{12}	Q'_{66}	Q'_{16}	Q'_{26}
Q_{11}	m^4	n^4	$2m^2n^2$	$4m^2n^2$	$-4m^3n$	$-4mn^3$
Q_{22}	n^4	m^4	$2m^2n^2$	$4m^2n^2$	$4mn^3$	$4m^3n$
Q_{12}	m^2n^2	m^2n^2	m^4+n^4	$-4m^2n^2$	$2(m^3n-mn^3)$	$2(mn^3-m^3n$
Q_{66}	m^2n^2	m^2n^2	$-2m^2n^2$	$(m^2-n^2)^2$	$2(m^3n-mn^3)$	$2(mn^3-m^3n)$
Q_{16}	m^3n	$-mn^3$	mn^3-m^3n	$2(mn^3-m^3n)$	$m^4-3m^2n^2$	$3m^2n^2-n^4$
Q_{26}	mn^3	$-m^3n$	m^3n-mn^3	$2(m^3n-mn^3)$	$3m^2n^2-n^4$	$m^4-3m^2n^2$

where $m = \cos\theta = \cos\gamma, \quad n = \sin\theta = \sin\gamma$

table A.8
"stiffness" transformation in multiple angle functions

	1	U'_2	U'_3	U'_6	U'_7
Q_{11}	U_1	$\cos 2\theta$	$\cos 4\theta$	$-2\sin 2\theta$	$-\sin 4\theta$
Q_{22}	U_1	$-\cos 2\theta$	$\cos 4\theta$	$2\sin 2\theta$	$-\sin 4\theta$
Q_{12}	U_4		$-\cos 4\theta$		$\sin 4\theta$
Q_{66}	U_5		$-\cos 4\theta$		$\sin 4\theta$
Q_{16}		$\frac{1}{2}\sin 2\theta$	$\sin 4\theta$	$\cos 2\theta$	$\cos 4\theta$
Q_{26}		$\frac{1}{2}\sin 2\theta$	$-\sin 4\theta$	$\cos 2\theta$	$-\cos 4\theta$

where

$$U'_1 = \frac{1}{8}[3Q'_{11} + 3Q'_{22} + 2Q'_{12} + 4Q'_{66}] = U_1$$

$$U'_2 = \frac{1}{2}[Q'_{11} - Q'_{22}] = U_2\cos 2\theta + 2U_6\sin 2\theta$$

$$U'_3 = \frac{1}{8}[Q'_{11} + Q'_{22} - 2Q'_{12} - 4Q'_{66}] = U_3\cos 4\theta + U_7\sin 4\theta$$

$$U'_4 = \frac{1}{8}[Q'_{11} + Q'_{22} + 6Q'_{12} - 4Q'_{66}] = U_4 \tag{A.9}$$

$$U'_5 = \frac{1}{8}[Q'_{11} + Q'_{22} - 2Q'_{12} + 4Q'_{66}] = U_5$$

$$U'_6 = \frac{1}{2}[Q'_{16} + Q'_{26}] = -\frac{1}{2}U_2\sin 2\theta + U_6\cos 2\theta$$

$$U'_7 = \frac{1}{2}[Q'_{16} - Q'_{26}] = -U_3\sin 4\theta + U_7\cos 4\theta$$

table A.9
"stiffness" transformation in invariant functions

	I	R_1	R_2
Q_{11}	U_1	$\cos 2(\theta+\delta_1)$	$\cos 4(\theta+\delta_2)$
Q_{22}	U_1	$-\cos 2(\theta+\delta_1)$	$\cos 4(\theta+\delta_2)$
Q_{12}	U_4		$-\cos 4(\theta+\delta_2)$
Q_{66}	U_5		$-\cos 4(\theta+\delta_1)$
Q_{16}		$\frac{1}{2}\sin 2(\theta+\delta_1)$	$\sin 4(\theta+\delta_2)$
Q_{26}		$\frac{1}{2}\sin 2(\theta+\delta_1)$	$-\sin 4(\theta+\delta_2)$

here

$$R_1 = \sqrt{U_2'^2 + 4U_6'^2}$$

$$\delta_1 = \frac{1}{2}\tan^{-1}\frac{2U_6'}{U_2'} \qquad \text{(A.10)}$$

$$R_2 = \sqrt{U_3'^2 + U_7'^2}$$

$$\delta_2 = \frac{1}{4}\tan^{-1}\frac{U_7'}{U_3'}$$

. transformation of S_{ij}, a_{ij}, d_{ij}, α_{ij}, β^*_{ij}, δ_{ij}, F_{ij}

A.10
pliance" transformation in power functions

	S_{11}'	S_{22}'	S_{12}'	S_{66}'	S_{16}'	S_{26}'
	m^4	n^4	$2m^2n^2$	m^2n^2	$-2m^3n$	$-2mn^3$
	n^4	m^4	$2m^2n^2$	m^2n^2	$2mn^3$	$2m^3n$
	m^2n^2	m^2n^2	m^4+n^4	$-m^2n^2$	m^3n-mn^3	mn^3-m^3n
	$4m^2n^2$	$4m^2n^2$	$-8m^2n^2$	$(m^2-n^2)^2$	$4(m^3n-mn^3)$	$4(mn^3-m^3n)$
	$2m^3n$	$-2mn^3$	$2(mn^3-m^3n)$	mn^3-m^3n	$m^4-3m^2n^2$	$3m^2n^2-n^4$
	$2mn^3$	$-2m^3n$	$2(m^3n-mn^3)$	m^3n-mn^3	$3m^2n^2-n^4$	$m^4-3m^2n^2$

ransformation equations are valid when β is symmetric.

table A.11
"compliance" transformation in multiple angle functions

	I	U_2'	U_3'	U_6'	U_7'
S_{11}	U_1	$\cos 2\theta$	$\cos 4\theta$	$-2\sin 2\theta$	$-\sin 4\theta$
S_{22}	U_1	$-\cos 2\theta$	$\cos 4\theta$	$2\sin 2\theta$	$-\sin 4\theta$
S_{12}	U_4		$-\cos 4\theta$		$\sin 4\theta$
S_{66}	U_5		$-4\cos 4\theta$		$4\sin 4\theta$
S_{16}		$\sin 2\theta$	$2\sin 4\theta$	$2\cos 2\theta$	$2\cos 4\theta$
S_{26}		$\sin 2\theta$	$-2\sin 4\theta$	$2\cos 2\theta$	$-2\cos 4\theta$

where

$$U_1' = \frac{1}{8}[3S_{11}' + 3S_{22}' + 2S_{12}' + S_{66}'] = U_1$$

$$U_2' = \frac{1}{2}[S_{11}' - S_{22}'] = U_2\cos 2\theta + 2U_6\sin 2\theta$$

$$U_3' = \frac{1}{8}[S_{11}' + S_{22}' - 2S_{12}' - S_{66}'] = U_3\cos 4\theta + U_7\sin 4\theta$$

$$U_4' = \frac{1}{8}[S_{11}' + S_{22}' + 6S_{12}' - S_{66}'] = U_4 \tag{A.11}$$

$$U_5' = \frac{1}{2}[S_{11}' + S_{22}' - 2S_{12}' + S_{66}'] = U_5$$

$$U_6' = \frac{1}{4}[S_{16}' + S_{26}'] = -\frac{1}{2}U_2\sin 2\theta + U_6\cos 2\theta$$

$$U_7' = \frac{1}{4}[S_{16}' - S_{26}'] = -U_3\sin 4\theta + U_7\cos 4\theta$$

table A.12
"compliance" transformation in invariant functions

	I	R_1	R_2
S_{11}	U_1	$\cos 2(\theta+\delta_1)$	$\cos 4(\theta+\delta_2)$
S_{22}	U_1	$-\cos 2(\theta+\delta_1)$	$\cos 4(\theta+\delta_2)$
S_{12}	U_4		$-\cos 4(\theta+\delta_1)$
S_{66}	U_5		$-4\cos 4(\theta+\delta_2)$
S_{16}		$\sin 2(\theta+\delta_1)$	$2\sin 4(\theta+\delta_2)$
S_{26}		$\sin 2(\theta+\delta_1)$	$-2\sin 4(\theta+\delta_2)$

where

$$
\begin{aligned}
R_1 &= \sqrt{U_2'^2 + 4U_6'^2} \\
\delta_1 &= \frac{1}{2}\tan^{-1}\frac{2U_6'}{U_2'} \\
R_2 &= \sqrt{U_3'^2 + U_7'^2} \\
\delta_2 &= \frac{1}{4}\tan^{-1}\frac{U_7'}{U_3'}
\end{aligned}
\tag{A.12}
$$

7. transformation of G_i

table A.13
"strength in strain" transformation in power functions

	G_1'	G_2'	G_6'
G_1	m^2	n^2	$-2mn$
G_2	n^2	m^2	$2mn$
G_6	mn	$-mn$	m^2-n^2

table A.14
"strength in strain" transformation in multiple angle functions

	p'	q'	r'
G_1	1	$\cos 2\theta$	$-\sin 2\theta$
G_2	1	$-\cos 2\theta$	$\sin 2\theta$
G_6		$\sin 2\theta$	$-\cos 2\theta$

where

$$p' = \frac{1}{2}(G_1' + G_2')$$

$$q' = \frac{1}{2}(G_1' - G_2') \tag{A.13}$$

$$r' = G_6'$$

table A.15
"strength in strain" transformation in invariant functions

	I	R
G_1	1	$\cos 2(\theta+\theta_o)$
G_2	1	$-\cos 2(\theta+\theta_o)$
G_6		$\sin 2(\theta+\theta_o)$

where

$$I = p = \frac{1}{2}(G_1 + G_2)$$

$$R = \sqrt{q'^2 + r'^2} \tag{A.14}$$

$$\theta_o = \frac{1}{2}\tan\frac{r'}{q'} = \frac{1}{2}\sin\frac{r'}{R} = \frac{1}{2}\cos\frac{q'}{R}$$

3. transformation of F_i, α_i, β_i

table A.16
"strength in stress" transformation in power functions

	F'_1	F'_2	F'_6
F_1	m^2	n^2	$-mn$
F_2	n^2	m^2	mn
F_6	$2mn$	$-2mn$	m^2-n^2

table A.17
"strength in stress" transformation in multiple angle functions

	p'	q'	r'
F_1	1	$\cos 2\theta$	$-\sin 2\theta$
F_2	1	$-\cos 2\theta$	$\sin 2\theta$
F_6		$2\sin 2\theta$	$2\cos 2\theta$

where

$$p' = \frac{1}{2}(F'_1 + F'_2)$$

$$q' = \frac{1}{2}(F'_1 - F'_2) \tag{A.14}$$

$$r' = \frac{1}{2}F'_6$$

table A.18
"strength in stress" transformation in invariant functions

	I	R
F_1	1	$\cos 2(\theta+\theta_0)$
F_2	1	$-\cos 2(\theta+\theta_0)$
F_6		$2\sin 2(\theta+\theta_0)$

where

$$I = p = \frac{1}{2}(F_1 + F_2)$$

$$R = \sqrt{q'^2 + r'^2} \tag{A.15}$$

$$\theta_o = \frac{1}{2}\tan^{-1}\frac{r'}{q'} = \frac{1}{2}\sin^{-1}\frac{r'}{R} = \frac{1}{2}\cos^{-1}\frac{q'}{R}$$

appendix B
unit conversion tables

1. $\sigma_i, Q_{ij}, A^*_{ij}, \ldots$	To convert Into	Nm^{-2}	$\#/in^2$	kgf/mm^2
$1\ Pa = 1\ Nm^{-2}$	Nm^{-2}	1	6.89; +3	9.81; +6
	$\#/in^2$	.145; −3	1	1.42; +3
	kgf/mm^2	.102; −6	.703; −3	1

2. $S_{ij}, F_i, \alpha^*_{ij}, \ldots$	To convert Into	$(Nm^{-2})^{-1}$	$(\#/in^2)^{-1}$	$(kgf/mm^2)^{-1}$
	$(Nm^{-2})^{-1}$	1	.145; −3	.102; −6
	$(\#/in^2)^{-1}$	6.89; +3	1	.703; −3
	$(kgf/mm^2)^{-1}$	9.81; +6	1.42; +3	1

3. F_{ij}	To convert Into	$(Nm^{-2})^{-2}$	$(\#/in^2)^{-2}$	$(kgf/mm^2)^{-2}$
	$(Nm^{-2})^{-2}$	1	21.0; −9	10.4; −15
	$(\#/in^2)^{-2}$	47.5; +6	1	49.4; −9
	$(kgf/mm^2)^{-2}$	96.2; +12	2.02; +6	1

4. N_i, A_{ij}	To convert / Into	N/m	#/in	kgf/mm
	N/m	1	.175; +3	9.81; +3
	#/in	5.71; −3	1	56.0; 0
	kgf/mm	.102; −3	17.8; −3	1

5. α_{ij}, a_{ij}	To convert / Into	m/N	in/#	mm/kgf
	m/N	1	5.71; −3	.102; −3
	in/#	.175; +3	1	17.8; −3
	mm/kgf	9.81; +3	56.0; 0	1

6. M_i, B_{ij}	To convert / Into	N	#	kgf
	N	1	4.45; 0	9.81; 0
	#	.225; 0	1	2.20; 0
	kgf	.102; 0	.454; 0	1

7. β_{ij}	To convert / Into	N^{-1}	$\#^{-1}$	kgf^{-1}
	N^{-1}	1	.225; 0	.102; 0
	$\#^{-1}$	4.45; 0	1	.454; 0
	kgf^{-1}	9.81; 0	2.20; 0	1

. D_{ij}	To convert Into	Nm	#in	kgf-mm
1 J = 1Nm	Nm	1	.112; 0	9.80; −3
	#in	8.85; 0	1	86.8; −3
	kgf-mm	102; 0	11.5; 0	1

. δ_{ij}, d_{ij}	To convert Into	$(Nm)^{-1}$	$(\#in)^{-1}$	$(kgf\text{-}mm)^{-1}$
	$(Nm)^{-1}$	1	8.85; 0	102; 0
	$(\#in)^{-1}$	.112; 0	1	11.5; 0
	$(kgf\text{-}mm)^{-1}$	9.80; −3	86.8; −3	1

0. k_i	To convert Into	m^{-1}	in^{-1}	mm^{-1}
	m^{-1}	1	39.4; 0	1; +3
	in^{-1}	25.4; −3	1	25.4; 0
	mm^{-1}	1; −3	39.4; −3	1

appendix C
general references

B. D. Agarwal and L. J. Broutman, *Analysis and Performance of Fiber Composites,* John Wiley & Sons, New York, 1980

J. E. Ashton, J. C. Halpin and P. H. Petit, *Primer on Composite Materials:* Analysis, Technomic Pub. Co., Lancaster, PA, 1969 Second Ed. 1984

J. E. Ashton and J. M. Whitney, *Theory of Laminated Plates,* Technomic Pub. Co., Lancaster, PA, 1970

L. J. Broutman and R. H. Krock, Eds., *Composite Materials,* Volumes 1 through 7, Academic Press, New York, 1974

L. J. Broutman and R. H. Krock, Eds., *Modern Composite Materials,* Addison-Wesley Pub. Co., New York, 1967

L. R. Calcote, *Analysis of Laminated Composite Structures,* Van Nostrand Reinhold Co., London, England, 1969

R. M. Christensen, *Mechanics of Composite Materials,* John Wiley & Sons, New York, 1979

C. Y. Chia, *Nonlinear Analysis of Plates,* McGraw-Hill, New York, 1980

A. G. H. Dietz, Ed., *Composite Engineering Laminates,* MIT Press, Cambridge, 1969

S. K. Garg, V. Svalbonas and G. A. Gurtman, *Analysis of Structural Composite Materials,* Marcel Dekker, Inc., New York, 1973

G. S. Holister and C. Thomas, *Fibre Reinforced Materials,* Elsevier Pub. Co., Amsterdam, Netherland, 1967

L. Holliday, Ed., *Composite Materials,* Elsevier Pub. Co., Amsterdam, Netherland, 1966

D. Hull, *An Introduction to Composite Materials,* Cambridge University Press, 1981

R. M. Jones, *Mechanics of Composite Materials,* Scripta Book Co., Washington, D.C., 1975

R. L. McCullough, *Concepts of Fiber-Resin Composites,* Marcel Dekker, Inc., New York, 1971

R. Nichols, *Composite Construction Materials Handbook,* Prentice-Hall, Inc., Englewood Cliffs, N.J., 1976

L. E. Nielsen, *Mechanical Properties of Polymers and Composites,* Volumes 1 and 2, Marcel Dekker, Inc., New York, 1974

N. J. Parrat, *Fibre-Reinforced Materials Technology,* Van Nostrand Reinhold Co., London, England, 1972

M. R. Piggott, *Load Bearing Fibre Composites,* Pergamon, Oxford et al, 1980

E. Scala, *Composite Materials for Combined Functions,* Hayden Book Co., Rochelle Park, N.J., 1973

V. K. Tewary, *Mechanics of Fibre Composites,* John Wiley & Sons, New York, 1978

S. W. Tsai, *Composites Design-1985,* Think Composites, Dayton et al, 1985

J. R. Vinson and T. W. Chou, *Composite Materials and Their Use in Structures,* John Wiley & Sons, New York, 1975

J. M. Whitney, I. M. Daniels, and R. B. Pipes, *Experimental Mechanics of Fiber Reinforced Composite Materials,* Society of Experimental Stress Analysis, Brookfield, Conn., 1982

index